GROUNDWATER CONTAMINATION

VOLUME 2

MANAGEMENT, CONTAINMENT, RISK ASSESSMENT & LEGAL ISSUES

GROUNDWATER CONTAMINATION

VOLUME 2

MANAGEMENT, CONTAINMENT, RISK ASSESSMENT & LEGAL ISSUES

Chester D. Rail

CDR — Environmental Regulations — Consultation and Research

CRC Press
Taylor & Francis Group
Boca Raton London New York

CRC Press is an imprint of the
Taylor & Francis Group, an **informa** business

CRC Press
Taylor & Francis Group
6000 Broken Sound Parkway NW, Suite 300
Boca Raton, FL 33487-2742

First issued in paperback 2019

© 2000 by Taylor & Francis Group, LLC
CRC Press is an imprint of Taylor & Francis Group, an Informa business

No claim to original U.S. Government works

ISBN-13: 978-1-56676-897-9 (hbk)
ISBN-13: 978-0-367-39869-9 (pbk)

Library of Congress Cataloging-in-Publication Data

Main entry under title:
Groundwater contamination: Management, Containment, Risk Assessment and Legal Issues, Volume 2

Full Catalog record is available from the Library of Congress

Library of Congress Card Number 1-56676-897-7

Visit the Taylor & Francis Web site at
http://www.taylorandfrancis.com

and the CRC Press Web site at
http://www.crcpress.com

To Joan Mary Stearley
May 19, 1967–July 12, 1997

Joanie, a Senior Veterinary Medicine student, whom we were very proud of,
was like a daughter to my wife, Carole, and me. She and I were also
marathon running partners. Her smile is missed by family, friends, and the
many pets she had. She was the "little sunshine" in our lives.

Preface   xi
Acknowledgements   xiii
Introduction   xv

**ONE/GROUNDWATER MANAGEMENT, INCLUDING LEGAL CONCEPTS THAT RELATE
TO PREVENTION OF CONTAMINATION** 1

Proper Management to Prevent Groundwater Contamination   1
The Groundwater Management Planning Process to Prevent Contamination   1
Selected Additional Information and References on Organization and Management of Groundwater Planning Concepts to
Prevent Contamination   3
  Organization and Management (Implementation and Feedback)   4
  Institutional Structure and Responsibilities   4
  Central and Regional Control   4
  Links to Other Planning Agencies   4
  Semi-Governmental Agencies and the Private Sector   4
  Selection, Definition and Approval Projects   5
  Presentation of the Plan   5
  License or Permit System for Well Drilling   5
  Update of Groundwater Development Plans that Prevent Contamination   5
Technical Aspects of Groundwater Management Related to Prevention of Groundwater Contamination   6
  Signs of Overpumping   6
  Recharge of Aquifers   6
  Indirect Recharge   7
  Artificial Recharge   7
  Outflow from Aquifers   7
  Groundwater Abstraction   7
Environmental Aspects of Groundwater Management Related to Contamination Concerns   7
  Objectives   8
  General Principles   8
  Conclusions   8
Methods and Techniques to Improve Water Resources Management Concerns in Groundwater Pumping Areas and
Relationship to Contamination   9
  Groundwater Quality Management of a Low Inertia Basin   9
California's New Groundwater Management Law   9
A Groundwater Management Model for Asian Developing Countries   10
Additional Selected Groundwater Management Plans (Methods and Techniques) Related to the Management
Planning Process   10

Protection of Public Water Supplies   10
Groundwater Contamination Control   11
Establishment of Groundwater Protection Programs at the State and Federal Levels   11
  Comprehensive State Groundwater Protection Programs   11
  Groundwater Contaminants and Their Sources (A Review of State Reports)   12
  Sources of Contamination in Groundwater (General)   12
  Discussion/Limitations/Conclusions   13
Programs for Corrective Action of Groundwater Contamination Problems   13
Focus of Programs for Prevention of Groundwater Contamination   14
U.S. National Policy Implications   14
Groundwater Technical Assistance to States and Local Governments   15
Research and Development Needs   15
Establishment of Groundwater Protection Programs at the Local Level   16
Protocol for Protection, Control, or Stabilization Programs to Prevent Groundwater Contamination   16
Recommended Steps   16
  Step 1: Select and Assemble Participants   16
  Step 2: Establish Purpose and a Development Plan   17
  Step 3: Technically Evaluate the Aquifer(s)   17
  Step 4: Select Computer Models for Data Summarization   18
  Step 5: Evaluate Water Law and the Responsibilities of Municipalities and Government   19
  Step 6: Prepare and Implement Plans and Programs   19
Overview of Groundwater Contamination—Summarization   20

TWO/**ECOTOXICOLOGICAL RISK ASSESSMENT (RISK ASSESSMENT STRATEGIES) AND
GROUNDWATER CONTAMINATION** 21

Ecotoxicology (Risk Assessment and Groundwater Interactions)   21
  *Reducing Uncertainty in Assessing the Risk of Environmental Contaminants and Relationship to Groundwater
  Contamination*   23
  Ecotoxicological Testing   23
  Process of Human Health Risk Assessment and Relationship to Groundwater Contamination   24
Risk Assessment and Aquifer Restoration   26
Risk-Based Management of Hazardous Waste and Groundwater   27
Pesticides and Fertilizer Toxicity and Groundwater Contamination   29
Nitrates as Fertilizers and Toxicity in Groundwater   29
Risk Factors and Radium   31
  Overview of Risk Assessment   31
Hazard Ranking Information (HRI) and Site Rating Methodology in Relation to Risk Assessment and Groundwater
Contamination   32
A Case History Hazard Ranking System (HRS) Example Taken from the U.S. Department of Energy Hanford Site   33
  The HRS Groundwater Route   33
  Application to Network Design   34
General Summary of Risk Assessment Strategies and Groundwater Contamination   34

THREE/**NONRADIOACTIVE HAZARDOUS WASTE AND GROUNDWATER
CONTAMINATION INTERACTIONS** 37

Use of Selected U.S. Department of Energy Facilities as Case Models   37
  Baseline Information on U.S. Department of Energy (DOE)   37
Hazardous Waste at DOE Facilities   38
General Information Related to Hazardous Waste Generation at U.S. DOE Facilities   39
  Waste Tanks   39
  Environmental Cleanup at U.S. DOE Facilities   39
Sources of Pollutants Related to Groundwater Contamination at U.S. DOE Facilities   40

Uranium Mining and Milling Wastes   40
Uranium Enrichment Wastes   40
Fuel and Target Fabrication Wastes   40
Reactor Irradiation Wastes   40
Chemical Separation Wastes   40
Fabrications of Weapons Components Wastes   40
Weapons Assemble and Maintenance Wastes   40
Research, Development, and Testing Wastes   40
Other Related Source Wastes   40
Environmental Remediation/Restoration at DOE Facilities and Relationship to Groundwater Contamination   41
Groundwater Contamination at U.S. DOE Sites   41
Treatment   42
Storage and Handling   42
Disposal   42
Regulatory Requirements for Groundwater Monitoring Networks at U.S. DOE Hazardous Waste Sites   42
Migration of Hazardous Waste Constituents to an Aquifer   43
U.S. Department of Energy Case History Studies and Evaluations Related to Groundwater Contamination   43
Los Alamos National Laboratory   43
U.S. Department of Energy Hanford Site   45
The Fernald Groundwater Concerns   46
The Savannah River Site (SRS)   47
Applicability of Land Disposal Restrictions to RCRA and CERCLA Groundwater Treatment Reinjections at U.S.
DOE Facilities   48
Need for Interpretation   48
Basis for the LDR Interpretation   48
Necessary Conditions   49
Groundwater Contamination As Defined by Federal, State, and Local Statutes   49
Groundwater Contamination and Analysis at Other Non-DOE Hazardous Waste Sites   50

FOUR/RADIOACTIVITY, INCLUDING OCCURRENCE/FATE/TRANSPORT AND REMEDIATION/ RESTORATION GROUNDWATER WITH CASE HISTORY EXAMPLE FROM U.S. DOE FACILITIES 53

General   53
U.S. Department of Energy (DOE)   53
Various Types of Radioactive Wastes   54
High-Level Radioactive Wastes, Transuranic Waste, Mining and Milling Wastes, Mixed Waste, and Low-Level Wastes,
within Department of Energy (U.S. DOE) Facilities   54
High-Level Radioactive Wastes   56
Transuranic Waste   57
Mining and Mill Tailings Waste   58
Mixed Waste   59
Low-Level Radioactive Waste   60
A Case History Example of some U.S. DOE Facilities within the Albuquerque Operations Office   60
DOE—Albuquerque Operations   60
Los Alamos National Laboratory, A Specific Case History Review   62
Occurrence, Fate, and Transport   65
Fate/Transport   65
Transport and Fate of Contaminants in the Subsurface   68
Groundwater Remediation/Restoration of Radioactively and Chemically Contaminated Sites   72
General   72
Additional Examples of Groundwater Remediation/Restoration Techniques/Methods for Radioactive and Hazardous
Wastes Sites   76
Sun Fuels Groundwater Remediation   76
Removing Groundwater Contaminants Through Irrigation   77

Use of Plants in the Remediation of Soil and Groundwater Contaminated with Organic Materials   77
Review of In Situ Air Sparging for the Remediation of VOC-Contaminated Saturated Soils and Groundwater   78
Groundwater Remediation Using the Simulated Annealing Algorithm   78
Remediation of Groundwater Polluted with Chlorinated Ethylenes by Ozone-Electron Beam Irradiation Treatment   79
Remediation of Contaminated Soil and Groundwater Using Air-Stripping and Soil Venting Technologies   79
Recovery of Toxic Heavy Metals from Contaminated Groundwaters   79
Cost Components of Remedial Investigation/Feasibility Studies   80
Superfund and Groundwater Remediation, Another Perspective   80

FIVE/ TECHNICAL EVALUATIONS OF GROUNDWATER AND GROUNDWATER PROTECTION PLANS RELATED TO CONTAMINATION 81
Technical Evaluations of Groundwater   81
The Mid Rio Grande Area of New Mexico   81
The Sandia National Laboratories Site-Wide Hydrogeologic Characterization Project   83
Los Alamos National Laboratory Groundwater Protection Plan   88

SIX/GROUNDWATER PROTECTION LAWS, REGULATIONS, STATUTES, AND A CASE STUDY GROUNDWATER PROTECTION PLAN FOR BERNALILLO COUNTY, NEW MEXICO 91
Federal, State, and Local Laws and Regulations   91
Federal Regulations   91
Bernalillo County Groundwater Protection Policy and Action Plan   97
Specific Policies   102
Protection Measures   104
Conclusions   116

BIBLIOGRAPHY 119
URL HYPERLINKS 127
INDEX 163

This book[1] on *Groundwater Contamination* [URL Ref. No. 169] is depicted in a series of two volumes designed to provide updated, integrated, interdisciplinary material in the form of bibliographic references and URL Internet WWW site information [URL Ref. No. 1–339, 336 {105}]. Information on hazardous (including radioactive) and nonhazardous waste [URL Ref. No. 145 {15}, 266, 279, 336 {68}, 337 {111}, 338 {17}] concerns, groundwater contamination in airports [URL Ref. No. 182, 249, 313] (Halm, 1996; Graham, 1997), radioactivity in groundwater [URL Ref. No. 285–286, 289–291, 301], occurrence/fate/transport [URL Ref. No. 302], remediation/restoration [URL Ref. No. 303], and mining and milling wastes concerns [URL Ref. No. 237, 246, 309–311], with specific case history examples from within the U.S. Department of Energy (U.S. DOE) complex [URL Ref. No. 61, 64, 88, 164, 165, 281, 282, 292, 294, 312, 338], are described and presented. Groundwater contamination concerns related to the Los Alamos National Laboratory [URL Ref. No. 97, 101, 288], Sandia National Laboratory [URL Ref. No. 73], and other U.S. DOE sites such as Hanford [URL Ref. No. 164–165, 292] and Fernald [URL Ref. No. 294], are also included along with basic technical information on high-level [URL Ref. No. 305], transuranic wastes [URL Ref. No. 289], mixed waste [URL Ref. No. 291], and low-level radioactive wastes [URL Ref. No. 290], and their relationships to groundwater issues. Risk assessment strategies [URL Ref. No. 261–262, 271–274] for groundwater contamination are also discussed, including the reduction of uncertainties [URL, Ref. No. 261] in assessing the risk of environmental contaminants, risk assessment [URL Ref. No. 235], and aquifer restoration [URL Ref. No.

263]. Specifically, the U.S. DOE Hanford site [URL Ref. No. 164–165, 292] and some of the groundwater contamination problems associated with it are presented as a case history and analysis example. Additionally, discussions and descriptions of military toxics [URL Ref. No. 259] [U.S. Department of Defense (DOD)] [URL Ref. No. 265], creosote [URL Ref. No. 339 {25}, transportation, Cryptosporidium oocysts (*Cryptosporidium parvum*) [URL Ref. No. 159] and Giardia [URL Ref. No. 111] are discussed. Information on important managerial and political implications [URL Ref. No. 318–321], which usually provide the impetus for program planning and implementation [URL Ref. No. 319–321], is also included. The Mid-Rio Grande River Basin Area of New Mexico [URL Ref. No. 54–55] and Bernalillo County [URL Ref. No. 89], including the City of Albuquerque [URL Ref. No. 160, 327], are included as case history examples to show how technical groundwater evaluations and groundwater protection plans [URL Ref. No. 105, 319, 321] can relate to groundwater contamination protection concerns [URL Ref. No. 120, 126, 298] and how specific measures to prevent such problems can be implemented [URL Ref. No. 188], reviewed, and studied. The prototype groundwater protection plan as presented can be used as a generic model to review/recognize, understand, and comprehend the pros and cons of implementing such a protection strategy. The groundwater protection plan prospectus as presented is meant to establish for the readers a framework and a basis for further discussion of the various points that are included within its contents and how they may be applied to any given related situation, specifically the readers' concerns.

Basic bibliographic literature references related to groundwater contamination (see also Rail 1989) [URL Ref. No. 169] provide basic and sound background material. And, pertinent and current bibliographical references (i.e., more than 1,300 total cited references) including substantial use of Internet URL (Universal Resource Locator) [URL Ref. No. 336 {105}] hyperlinks (i.e., more than

---

[1] This book is designed to complement the use of Internet URL Reference Hyperlinks (URL) web pages, which are presented to the reader as the text is read. Internet WWW access to the URLs referenced is necessary, if readers (or participants) are to absorb the full content and comprehensive dimension of groundwater contamination that is presented in this two volume series. The reader or participant is expected to have a basic comprehension of Internet WWW use, literacy, and potential liability if used improperly.

2,300 total groundwater-related WWW URL sites)[2] are included.

As indicated, World Wide Web (WWW) URL hyperlink site reference numbers [URL Ref. No. 1–339] are depicted within the text at appropriate locations of the reading boundaries, as specific subject matter relates to them. And by using, listing, and showing specific Internet WWW URL hyperlink references within text areas, I hope that these sites will be systematically accessed and reviewed as part of the reading process and that they will enhance and supplement the knowledge base of the readers. Also, since active and most pertinent WWW URL sites are generally updated daily, periodically, or systematically, this two-volume series in essence becomes a "living, evolving book and URL WWW related continuous document" that is modified and updated at intervals, specifically, as groundwater contamination related URL WWW pages are changed, modified, or updated by their appropriate home web page host(s) or webmaster.

The bibliographic references cited in this two-volume set and the Internet URL WWW reference hyperlinks that are presented, encompass an extremely integrated, comprehensive, diversified, and interdisciplinary overview of groundwater contamination issues, problems, and concerns, which is the priority and intent of this series that is based on *Groundwater Contamination* (Rail, 1989). In fact, this integrated overview in some instances will appear mind boggling and incomprehensible [URL Ref. No. 45, 61, 88], but after one studies the text, one's mind will then eventually flow with both the bibliographic reference and the URL web page reference material that is presented. Thus, a manageable, integrated, comprehensive perspective of information about groundwater contamination from bibliographic references and Internet URL web pages then emerges and makes sense, especially now that a tremendous amount of groundwater-related information is available on the WWW.

Additionally, many of the problems related to groundwater contamination that are discussed relate directly to and are the result of the wartime Manhattan Project of the Atomic Energy Commission [URL Ref. No. 330] and Department of Defense (DOD) [URL Ref. No. 265] activities, both of which have supported a wide array of nuclear weapons production and radiological experiments that have used many hazardous chemicals [URL Ref. No. 266, 279] and toxic substances [URL Ref. No. 256, 295] (i.e., known and unknown) in various related nuclear weapons experiments [URL Ref. No. 61, 88, 164, 232] since the early 1940s. And, in essence, as we know, some of these experiments have included nuclear facility work [URL Ref. No. 284, 289, 290, 293, 304, 305, 310] that is now managed by the U.S. Department of Energy (DOE) [URL Ref. No. 61, 88] and various subcontractors such as those at the Los Alamos National Laboratory [URL Ref. No. 97, 101, 288], the Sandia National Laboratory [URL Ref. No. 73], and the facilities at Hanford [URL Ref. No. 164–165, 281, 292] and Fernald [URL Ref. No. 45, 294].

The use of the WWW for groundwater contamination related electronic literature searches, including the use of web-based indexes such as First Search, SciSearch, CARL/ Uncover Web, Cambridge Scientific Abstracts (CSA), Water Resources Abstracts, USGS Selected Water Resources Abstracts, WorldCat, Department of Energy (DOE) Reports Bibliographic Database, Dissertation Abstracts, Environmental Science and Pollution Management, Applied Science & Technology Index, ASCE's Civil Engineering Database, GeoBase, GeoRef, Books in Print, NTIS, Books in Print, Pub Science, Kluwer Online, Library of Congress, and other related index searches [URL Ref. No. 335 {6–7, 10, 13, 15, 17, 22}] and the general use of Internet search engines [URL Ref. No. 335 {106, 107}] have all been used to obtain and review technical and general information that is presented.

Local, state, and federal groundwater personnel, specialists, scientists, water resource managers, regulators, water-supply operators, environmental health professionals [URL Ref. No. 166], consultants, university professors and students, librarians, and individuals interested in an integrated comprehensive doctrine or treatise on groundwater contamination, will find this two-volume set interesting, educational, timely, and appropriate. Water resource management personnel and individuals that prepare groundwater protection plans, will find the information germane, useful, timely, and necessary as a reference.

The term *groundwater* is used as one word throughout the text in continued recognition of its importance and status as a single concept and phenomenon as in the book on which this series is based (Rail, 1989). *Contamination* is meant to be analogous to pollution throughout the text and does not necessarily imply infectious organisms or a given state thereof. *Stabilization* is described as maintaining contamination/ pollution in place through time and not necessarily improving or diminishing the quality of groundwater or usage of such for any given specific purpose or use. The use of the term *integrated*, implies the continuous commitment to keep an open mind to the subject of groundwater contamination and to always think interdisciplinary in nature The use of the term "Internet URL reference hyperlink" (i.e., and in the *format* used by me, including the way it is presented and depicted throughout this two-volume series), is basically used to reference an Internet hyperlink WWW site address that provides direct and/or supplementary information as related to the material being shown, described, or discussed within the text itself. Therefore, the viewing, interpretation, and use of any referenced URL WWW site hyperlink addresses remains the full responsibility of the reader(s), reviewer(s), or users.

---

[2] This does not include the additional number of URL hyperlinks that can be accessed by each one of these sites when they are viewed. The actual theoretical number estimated by me that logically can be accessed could be over 100,000 sites.

ACKNOWLEDGEMENTS

I would like to acknowledge the numerous contributions made in the area of *Groundwater Contamination* by individuals from the U.S. Geological Survey, the U.S. Environmental Protection Agency, the U.S. Department of Energy, the U.S. Department of Defense, various universities both local and abroad, professional groundwater consultants and associations, various states, and local municipalities. Truly, all have conducted and coordinated many substantial studies on groundwater and reviews dealing with this subject area, and are the real experts.

I also am grateful for the assistance rendered to me by the University of California Library system at Los Alamos National Laboratory, Los Alamos, NM, especially Susan Miller, Research Associate Librarian. I also appreciate the assistance I have received from the University of New Mexico Zimmerman Library, Albuquerque, N M, their Office of Government Documents, and the UNM Centennial Science and Engineering Library, especially Dena R. Thomas, Associate Professor and Civil Engineering Librarian, and Linda "Kash" Heitkamp, Library Information Specialist Librarian for Earth and Planetary Sciences.

I am also grateful to the following individuals for their assistance in obtaining information related to this book, especially, Della Gallegos, Manager Albuquerque City Council Affairs; Jean Weatherspoon, City of Albuquerque Water Resources Department, Water Conservation Director; and Debra James, Public Relations Manager, City of Albuquerque Water Resources Department.

I also acknowledge the technical and continued support from Professor and Dean William M. Hadley, College of Pharmacy, University of New Mexico; Professors Bruce Thomson and Stephen P. Shelton, Civil Engineering Department, University of New Mexico; Professor D. Matthews, Chair of the University of New Mexico Geography Department; and Bruce M. Gallaher of the Los Alamos National Laboratory EM-18 Group; although I am solely responsible for the full content of this book.

Also, I fondly appreciate the fine work and relationship that Susan G. Farmer, Managing Editor; Dr. Joseph L. Eckenrode, Vice President and Publisher; Lori A. Eby, Project Manager; Teresa Wiegand, Managing Editor; and Stephen C. Spangler, Production Manager; at Technomic Publishing Co., Inc., and I have shared during the preparation and final revisions of this two-volume series of *Groundwater Contamination*. I am also very appreciative of the "unknown" technical editors and typesetters from Technomic Publishing Co., Inc. for their patience and understanding when working with this two-volume manuscript series on groundwater contamination. I also appreciate the support and encouragement I received from my first graduate school Professor, Dr. Doug Jester. I also acknowledge the continued positive support and words of encouragement that I have always received from my good friend, mentor, and coffee-drinker buddy, Simon O. Santillanes.

New Materials and Information
in this Two-Volume Series

As mentioned in the *Preface,* updated groundwater contamination related interdisciplinary material in the form of
bibliographic references (i.e., more than 1,300 literature references) and URL[1] Internet WWW site information and content (i.e., more than 2,300 Internet web sites) are presented
(i.e., an integrated approach). By using Internet WWW
URLs within various text areas along with cited literature
references, it is hoped that these Internet WWW referenced
sites will be systematically and periodically accessed in an
integrated format as part of the reading process, so as to enhance and supplement the groundwater knowledge base of
any reader. Also, since active and most pertinent URL
WWW sites related to groundwater are generally updated on
a daily or periodic basis, this is in essence, designed to be a
"living, evolving book set" that is timely and completely integrated with the Internet and WWW.

The reader can be kept "Abreast of the Art" concerning
groundwater contamination concerns as long as the various
webmasters or web sites keep the Internet WWW sites that
are referenced in this series updated at regular intervals.
Since the majority of Internet WWW sites that are referenced are U.S. Government [URL Ref. No. 7–8, 19–20, 22,
37, 40, 45, 137, 165, 265, 276, 285, 332], state government
[URL Ref. No. 41, 138, 146, 228, 333], or local government
managed [URL Ref. No. 89, 160, 327], include professional
groundwater-related organizations [URL Ref. No. 1, 78,
105, 107, 166, 171, 306], and/or universities [URL Ref. No.
43, 66, 83, 91, 102, 147], the probability is high that if one
views the Internet WWW information sites as addressed,
one can maintain a timely grasp on groundwater contamination matters and issues.

[1] URL [URL Ref. No. 336 {107}] is a standard for specifying the location
of an object on the Internet, such as a file or a newsgroup. URLs are used on
the World Wide Web (WWW) to specify a hyperlink that is often another
document stored on another computer.

Groundwater Contamination Concerns
Including Philosophical Issues

One major issue that is discussed in the groundwater legal
arena [URL Ref. No. 325] is the use of water for irrigation
and industries competing with supplies for domestic use and
consumption. Another is the degree to which groundwater
contamination should be defined and rigidly controlled.
Maintaining reasonable environmental rules [URL Ref. No.
68–69, 258, 279–280, 287, 291, 300, 328] is acceptable, but
using discretionary judgment and degrees of toleration, getting optimal value from data and analyses [URL Ref. No. 8,
23, 26–27, 32, 56, 61, 63–64], and debating some issues in
the marketplace are paid more attention to in this series than
in the book on which it is based (Rail, 1989).

Additionally, however, as explained in *Groundwater Contamination* and in agreement with LeGrand (1995), public
awareness of problems that arise from groundwater contamination will continue to prompt new regulations [URL Ref.
No. 66–69] and bureaucratic procedures (Colten, 1998). Although the intentions of the regulatory bodies are always
meant to be good, many regulations are still severe [URL Ref.
No. 68–69] and do not fit uncertainties [URL Ref. No. 261,
271–274] or the great variety of possible situations that
could realistically arise now and in the future (Hall, 1999).
Consequently, the philosophy that is depicted in this text,
compared to the book on which it is based (Rail, 1989),
agrees that uneven degrees of effort in work programs will
follow and eventually lead to projects with excessive costs in
some cases and neglect in others. It will be necessary to get
more value from imprecise data, now too often discarded.

LeGrand (1995) also mentions that the number of workers
in groundwater programs should not decrease in the future,
but a shift or emphasis in type of work activities is likely.
The current straightforward dedication to complying with
narrowly defined prescriptive rules and procedures, now
overwhelming, could become a smaller part of the overall
work. More discretionary power will likely be given to

skilled hydrogeologists [URL Ref. No. 6, 13, 14] or ground-water contamination generalists [URL Ref. No. 35, 58, 139, 182] who must use a commonsense approach toward issues and particular characterization/methods/techniques assessment [URL Ref. No. 36, 48, 157, 252, 263] and other groundwater site evaluation [URL Ref. No. 15, 132, 134, 145, 152, 292, 294, 307] problems.

As a sensible basis for impending changes, the following subject areas in the groundwater arena are likely to be considered at international, national, and state levels, and revamped and amended rules will always come from them:

(1) The philosophical basis for groundwater protection and management

(2) The hydrogeologic experience at policy and regulatory levels

(3) Discrimination among the importance of subjects to be considered and proper priorities designated

(4) The approaches to make optimal use of existing hydrogeologic experience and data.

One cannot pinpoint or definitely narrow cost-effective changes that will affect groundwater work, but the forefront of activities will involve using hydrological skills and exercising good judgment and substantial will and perseverance.

Additionally, individuals that conduct research (Glaze, 1996) or investigate certain subject areas (i.e., such as water resources and groundwater contamination) [URL Ref. No. 7, 10, 11, 25, 29, 37, 40–41, 43, 45, 50, 83–84, 91, 100–101, 139, 152–153, 178, 265, 333, 335–339] must recognize that a properly conducted library or literature electronic search (i.e., Internet subject WWW search) is necessary and has never been more valuable because, as the volume of published literature increases and overwhelms most of us, there is a direct need for a quick reliable, readable, concise, and adequate review of important groundwater, engineering, and other water quality/quantity-related issues and concerns that have been published (Water Pollution Control Federation, 1989, 1990, 1991; Water Environment Research Journal, 1992, 1994, 1995, 1996, 1998; Weber, 1994) and are available on the WWW [URL Ref. No. 1, 16–17, 30, 38, 77–78, 85–86, 90–91, 103, 107, 135, 144, 147, 150, 154, 156, 169, 335 {2–4, 6, 9–14, 22}, 336 {7}].

Groundwater Management Concepts as Described in Volume 2

Proper management of groundwater aquifers [URL Ref. No. 145, 318–319] to eliminate, control, or stabilize groundwater contamination, requires appropriate credible institutional structures that implement and follow through on strategies of significance that are proactive in nature. In Volume 2 information is presented that involves the implementation of short- or long-term groundwater management objectives such as maintaining, controlling, or stabilizing groundwater levels so as to minimize the opportunity for contamination of infiltrate from surface sources; maintaining, controlling, or stabilizing groundwater levels to prevent upward movement of more saline [URL Ref. No. 244, 255] and warmer water into the aquifer; regulating the quality of water used to artificially recharge [URL Ref. No. 322] the aquifer-storm runoff [URL Ref. No. 249] that is collected in upstream reservoirs, stored, and then released into spreading areas; preventing saltwater intrusion [URL Ref. No. 255] and inflow of poor quality natural waters [URL Ref. No. 186, 189] from adjacent surface areas and aquifers, with poor quality water from underground sources usually being excluded by many pumping wells installed in a line, while surface waters are intercepted by drainage ditches and diverted from the area; regulating the drilling, completion, and operation of all types of wells penetrating the aquifer in question; reducing salt loads by exporting groundwater's, wastewater's, or brines high in salinity; systematically and comprehensively monitoring the quality of groundwater [URL Ref. No. 48, 252] throughout the aquifer system to identify and locate contamination sources, including leaking underground fuel tanks, radionuclides, etc., or to verify if corrective or stabilization or control measures have been successful or implemented; and properly implementing comprehensive planning programs [URL Ref. No. 145, 318–319] aimed at controlling, stabilizing, or abating groundwater contamination.

Technical Hydrogeological Evaluations and Groundwater Protection Plans

An example (case history analysis) of Technical Hydrogeological Evaluations [URL Ref. No. 13, 187] and a Groundwater Protection Policy and Action Plan along with a summary of existing government protection laws and regulations is also presented in *Volume 2*. The Albuquerque-Belen Basin [URL Ref. No. 54–55] located in central New Mexico [URL Ref. No. 339 {42}] and Bernalillo County [URL Ref. No. 89] is used as the geographical area model to present technical and groundwater protection plan development information because within the broad boundaries of the area are included the U.S. DOE Sandia [URL Ref. No. 73] and Los Alamos National Laboratories [URL Ref. No. 97], the City of Albuquerque (Bernalillo County) [URL Ref. No. 89, 327], and various Indian Reservation centers [URL Ref. No. 339 {41}]. Included also are many references taken from the U.S. Geological Survey [URL Ref. No. 8] that provide in-depth groundwater-related information and descriptions of the area in question, including site-wide hydrogeological characterization reports. A comparison of what is being done at the local, state, and federal levels in relationship to groundwater contamination prevention within the Mid-Rio Grande area of New Mexico is also presented in *Volume 2*.

Additionally, what is also new to this text involving eco-toxicological risk assessment [URL Ref. No. 260] and groundwater interactions (*Volume 2*) includes the following: (i.e., information concerning distribution of dose-response exposures to various human and animal populations, intrinsic sensitivity among the populations exposed, and translation of animal effects data into estimates of consequence to humans [URL Ref. No. 273]. The definition of *risk analysis* [URL Ref. No. 235] is considerably expanded in this series, compared to the book on which it is based (Rail, 1989), to include its usefulness as a relative tool in ranking, priority setting, allocating resources, and assessing research needs to improve the understanding of environmental problems. *Risk* as presented is also meant to involve a geographical component (Lantzy et al., 1998), even at locations in space where receptors (i.e., human or environmental) and hazards come together. Information is also presented that discusses quantitative risk assessment [URL Ref. No. 235] and its use as a tool at hazardous waste sites [URL Ref. No. 266, 279, 328] for evaluating the need for treatment and determining which control strategies should be implemented (Batchelor, 1997). The conduction of quantitative risk assessments at contaminated sites is also reviewed and a framework for evaluating risk associated with groundwater contamination by disposal of materials treated by solidification and stabilization is discussed. Groundwater contamination and risk factors related to pesticides [URL Ref. No. 239] and groundwater ecosystems [URL Ref. No. 182, 260] that have been conducted are also presented in *Volume 2* along with a discussion concerning a series of risk factors of groundwater contamination related to livestock holding pens [URL Ref. No. 240], wellhead management, hazardous waste, fertilizer storage and handling, petroleum product storage [URL Ref. No. 253], milking center wastewater, and livestock manure storage areas [URL Ref. No. 238]. Risk-related information is also presented that evaluates the necessity of bringing groundwater up to drinking water standards based on health-risk-based criteria [URL Ref. No. 262]. A discussion on the real lack of knowledge of toxic effects of contaminants, especially at low levels for prolonged periods and whether the health-risk-based approach is effective is also discussed. An in-depth discussion on reducing uncertainty [URL Ref. No. 261] in assessing the risk of environmental contaminants in groundwater is presented. The strategy for reducing uncertainty also recognizes that the cost of building new models and collecting data must be balanced by the value of the information obtained. Also, that it is possible to use the analytical framework of statistical decision analysis [URL Ref. No. 339 {31}] to determine when additional information is beneficial. Human Health Risk Assessment [URL Ref. No. 262] is also discussed and presented, and the process of human health risk assessment is judged by its ability to predict adverse outcomes of particular environmental contaminants or exposures for individual humans (Burger, 1994). It

is surmised that views and the knowledge of environmental risk assessments and ecotoxicological concepts are still evolving and assessments that can apply meaningful information during aquifer restoration [URL Ref. No. 263] are not always easy to explain.

U.S. Department of Energy (U.S. DOE) Environmental Management Program and Groundwater Contamination Concerns

The U.S. Department of Energy Environmental Management Programs [URL Ref. No 45, 61, 88] and how they contend with groundwater contamination concerns are emphasized and discussed (McDonald, 1999). Specific case history examples from the U.S. DOE are used to discuss hazardous and nonhazardous materials related to groundwater contamination. Information is also presented on how the work that was conducted concerning the development of nuclear weapons has impacted groundwater contamination problems and their relationship to radioactive isotopes [URL Ref. No. 232, 301, 304]. Specific examples and information about what is being done at various U.S. Department of Energy Laboratories, such as the Los Alamos National Laboratory [URL Ref. No. 97, 288], Sandia National Laboratory [URL Ref. No. 73], Hanford [URL Ref. No. 292], Fernald [URL Ref. No. 294], and Savannah River site [URL Ref. No. 45], are presented in *Volume 2*. Basic information is presented concerning the "nuclear weapons complex" [URL Ref. No. 45] that includes thousands of large industrial structures such as nuclear reactors, chemical processing buildings, metal machining plants, maintenance facilities, and their relationships to groundwater contamination.[2] Concerns including those related to low-level radioactive waste [URL Ref. No. 290], mixed waste [URL Ref. No. 291], high-level radioactive waste [URL Ref. No. 305], and transuranic radioactive waste [URL Ref. No. 289], are also discussed. How the U.S. DOE interrelates with hazardous waste [URL Ref. No. 266, 279], radioactivity [URL Ref. No. 304], and restoration/remediation [URL Ref. No. 303] groundwater contamination concerns are also presented. Various Internet WWW sites related to the U.S. DOE and other related agencies [URL Ref. No. 2, 8, 19, 265] that address radioactivity, restoration, remediation, fate and transport interactions, hazardous and nonhazardous waste disposal, including mixed waste, are also presented.

The reader of this text will have an understanding of the Environmental Management Program [URL Ref. No. 61, 88] that is in place at the U.S. DOE (i.e., Office of Environmental Management), including present and proposed future strategies of the program.

---

[2] U.S. DOE. "Accelerating Cleanup: Paths to Closure," Office of Environmental Management, Washington, D.C., DOE/EM-0362 (1998).

The relationship that the U.S. DOE has with the U.S. Environmental Protection Agency (U.S. EPA) [URL Ref. No. 19] and the application of various State, Federal Regulations and U.S. DOE Orders [URL Ref. No. 275] and other mandated Directives are also presented. Activities that revolve around hazardous waste management units (HWMUs) [URL Ref. No. 299] at U.S. DOE are complex and involve interaction among DOE Operations Offices, [URL Ref. No. 45] EPA Regions, [URL Ref. No. 19, 124] and states demonstrate that no single document can contain all the information relevant to hazardous waste management facilities. The relationship of nuclear wastes to groundwater contamination is a complex issue, but an attempt to present consistent information is provided.

Environmental Remediation/Restoration at U.S. DOE Facilities

Environmental remediation/restoration at DOE [URL Ref. No. 303] encompasses activities at all sites within the Environmental Management Program [URL Ref. No. 61, 88], and these sites at one time involved 10,500 potential release sites. To establish the case for environmental restoration, the DOE is depending on ongoing efforts, which are discussed in *Volume 2,* directed at containing contaminants to prevent them from migrating from the source and eliminating the initial source. How radionuclides [URL Ref. No. 301] and other contaminants such as heavy metals [URL Ref. No. 207, 317] that cannot be destroyed and how they relate to remediation are also discussed. Information concerning technology development activities such as oriented technology to support environmental restoration, nuclear material and facility stabilization, and waste management activities, are also presented.

Regulatory Requirements for Groundwater Monitoring Networks at U.S. DOE Hazardous Waste Sites

Information concerning the *integrated approach* used by the U.S. Department of Energy [URL Ref. No. 45] to protect groundwater through the various standards and classifications, which are based on a comprehensive regulatory and policy analysis (Keller, 1990), are presented. Groundwater monitoring programs [URL Ref. No. 48] at hazardous waste sites (i.e., active and inactive) as depicted by the U.S. DOE are *integrated* with site-specific requirements and regulatory requirements. Statutes and regulations that are related to this concern such as the Resource Conservation and Recovery Act (RCRA) [URL Ref. No. 280], the Comprehensive Environmental Response Liability and Compensation Act (CERCLA) [URL Ref. No. 264], the Safe Drinking Water Act (SDWA) [URL Ref. No. 258], the Clean Water Act (CWA) [URL Ref. No. 287], the Low Level Radioactive Waste Policy Act (LLRWA) [URL Ref. No. 290], and the Nuclear Waste Policy Act (NWPA) [URL Ref. No. 278], are also examined in detail.

U.S. Department of Energy Case History Studies and Evaluation Examples

Environmental Restoration and Waste Management Programs (i.e., past, present, and future missions) are discussed in *Volume 2* as these programs pertain to the Los Alamos National Laboratory [URL Ref. No. 97] that was established in 1943 for the design, development, and testing of nuclear weapons. Information is presented on how the environmental restoration program [URL Ref. No. 288] has identified approximately 2,100 potential release sites and on how potential release sites (e.g., field units) and releases could pose a health risk [URL Ref. No. 235, 262] to surrounding communities via groundwater contamination.

The waste types that continue to be generated at the Laboratory include radioactive waste (transuranic waste and mixed transuranic waste [URL Ref. No. 289], low-level radioactive and low-level mixed waste [URL Ref. No. 290], and accelerator-produced radioactive materials), hazardous chemical waste, biological waste, medical waste, and sanitary solid and liquid waste.

The U.S. DOE Hanford Site

The U.S. Department of Energy Hanford site [URL Ref. No. 164, 292] is presented as a case history study. More than 1,500 waste disposal sites have been identified at this site (Sherwood, 1990). At the request of the U.S. EPA [URL Ref. No. 19], these sites were aggregated into four administrative areas for listing on the National Priority List. Within the four aggregate areas, 646 inactive sites were selected for further evaluation using the Hazard Ranking System (HRS) [URL Ref. No. 274]. Evaluation of inactive waste sites by the HRS provided valuable insight for designing a focused radiological and hazardous substance monitoring network. The Hanford Site [URL Ref. No. 164, 292] groundwater monitoring program was expanded to address not only radioactive constituents but also hazardous chemicals. Information on how this was done is also presented (Valenti, 1993; Illman, 1993).

U.S. DOE Fernald Environmental Project

A discussion of the U.S. DOE Fernald Environmental Management Project's [URL Ref. No. 294] groundwater concerns is also presented in *Volume 2*. The Fernald Environmental Management Project is a 1,050-acre facility located 18 miles northwest of downtown Cincinnati near the farming community of Fernald, Ohio (Nelson and Janke, 1995). While in active operation from 1952 until 1989, the Feed Material Production Center (FMPC) [URL Ref. No. 294], as it was then called, produced highly purified uranium metal [URL Ref. No. 283] for ultimate use in the manufacture of nuclear weapons. In 1986, the U.S. EPA [URL Ref. No. 19] and the U.S. DOE [URL Ref. No. 45] entered into a Federal Facility Compliance Agreement (FFCA) covering environmental impacts associated with the FMCP. Information is presented concerning the site-wide Remedial Investigation/Feasibility Study (RI/FS) initiated pursuant to the Comprehensive Environmental Response and Liability Act (CERCLA) [URL Ref. No.264] as amended by the Superfund Amendment and Reauthorization Act.

U.S. DOE Savannah River Site

Information on how the U.S. DOE Savannah River Site (SRS) [URL Ref. No. 45] relates to potential groundwater contamination concerns is also presented in *Volume 2* The SRS groundwater remediation program that is removing industrial solvents from groundwater, is the largest such program within the U.S. DOE complex and is among the largest groundwater cleanup programs in the nation. How the remediation program was initiated at this site with a groundwater extraction and treatment system using an air stripper is discussed.

Occurrence/Fate/Transport and Remediation/Restoration with some Examples from U.S. DOE Facilities

An in-depth discussion on radioactivity and remediation concerns at U.S. DOE facilities is presented in *Volume 2*. A general summary of information concerning subsurface Transport/Fate Processes [URL Ref. No. 302] and Groundwater Remediation/Restoration [URL Ref. No. 276, 285] is also presented. The U.S. DOE facilities (i.e., with emphasis on the Los Alamos National Laboratory) [URL Ref. No. 97] are used as a case history model so that a discussion that can be extrapolated to non-DOE areas can be presented (Knox, et al., 1993; Gray, 1990).

U.S. DOE Mining and Milling Wastes

The processes of mining and milling [URL Ref. No. 310] uranium [URL Ref. No. 283] and thorium [URL Ref. No. 339 {35}] ores have generated large quantities of rock, sludge, and liquids. These wastes contain daughter nuclides such as radium, polonium, bismuth, and lead (Gershey et al., 1990; U.S. DOE, 1994), and they are generated during the exploratory and operational phases of mining and consist of large amounts of rock from excavations and liquids from surface drainage, seepage, and *in situ* [URL Ref. No. 246] leaching. How groundwater contamination related to uranium [URL Ref. No. 283] mill tailings and the remedial action that has been conducted by the U.S. DOE is discussed in *Volume 2* along with an in-depth discussion on how the U.S. DOE UMTRA Project Program (Uranium Mill Tailings Remedial Action Program) [URL Ref. No. 310] works.

ONE

Groundwater Management, Including Legal Concepts That Relate to Prevention of Contamination

Proper Management [ULR Ref. No. 145, 318] to Prevent Groundwater Contamination

Proper management of groundwater aquifers to eliminate (IWRA, Water International, 1999), control, or stabilize groundwater contamination, requires appropriate credible institutional structures that allow for implementation and follow-through on strategies of proactive significance (Archey and Mawson, 1984; McCabe et al., 1997; Burke et al., 1999; Sun and Zheng, 1999; Petts et al., 1999). Groundwater management criteria [URL Ref. No. 318] should include the following as a minimum:

- maintaining, controlling, or stabilizing groundwater levels so as to minimize the opportunity for contamination of infiltrate from surface sources
- maintaining, controlling, or stabilizing groundwater levels to prevent upward movement of more saline and warmer water into the aquifer
- regulating the quality of water used to artificially recharge the aquifer storm runoff [URL Ref. No. 249] collected in upstream reservoirs, stored, and then released into spreading area; this could be of a higher quality than groundwater, however, imported and reclaimed waters may not be
- preventing saltwater intrusion [URL Ref. No. 255] and inflow of poor quality natural waters from adjacent surface areas and aquifers, with poor quality water from underground sources usually being excluded by many pumping wells installed in a line, while surface waters intercepted by drainage ditches are diverted from the area
- regulating the drilling, completion, and operation of all types of wells penetrating the aquifer in question
- reducing salt loads by exporting groundwater, wastewater, or brines [URL Ref. No. 237] that are high in salinity

- systematically and comprehensively monitoring the quality of groundwater throughout the aquifer system to identify and locate contamination sources, including leaking underground fuel tanks (McKee et al., 1972) [URL Ref. No. 251], radionuclides [URL Ref. No. 304], etc., or to verify if corrective or stabilization or control measures have been successful or implemented
- properly implementing comprehensive planning programs [URL Ref. No. 319] aimed at controlling, stabilizing, or abating groundwater contamination (i.e., such as comprehensive pretreatment programs for industrial waste discharge) [URL Ref. No. 227]

The Groundwater Management Planning Process to Prevent Contamination

General groundwater management [URL Ref. No. 318] and development has long been directed to satisfying demands without taking into account the scarcity of this natural resource (Leusink, 1992; American Society of Civil Engineers, 1987). As a result, groundwater resources are overexploited in many areas, especially in arid and semiarid regions. Sustained groundwater resources development to prevent contamination, therefore, requires a broader scope and an integrated approach to water resources management and planning [URL Ref. No. 34, 320]. Relevant elements within integrated water resources management (Rushton, 1999; Lee, 1999; Berg et al., 1999) include water conservation [URL Ref. No. 320], the role of surface water, water quality, demand and supply management, institutional credible organization, and the role of beneficiaries.

Groundwater resources management strategies [URL Ref. No. 320] seem to have focused on the development of groundwater resources (Elliot et al., 1999; Psilovikos, 1999) to satisfy the increasing demands of society for water, and these groundwater resources were developed to survive

1

national disasters such as droughts. The effects of groundwater utilization also have medium and long-term impacts that have affected future availability and access to the source, and it is very likely that in the coming decades, a different kind of water resources scarcity will be encountered, including one that cannot be solved by countermeasures to cancel the adverse side-effects of exploitation. Consequently, the approach to formulating groundwater resources management strategies [URL Ref. No. 320] will continue to change (Barrett, 1999), and principles such as water conservation [URL Ref. No. 110], demand management, and institutional organization will always be starting points.

General Principles

Proper allocation of available groundwater resources, especially in semiarid areas, requires planning because competitive users apply for limited resources. In relation to time decisions, they have to be made on long-term developments and on the daily operational basis (Hinds et al., 1999; Pohll et al., 1999). Planning then is primarily related to decisions with long-term effects, with a time horizon that varies from five to 20 years (i.e., although it can have a longer time frame when various radionuclides are involved). Planning along these lines then needs three basic elements: anticipating decision making about a course of action to follow, creating coherence between or among decision-making authorities, and aiming at a desired situation and eliminating wrong courses of action (i.e., wrong in terms of defined objectives). Planning and subsequent decisions then can generate consistent, coherent, courses of action focused to obtain a desired situation in the future, and as such, proper planning tries to *reduce* or *eliminate uncertainties* [URL Ref. No. 261] due to developments in the planning area.

At present in most water resources management situations, some form of planning exists (De Sena, 1999; Baca, 1999), but the subjects and the approaches vary widely. Four characteristics of the complex planning process can include an understanding of the following:

(1) Planning and decision-making processes are dynamic processes due to conditions and circumstances and all inputs to the water resources system being subject to changes (i.e., population growth, economic development, land use, quantity and quality change due to natural developments, and human interactions), including technological options, and social and political preferences regarding objectives.

(2) Groundwater resources planning processes are part of the water resources system and are strongly interrelated with other planning areas (i.e., agricultural development, industrial development, urbanization, and family planning) with the multiplicity of objectives formulated from those perspectives having strong implications for

the planning and development of groundwater contamination prevention projects.

(3) Planning processes take place at different administrative levels (i.e., national, regional, and local) that can be hierarchically linked; consequently, when hierarchical relations exist, it is important to indicate the framework of the interdependent levels to facilitate integration and consistency of formulated plans.

(4) Planning processes must take into account a variety of constraints that play an important role in prescreening and evaluating various alternative strategies.

Additionally, however, sometimes strategies may be adjusted to the constraints, but other ones then become lethal factors (Gibbons et al., 1999). These factors that must be included in an early stage of the screening process in groundwater planning include the following:

(1) Natural factors such as safe yield and total capacity of an aquifer, geohydrological regime [URL Ref. No. 188], land use destination, and geographical location

(2) Socioeconomic factors such as water use patterns, water demand characteristics, price elasticity, and available funds

(3) Administrative factors such as existing laws and regulations [URL Ref. No. 68–69, 89, 333], the institutional organization, decision-making procedures, and conditions set by other sectors

(4) Technical factors such as the capacity of wells and well fields

The actual application of groundwater management practice faces serious difficulties, and, wherever applied, its success is rather limited (Anzzolin et al., 1999; Schintu and Robert, 1999). The causes of these difficulties are twofold and include deficiencies in the information and in the use of the information. Analysis for the groundwater management planning process, therefore, requires information on the follow-ing components, according to Leusink (1992), the American Society of Chemical Engineers (1987), and Tuinhof (1992):

(1) Anticipated demand for groundwater or groundwater-related conditions (e.g., price, quantity, and quality as functions of location and time) over the given planning time frame

(2) The configuration of the system, its present state (e.g., water levels, water quality, volume in storage, and discharges), and its trend relative to the demand

(3) Feasible controllable measures and corresponding actions that can aid in closing the gap between supply and demand and available resources

(4) Anticipated exogenous inputs to the system (water and substances), natural and anthropogenic, that may affect the supply of groundwater

(5) Anticipated state and supply of the groundwater as a result of alternative courses of action

(6) Resulting payoffs and losses in each alternative

It is also difficult to predict the vulnerability of passive objectives for groundwater abstractions [URL Ref. No. 323]. This is due to the lack of information on the previously listed components [(1)–(6)], and the incompleteness of knowledge about cause-effect relationships and of systematic and comprehensive methods to visualize the interrelation between abstractions and consequences.

Another key segment of the groundwater planning cycle shows the formulation and analysis of groundwater resources management strategies [URL Ref. No. 318], including a step in which policy options that are clearly unattractive are screened out. The result is a limited list of promising methods that deserve a more thorough evaluation. Screening can also significantly reduce the cost and time of carrying out policy method/technique studies involving large numbers of alternatives.

The procedure for formulating and analyzing groundwater management strategies must be strong and reliable (Vogel, 1999; Barcelona, 1999). In this respect, analytical methods exist to contribute to the evaluation of strategies, such as cost-benefit analysis and multicriteria analysis. Besides technical and economic criteria planning of groundwater concerns, planning must also involve social, environmental, institutional, and political impacts. Decision makers should use multiple criteria in making decisions.

The concept of integrated groundwater resources planning will become more important in the coming decades, and, it is, therefore, necessary to be more specific about the word *integrated,* because it is sometimes used in many different contexts (Nyler et al., 1999). The following are, therefore, considered relevant when using the term:

(1) *Integration of issues:* The scope and objectives of water resources planning include surface water and groundwater and quantity and quality issues. Criteria for these issues must be developed against the background of environmentally sound concerns. Water resources planners should look at the carrying capabilities of resources first, rather than just plan and summarize adverse impacts later. Sustainable development can be the leading principle for planners and decision makers who have not constantly dealt with the *scarcity phenomenon* of water resources that may be expected more and more into the future.

(2) *Integration of disciplines:* In groundwater resources planning, coordination of concerns is required with other sectors of the economy. Among a team of planners, different expertise should always be available to incorporate all aspects in the plan (i.e., engineering, economic, ecological, legal, and social aspects, etc.). The various disciplines should cooperate and communicate. Different sectors, such as public water supply, agriculture, natural environment, and recreation, should be strengths that support pertinent coordination efforts during policy preparation and implementation.

(3) *Integration of supply and demand:* Generally, water should be provided at low cost in amounts and qualities desired. Nowadays, water conservation techniques [URL Ref. No. 110] and water demand management are still not generally applied, although it is expected that these concepts will become very important and must be given more attention in the future. Demand management concepts should include the formulation and application of implementation incentives (i.e., economic instruments, such as charges for the utilization of water resources often directly related to the production costs; legal instruments, such as licenses for water abstraction [URL Ref. No. 323]; and effluent charges that can reduce industrial water use).

(4) *Integrated plan of activities:* The groundwater planning cycle can consist of several steps including policy formulation, planning and analysis, implementation, operation and maintenance, monitoring, auditing, and quality assurance. It is less effective to focus the main attention on only one step of the cycle and forget the others. Different steps in the planning process should receive balanced attention in order to achieve improvement in the whole process of planning adequately within a defined groundwater resources area.

(5) *Integration of consumers:* Water users can play a dual role in water resources planning, since they are the ultimate beneficiaries. Cooperation between consumers and community-based managers is required for virtually every water resources development project, and it is desirable to involve all the beneficiaries at an early stage.

Selected Additional Information and References on Organization and Management of Groundwater Planning Concepts to Prevent Contamination

Groundwater development plans are based on hydrogeological consequences of simulated scenarios with time-dependent elements (Tuinhoff, 1992; Lee et al., 1999), and control and monitoring of the implementation of a plan is essential in order to verify the assumptions on which it is based and to collect the information needed to update and improve any groundwater resources plan after a number of years. A system of feedback is needed to guarantee that the information is returned to the planning agency for continuous update of the plan (Hall, 1999).

Organization and Management (Implementation and Feedback)

The decision on a certain groundwater development project should first depend on a regional plan. A regional plan provides the outline for the long-term allocation of groundwater resources (10–20 years) and generally includes the operational plan with shorter time horizons (1–10 years). Some of the steps of a regional plan can include selection and approval of projects, design of projects, tendering and construction, and operation and maintenance of schemes.

By its nature, a groundwater development plan cannot be static because it is based on data that is available at the initial planning stage and on certain scenarios. Hence, the following components also become important factors that should be integrated into the implementation of any groundwater development plan: monitoring and control, feedback to the planning agency, and update of the groundwater development plan. These then are relevant to the implementation of any groundwater plan.

Institutional Structure and Responsibilities

Implementation of water resources development plans requires an institutional structure that in turn, defines roles and responsibilities of the involved parties (i.e., water resources experts, planners, decision makers, users, and implementing agencies, etc.) and the communication between them. Governmental and semigovernmental agencies must play a substantial role in the implementation of plans together with representatives from the private sector. For example, the United Kingdom has relatively no international boundaries and relatively short rivers (Tuinhoff, 1992) that have permitted the creation of simple administrative units based on a group of adjacent catchments. And, in France, the planning of water resources lies in the six basin agencies, four of which are wholly within France and are based on their respective catchment units, with two basins being international, the first forming part of the Rhine system (*Agence Financiere de bassin Rhine-Meuse*) and the second representing a number of small rivers in the northwest flowing into the English Channel. Holland, as another example, is a small country occupying the delta region of a major international basin, and it is mainly dependent on an external water supply. For arid countries like those in the Middle East, there is little reason to create water management units based on hydrological or geographical boundaries. The few rivers (e.g., Nile, Euphrates, Jordan, and Tigris) are international rivers that originate from more humid areas and are not recharged in the downstream countries such as Egypt (Nile), Jordan (Jordan), and Syria/Iraq (Euphrates and Tigris). Sharing of the water in these areas requires international consultation, and within these countries, the main water management issues concern distribution of scarce water resources.

Central and Regional Control

An important consideration in the management of groundwater resources is the question of the balance between a central and a regional control area. This is because the extent to which the control of water should rest with the central government or regional bodies is governed by availability of water, water demands, political structure and political philosophy of the nation, and economic state of the nation. Water resources planning [URL Ref. No. 320] in arid regions is implemented at a central level because of the scarcity and uneven distribution of the available water resource, making it a national interest to allocate water to different sectors of the area. Although these decisions go beyond the authority of regional administrators, they are necessary, if water is to be conveyed between regions.

Surface water resources, such as rivers in arid regions, are almost never internal to a single region or state and do not originate and are not recharged in the country in question. Replenishment of groundwater in many areas is linked to rivers, as they are one of the main sources of groundwater recharge, either natural (i.e., from the rivers) or artificial (i.e., irrigated fields, canals, reservoirs, etc.). In the absence of rivers or surface waters, aquifers receive little direct recharge [URL Ref. No. 322], and effects of overpumping may propagate and be demonstrable over long distances (i.e., especially in confined aquifers with little recharge).

The role of regional groundwater planning bodies lies primarily in the implementation of plans developed through the planning process, and the extent of regional input should be part of the initial groundwater development plan. Input from the region should be limited to controlling and monitoring projects depending on the political structure and philosophy of the nation or areas in question (Eiontek, 1999).

Links to Other Planning Agencies

The scarcity of fresh water in arid and semiarid countries puts a heavy burden on different user groups and the Ministries at the regional and national levels. Coordination and cooperation between the Ministry responsible for water resources and other Ministries, like Public Health, Agriculture, Public Works, Housing, Industry, Energy, Transportation, and Tourism, are needed to draw up and implement an adequate water resources development plan in accordance with national policies and priorities.

Semi-Governmental Agencies and the Private Sector

Both private sector and semigovernmental agencies may play a role in the implementation of any groundwater resources development plan. However, their role should be limited to the design, construction, and operation of groundwater projects, since the control and monitoring of these projects should remain in the hands of the governmental agency responsible for the implementation of the groundwater development plan. Additionally, the planning agency is responsible for the preparation of the groundwater development plan and is generally a central-level body that is also responsible for the implementation.

Selection, Definition and Approval Projects

Project definition and selection is one of the first steps in the implementation of a groundwater contamination and prevention plan in which a planning agency and an implementation agency can both be involved. The selection of large projects that prevent groundwater contamination are generally controlled at a central level, not only because of the complexity but also because the financial means for such projects generally form part of the central government budget. Selection and definition of projects that prevent groundwater contamination would be based on information from groundwater development plans. The plans would provide the overall guidelines and framework for project selection that also include drinking water supply or industrial allotments as a first priority. The overall guidelines would then include maximum allowable extraction as a function of the area in question, water quality distribution and affects on water use, and environmental guidelines that are relevant and pertinent to water use in the area in question. Given these previously listed constraints and priorities, the decision makers would then select and define projects, although the request for such projects may originate from the private sector, the public sector, or from government.

Presentation of the Plan

The hydrogeological information [URL Ref. No. 13–14, 187] in the groundwater development plan to prevent contamination is not always easily understood by planners (Arnade, 1999) and decision makers, as they speak and understand a different technical language for the planning process in most instances. But, both benefit if information is presented to them in a simple, systematic, and credible way, including use of statistics, figures, maps, and other related graphic transfers. Maps may also be produced with the help of a Geographic Information System (GIS) [URL Ref. No. 163], and time-dependent changes, such as water level lowering and water quality, should be presented in hydrographs.

License or Permit System for Well Drilling

Effective control of groundwater development to prevent contamination cannot be achieved without a licensing or permitting system. A license or permit provides not only a legal basis for registration and administration of approved projects, but it also enables the opportunity to prescribe conditions for design and implementation that are needed to monitor and control groundwater development. These should include general guidelines with respect to well depth, maximum pumping rate, and well spacing; required site investigations, including the installation of observation wells; and requirements and specifications for implementation that involve evaluation of geological sampling, geophysical borehole logging, land leveling of wells, well testing, water sampling and analysis, obligation to drill a pumping well to a larger depth, and installation of flow meters.

Update of Groundwater Development Plans that Prevent Contamination [URL Ref. No. 321]

State-of-the-art continuous systematic monitoring and surveillance of water levels, water quality, groundwater abstraction [URL Ref. No. 323], and water law [URL Ref. No. 325] should continue in order to provide important information to verify whether or not groundwater development proceeds in agreement with the initial plan to prevent contamination. This information can then be used by the implementing agency in question to take immediate action if necessary (i.e., if the groundwater abstraction increases more rapidly than predicted and causes detrimental or negative effects that require a quick response). Predicted effects from monitoring networks may also be verified and adjusted with new information leading to an update of the allocated amounts of groundwater that can be abstracted to prevent groundwater contamination.

The reliability of groundwater contamination and prevention models will continue to be highly dependent on the availability of the necessary input, although some sophisticated models for regional groundwater evaluation require large amounts of data. Unfortunately, these data are not generally available, especially when a groundwater evaluation for a region is carried out for the first time. A recommended approach, therefore, is to design flexible groundwater models for any area in question in accordance with available input rather than to provide missing input with subjective information. This type of groundwater model can later be extended and refined as more reliable and consistent data become available.

A regular systematic update of a groundwater development plan related to prevention of groundwater contamination is not only needed to incorporate new acceptable hydrogeological information, but also to emphasize that the plan [URL Ref. No. 321] is one link in the dynamic process of water resources management that is subject to continuous change and adaptation. This means that inputs to the plan should be reviewed and incorporated from time to time as objectives and groundwater contamination conditions warrant.

Conclusions

The following are listed as areas that should be considered in designing or implementing any groundwater development or implementation plan that relates to prevention of groundwater contamination:

(1) Preparation of groundwater development plans (i.e., as part of the groundwater resources management system) in arid and semiarid regions with scarce water resources should be the responsibility of a central office with criteria and elements defined by the geographical area in question.

(2) Implementation of any groundwater contamination prevention plan should also focus on delegation of responsibilities to adequately qualified lower administrative levels to promote effective monitoring, surveillance, and control, and to stimulate the involvement of present and potential water users.

(3) Presentation of the information in the plan should be in the language of decision makers and water users and should be tailor-made to the geographical area.

(4) Evaluation of a system of permits and licensing should be used as an effective control tool to ensure that groundwater development proceeds according to the plan and that present or potential water users contribute supplementary data that systematically and periodically updates the plan.

(5) Adequate monitoring of water levels, water quality, and environmental impacts should be routine, and the central level and lower administrative levels should play a significant role in harmony with water users and their needs.

(6) Presentation of credible feedback of hydrogeological and environmental information from the field to the planning agency should be done to ensure that groundwater development plans related to prevention of contamination remain up to date and that health, safety, and environmental concerns always be kept at the forefront and adequately considered.

Technical Aspects of Groundwater Management [URL Ref. No. 145, 318] Related to Prevention of Groundwater Contamination

Water balance forms the basis for groundwater management (Van der Molen, 1992), and inputs and outputs should always be analyzed according to their nature and quantified as accurately as possible. Overexploitation of aquifers inevitably leads to a decline in water levels and often to an increase in salinity [URL Ref. No. 244] or changes in water quality parameters. Overcharging of aquifers may also cause

waterlogging in semiarid regions. Therefore, strong points to be considered in planning groundwater scenarios related to prevention of groundwater contamination include overpumping and overcharging concerns.

Signs of Overpumping

The hydraulic head in aquifers will show a continuing decline in areas that are overpumped. Another sign of overpumping is a gradual decline in water quality, caused by an increasing salt content (i.e., change is due to the upconing of more saline water [URL Ref. No. 244] from greater depths, attracted by the pumping).

Unconfined aquifers underlying large irrigated areas [URL Ref. No. 241] often receive more water than they lose and as a consequence, a gradual rise in groundwater levels occurs that finally results in waterlogging of irrigated fields. When the water table comes within reach of plant roots and capillary rise sets in, salinization of the soils then becomes a problem.

Additionally, to understand the mechanisms involved in overpumping and overcharging, it is necessary to review the concept of *water balance* [URL Ref. No. 323] as presented by Van der Molen (1992):

recharge = outflow + abstraction + increase in storage

or

$$R = O + A + DST$$

According to Van der Molen (1992), abstraction leads to a decrease in storage (negative DST) that can be established by measuring the fall in groundwater levels. Recharge [URL Ref. No. 322] and outflow are far more difficult to quantify, although both may be influenced by abstraction [URL Ref. No. 323]. Abstraction influence then obeys the same rules as the Van't Hoff Law in chemistry that describes any enforced change in a system inducing other changes, such that its effects will be diminished and a new equilibrium will be established. Thus, groundwater abstraction [URL Ref. No. 323] will tend to cause an increase in recharge and a decrease in outflow, and the aquifer system will attempt to restore equilibrium at a lower level.

Recharge of Aquifers [URL Ref. No. 322]

Recharge can be natural, but it can also be induced by human activities (i.e., artificial recharge). Recharge can be *direct,* by rainfall and snowmelt at the same site, or *indirect,* by surface runoff that has concentrated in a certain area where it infiltrates and eventually reaches the aquifer.

Direct natural recharge is the main process in humid and subhumid areas where there are seasons when precipitation exceeds evaporation. If the excess generates a surplus, this surplus may recharge underlying aquifers.

Indirect Recharge

Indirect recharge is the main process that occurs in semi-arid to arid climates. It may be caused by infiltration of river water originating from regions with a different climate, such as mountain areas. If the hydraulic head in the underlying aquifer is lower than the water level in the river, recharge will occur. Pumping such aquifers that have indirect recharge will increase infiltration and induce it in places where it was not present in the past.

Temporary rivers (e.g., arroyos) provide indirect recharge. This is because in their head waters, rocky areas occur that are impermeable, and runoff will commence after slight rains. When this runoff reaches the plains, it seeps into the sandy bed of these streams. In this way, recharge can take place around rocky outcrops, even in deserts. Indirect recharge can be enhanced by building small dams or constructing infiltration basins that promote artificial recharge.

Artificial Recharge

Artificial recharge is mostly of the indirect type (i.e., seepage from stored surface waters into aquifers). Inadvertent artificial recharge is mainly due to losses from irrigated areas, irrigated fields, leaking canals, and seepage from storage reservoirs. This inadvertent recharge will often lead to overcharging and, consequently, to waterlogging and soil salinization.

Outflow from Aquifers

Natural outflow of groundwater from an aquifer can take place as visible springs or seeps, but it can also be invisible in the form of evaporation from bare soil or as uptake by plants. In the latter case, it is often indicated by a different or more luxurious vegetation; in the former, it often leads to an accumulation of salts, at least in dry climates. Springs and seeps can be a spectacular phenomenon, especially as the climate becomes more arid. These areas have been used from time immemorial for watering cattle and for irrigation, and water rights [URL Ref. No. 325] are usually attached to these sources.

Additionally, salt flats in deserts are outlet areas for groundwater, and around such places, where the water table is still shallow but the salinity is less, natural vegetation is established and humans can take advantage of the water for irrigation or various purposes. Oases often abound around salt flats or salt lakes.

Probably some shrubs and trees are also able to reach and use this groundwater. Where the water table is shallower and where it is replenished by flow from elsewhere, swamp vegetation will develop. Such places, as previously listed, can be identified by remote sensing methods, especially by false color infrared photography (i.e., near infrared from sunlight is strongly reflected by actively growing vegetation and can be photographed with camera and film).

Groundwater Abstraction [URL Ref. No. 323]

Artificial outflow can be achieved by construction of wells, surface or subsurface drains, galleries, or qanats [URL Ref. No. 339 {16}]. The latter spontaneously collect phreatic water, whereas wells can also abstract water from greater depths and from confined aquifers. In this respect, there exists a considerable difference between confined and unconfined aquifers. Abstraction from a confined aquifer has little local influence apart from falling hydraulic heads and increased pumping costs (i.e., any consequences are felt far away, even in adjacent areas). On the other hand, pumping from unconfined aquifers may give rise to drought phenomena within a neighborhood (i.e., springs or qanats falling dry, desiccation of existing agricultural lands, and deterioration of wetlands) [URL Ref. No. 128].

A second difference between both types of aquifers is the fraction of the soil involved in effective storage, as expressed by the storage coefficient. It is of the order of 20 percent for unconfined aquifers, but only a few per thousand for the confined type. Still, if extensive enough (i.e., like the confined aquifers under the Sahara), large amounts can be mined from such formations. This is now the case in many countries in North Africa, where fossil water is being abstracted from very extensive confined aquifers (i.e., sometimes the depth from which this fossil water is taken is so great that the water has to be cooled before it is applied to crops).

Environmental Aspects of Groundwater Management Related to Contamination Concerns

Like surface water, groundwater has many implications if it is to be managed in an environmentally sound manner to prevent contamination. The three main considerations that need to be evaluated for environmentally sound groundwater management concerns include the following (Biswas, 1992):

(1) Groundwater development must be sustainable on a long-term basis. This means that the rate of abstraction [URL Ref. No. 323] should be equal to or less than the rate of recharge. If the rate of abstraction is higher than the rate of recharge [URL Ref. No. 322], it will result in

groundwater mining, which can be carefully considered for some specific cases. If mining occurs, groundwater levels will decline, which will steadily increase pumping costs, and then it would no longer be economical to pump it for its various uses.

(2) Human activities that impair the quality of groundwater for potential use should be controlled. This would include leaching of chemicals like nitrates [URL Ref. No. 201] and phosphates [URL Ref. No. 202] from extensive and intensive agricultural activities [URL Ref. No. 238, 240], contamination by toxic [URL Ref. No. 256, 317] and other undesirable chemicals from landfills [URL Ref. No. 220] and other environmentally unsound waste disposal practices (i.e., including bacterial [URL Ref. No. 111] and viral [URL Ref. No. 111] contamination due to inadequate sewage treatment and wastewater disposal practices), and increased salinity content due to inefficient or improper irrigation practices.

(3) Improper groundwater management often contributes to adverse environmental impacts. Among these are land subsidence in certain urban centers due to a high rate of groundwater abstraction, as in Bangkok, and sudden strict control of groundwater abstraction that allows the groundwater table to rise steadily over its recent long-term levels, which could contribute to structural damage as in London and Birmingham in the UK.

Also, according to Biswas (1992), an Environmental Impact Assessment (EIA) [URL Ref. No. 308] can be considered to be a planning tool that assists planners in anticipating potential future impacts of alternative groundwater development activities (i.e., both beneficial and adverse), with a view to selecting the optimal alternative that maximizes beneficial effects and mitigates adverse impacts on the environment. EIA procedures have been successfully used in several developed countries during the past two decades, but only within the past 10 years or so has the EIA process been introduced in several developing countries. While most EIAs carried out in the past, and being carried out at present, have dealt with the potential environmental implications of proposed development projects, there is an urgent need to monitor environmental changes once a project is operational. Such monitoring is necessary not only for those projects in developing countries for which EIAs were carried out during planning stages, but also for the vast majority of currently existing development projects that received very little environmental attention during their planning and construction phases.

Objectives

The objectives for applying an EIA to groundwater management could include the following:

(1) To identify adverse environmental problems that may be expected to occur

(2) To incorporate into the development action appropriate mitigation measures for the anticipated adverse problems

(3) To identify the environmental benefits and disbenefits of the project as well as its social and environmental acceptability to the community

(4) To identify critical environmental problems that require further studies and/or monitoring

(5) To examine and select the optimal alternative from the various relevant options available

(6) To involve the public in the decision-making process related to groundwater management

(7) To assist all parties involved in the specific development project in understanding their individual roles, responsibilities, and overall relationships with one another

General Principles

Human activities and natural phenomena can cause groundwater deterioration, but as a general rule, *human activities contribute to maximum damage through overexploitation and irrational use.* EIAs can be used successfully to identify beneficial as well as adverse consequences of human activities, and, thus, are of prime importance to all parties involved in development planning and implementation of groundwater projects. It is equally applicable to all new development actions as well as to the expansion or modification of currently existing actions. Furthermore, in most developing countries, few environmental considerations have been incorporated in past development actions. Consequently, many of the benefits that were originally anticipated by the planners are either not occurring or are being negated by unanticipated adverse side effects. Thus, there is a need to carry out environmental reviews of existing projects so that the major problems can be rectified or resolved.

Conclusions

Good environmental impact assessments [URL Ref. No. 308] have to be at the center of any sound groundwater management plan for prevention of contamination. However, because of the complexities and uncertainties [URL Ref. No. 261] that are invariably associated with groundwater regimes, it has generally not been possible to carry out proper environmental impact assessments of groundwater development projects in developing countries or in many situations (i.e., many such projects have shown to be neither sustainable nor environmentally acceptable on the long-term basis).

With improvements in expertise on groundwater management [URL Ref. No. 318], and with increases in interest in regular monitoring and surveillance of the quality of groundwater, more geographical areas should now be in a position to initiate and implement rational groundwater management plans [URL Ref. No. 319] related to prevention of groundwater contamination. Also, as environmental impact assessment becomes an integral part of the planning and management process, there is no doubt that use of EIAs can be considered to be a beneficial development in the production of adequate groundwater management plans.

Methods and Techniques to Improve Water Resources Management Concerns in Groundwater Pumping Areas [URL Ref. No. 145, 318] and Relationship to Contamination

De La Cruz and Pena (1994) describe the methodology used to obtain and analyze information concerning the operating conditions of wells, pumping equipment, and irrigation systems to develop a sustainable groundwater management program [URL Ref. No. 145, 318] in the main aquifers of Mexico. Low-efficiency pumping equipment, inefficient water use in irrigation systems, and the high cost of power combined with drawdown have produced rising production costs along with falling returns for farmers. The dimensions of this problem can best be shown in the following example, which includes the irrigation unit of Costa de Hermosillo, in northwestern Mexico. In this area, from 1968 to 1988, the drawdown registered was 19 m, approximately 0.95 m per year. In nearby zones, the fall in the water level was from 5 to 35 m. From 1989 to 1993, some wells were eliminated, and others were relocated. This action, together with the extraordinarily high rainfall, resulted in a reduction in the drawdown to 0.18 m per year. The problem of drawdown is generalized in Mexico and as such, methods that can assist in controlling groundwater abstraction [URL Ref. No. 323] must be found.

Irrigation districts that have experimented with aquifer drawdown can usually be characterized by pumped volumes in excess of natural aquifer recharge quantities, water being pumped from greater depths generating a greater demand for electricity, using pumping equipment that is often old and electromechanically inefficient, and poorly maintaining channels and low-efficiency application systems that combine to increase the volume of irrigation water abstracted.

Groundwater Quality Management of a Low Inertia Basin

A two-dimensional finite element model was applied to the San Mateo Basin, California, in order to investigate feasible and efficient management alternatives to enhance basin yield and reserve basin water quality (Bagtzoglou et al.,

1993). The model utilized lumped approximation methods for the determination of its subsurface boundary conditions and incorporated a variety of hydrological processes. The model solved uncoupled flow and transport equations [URL Ref. No. 302] by use of a nodal domain integration technique for the flow model and an integrated finite difference method for the transport model. Modeling results additionally indicated that sustained yield may be maximized by interception of ocean outflow from the basin with an improvement of about four times the historical sustained yield being achieved.

Although the study by Bagtzoglou et al. (1993) initially investigated the San Mateo Basin to determine the feasibility of using it as a groundwater storage element and to determine ways to enhance sustained yield of the basin without introducing adverse effects, it was determined that the primary adverse effect was subsurface saltwater intrusion [URL Ref. No. 255] due to increased pumping, with some environmentally sensitive freshwater plant species that provided habitats to severely threatened or endangered bird species [URL Ref. No. 339 {26}] being affected.

California's New Groundwater Management Law [URL Ref. No. 324–325]

Local agencies are implementing California's new groundwater legislation (Horseley, 1995a, 1995b). The law was drafted to let local governments manage their own groundwater concerns while coordinating their efforts with state and federal agencies. Three regional workshops on *A.B. 3030* were sponsored by the U.S. Environmental Protection Agency [URL Ref. No. 19] to help local officials develop these groundwater management plans [URL Ref. No. 319]. The law sets up a planning process for local water suppliers to develop a consensus among water users within a groundwater basin. It also allowed the enabling authority to implement a plan that involved regulatory and nonregulatory management strategies. Because of this law, local agencies were then able to form cooperative agreements with neighboring agencies and private water suppliers.

The groundwater management plans developed in this instance included the following elements:

(1) Control of saltwater intrusion [URL Ref. No. 255]
(2) Identification and management of wellhead protection areas [URL Ref. No. 106, 108] and recharge areas [URF Ref. No. 322]
(3) Regulation of the migration of contaminated groundwater
(4) Administration of a well abandonment and well destruction program
(5) Mitigation of overdraft conditions

(6) Replenishment of groundwater extracted by water producers

(7) Monitoring of groundwater levels and storage [URL Ref. No. 48]

(8) Facilitation of conjunctive use operations

(9) Identification of well construction policies

(10) Construction and operation by the local agency of groundwater contamination cleanup [URL Ref. No. 263, 276, 303], recharge [URL Ref. No. 322], storage, conservation [URL Ref. No. 110], water recycling, and extraction projects

(11) Development of relationships with state and federal regulatory agencies [URL Ref. No. 8, 19, 33, 45, 89]

(12) Review of land use plans and coordination with land use planning agencies to assess activities that create a reasonable risk [URL Ref. No. 235] of groundwater contamination

The previously listed groundwater planning elements can be divided into *quantity* and *quality* issues, although from a management perspective, it is difficult to distinguish surface and groundwater since, basically, it's all the same resource, water.

California water law [URL Ref. No. 324] is overwhelmingly complex, although it is not the only one (Blair and Wood, 1999; Swenson, 1999; Colten, 1999; Jones, 1999). Many of the groundwater basins have been adjudicated resulting in a myriad of unique regulatory requirements and overlapping water rights. While surface water law has been somewhat clearly addressed in their Water Code, groundwater law has generally been left up to lawyers and courtrooms. *A.B. 3030, the Water Code,* could provide a renewed opportunity to untangle some of this.

A Groundwater Management Model for Asian Developing Countries

Gupta and Onta (1994) provided an overview of groundwater management modeling, with particular reference to the situations prevalent in the developing countries of Asia. A state-of-the-art review of mathematical models [URL Ref. No. 339 {31}] developed primarily in the advanced countries is first made, and different approaches and techniques of systems analysis are then considered. Also, due to their increasing importance, water quality aspects are included and emphasized. Some recommendations are made on appropriate modeling strategies for groundwater management in Asian developing countries.

According to Gupta and Onta (1994), each country must develop its own modeling strategy [URL Ref. No. 339 {31}] that considers specific situations and constraints. Applied research considering the existing knowledge gaps and new developments and emphasizing real-world applications of management models involving important problems facing the developing countries can lead to greater understanding of the actual system as well as result in much improved planning and management decisions.

Additional Selected Groundwater Management Plans (Methods and Techniques) Related to the Management Planning Process
[URL Ref. No. 145, 318]

Protection of Public Water Supplies
[URL Ref. No. 85 {21}, 104–108, 118–121, 127, 337 {5}]

Protecting groundwater resources from contamination requires many different best management practices (BMP), including the continued development of plans at the local level to control activities that threaten the resource (Josephson, 1980; Wise, 1977; Rail, 1985a). Adequate protection of groundwater resources used for public water supply is dependent on an adequate understanding of the fundamentals of groundwater hydrology (U.S. EPA, 1985). However, the subsurface environment is a complex system subject to contamination from a host of sources. Furthermore, the slow movement of most contaminants, when they do occur through the groundwater environment, results in longer residence time and little diffusion. The restoration of groundwater quality after contamination problems occur, therefore, becomes difficult and can be expensive (Anonymous, 1981). Restoration costs usually exceed short-term value concerns of the groundwater resource, and the most viable approach to water quality protection must be one of prevention and not cure (Morrison, 1981).

And, since adequate prevention, planning, and emergency response (U.S. EPA, 1983) mechanisms are usually beyond the resources of water utilities, it therefore becomes the responsibility of the local community, if possible, to carry out planning and preventive programs in association with state, federal, or regional concerns that assess drinking water needs and protect present and future water supplies. Suitable aquifer protection controls must be adopted that take into account land use, industrial development, health, housing, and agriculture. Since no one approach successfully protects all aquifers, the entire community, including local, state, and federal government, is responsible for balancing risks [URL Ref. No. 235], costs, and benefits involved in protecting the groundwater supply [URL Ref. No. 89].

Community planning has been evident in some states [e.g., Arizona, California, Massachusetts and Cape Cod, Colorado, Connecticut, Florida and Dade County, Kansas, New York, Long Island (New York), New Jersey, and Wisconsin] (National Research Council, 1986b [URL Ref. No. 171]; Moorehouse, 1985). The National Research Council (NRC) (1986b) has focused on state and local groundwater protection programs, identifying prevention of groundwater contamination problems with respect to their scientific bases, performance over time, administrative requirements, and legal and economic frameworks. The resulting report by NRC (1986b) summarizes the committee's review of case studies and identifies those significant technical and institutional features that show progress and promise in providing protection of groundwater quality.

In essence, in the reports of the National Research Council (1986a, 1986b), the following information was presented: background information on groundwater protection strategies, groundwater quality standards and contamination sources, summaries of the state and local groundwater programs reviewed, and state and local strategies to protect groundwater.

Groundwater Contamination Control

Various methodologies for groundwater quality protection or treatment additionally depend on whether contamination problems are *acute* or *chronic* (Canter and Knox, 1986). Acute contamination problems may occur, for example, from inadvertent spills of chemicals or releases of undesirable materials and chemicals during a transportation accident. Acute contamination events are unplanned and are characterized by their emergency nature. Chronic aquifer contamination may occur from numerous point and area sources and may involve traditional contaminants such as nitrates [URL Ref. No. 201] and bacteria [URL Ref. No 111], or unique contaminants such as petroleum fuels (benzene) [URL Ref. No. 247, 316], metals [URL Ref. No. 317], organic chemicals [URL Ref. No. 210], and other contamination substances.

It should be recognized that a given aquifer cleanup project may involve usage of several methodologies in combination, such as excavation, backfilling, transportation of wastes to hazardous waste disposal sites, contaminant removal wells, treatment of contaminated water, discharge to municipal or surface drainage, surface capping, subsurface barrier installation, and *in situ* chemical treatment. Cantor and Knox (1986) discuss many basics of groundwater control contamination in their pollution control text (i.e., physical control measures, treatment of groundwater, *in situ* technologies, aquifer restoration, decision making, risk assessment, public participation, case studies, and other subjects).

Establishment of Groundwater Protection Programs at the State and Federal Levels
[URL Ref. No. 108, 118–121, 127, 141–142]

State and federal programs that protect groundwater quality, including detecting, correcting, and preventing contamination, seem to be on the right track (Bacon and Oleckno,

1985; Horseley, 1995b). Consequently, in some states, major sources of groundwater contamination have been identified, inventories conducted, incidents documented, and advances made in understanding the groundwater hydrogeology of their respective areas.

At the federal level, many regulatory statutes [URL Ref. No. 68–69] now authorize programs relevant to groundwater protection, and more than two dozen agencies and offices are involved with groundwater contamination related activities. Most states, however, are now concerned about ongoing contamination and have programs, at varying stages of development, to protect their groundwater or *stabilize* known contaminated areas. However, despite expanding local, federal, and state efforts, programs are still limited in their ability to protect many areas against groundwater contamination. This is because there is no one explicit national legislative mandate to protect groundwater quality, and although the groundwater protection strategy of the U.S. EPA [URL Ref. No. 19] acknowledges the need for comprehensive resource management at the federal level, and continues to move in that direction, details have not been provided. Significant efforts, however, are being made with the amendments to the Safe Drinking Water Act (SDWA) [URL Ref. No. 258] and the Clean Water Act (CWA) [URL Ref. No. 287] that have included Wellhead Protection grants [URL Ref. No. 106, 108]. Authorized programs seem to be in early stages, although some are still a long time from being fully implemented. As a result, many states, in addition to federal legislation, have their own standards for drinking water and groundwater quality (U.S. EPA, 1976a, 1976b).

Comprehensive State Groundwater Protection Programs

At least one state in each of the 10 EPA regions has embarked upon the development of a Comprehensive State Groundwater Protection Program (CSGWPP) (Horseley, 1995b). Two northeastern states (New Jersey and Connecticut) have completed documents that provide indications of where this program may be heading.

Like pilot states in other EPA regions, Connecticut (EPA Region 1) and New Jersey (EPA Region 2) were selected as candidates for the first round largely based upon their past accomplishments in groundwater management and their interest and willingness to participate in the GSGWPP process. Both states have EPA approved Wellhead Protection Programs [URL Ref. No. 106, 108] and have been pioneers in many other aspects of groundwater management.

Connecticut developed an aquifer/groundwater classification program in 1980, and this program has been used as a model for other states that have developed classification programs. The classification program divides groundwater into four classes, including the following:

(1) Public water supplies

(2) Private water supplies/potential public water supplies

(3) Commercial/industrial water supplies

(4) Groundwater impacted by landfills [URL, Ref. No. 220]

This program serves as a core for Connecticut's other groundwater programs. Groundwater withdrawal permits are not issued where water quality standards would be violated.

New Jersey has also developed a groundwater classification system. It divides groundwater into three classes:

(1) Ecologically significant groundwater

(2) Potable water supplies

(3) Groundwater other than potable

This classification system integrates groundwater and surface water quality by recognizing the value of high-quality groundwater discharge to down-gradient surface waters and wetlands.

Another key component reflected in Connecticut's and New Jersey's GSGWPPs is their vision regarding a fully integrating GSGWPP. In both cases, two principles emerge: data management and an increasing emphasis on pollution prevention. Both states are aiming for streamlining the remediation process and investigating-organizing more resources into pollution prevention/management activities. Their wellhead protection programs have ambitious goals of delineating all of the wellhead protection areas throughout the state, conducting inventories of potential contamination sources, and developing and implementing wellhead protection strategies (i.e., such as land use regulations and non-regulatory measures).

Groundwater Contaminants and Their Sources (A Review of State Reports)

Pursuant to section 305(B) of the Clean Water Act of 1987 [URL Ref. No. 287], each state is required to submit biennially a water quality report to the U.S. Environmental Protection Agency [URL Ref. No. 19]. States were also requested to identify the contaminants and their associated sources in public groundwater supply systems and to delineate the frequency of occurrence. Forty-two water quality reports were received and reviewed by the U.S. EPA; not included were the reports from Kentucky, Louisiana, Maryland, Massachusetts, Nevada, New Jersey, Utah, and West Virginia. Information on sources of contaminants and the detection frequency of specific physical, chemical, and/or bacteriological/virological [URL Ref. No. 111] contaminants can be used in planning and implementing groundwater quality protection and remediation programs [URL Ref. No. 130, 132, 145, 155]. Canter and Maness (1995) include sections on sources of contamination and contaminants found in pub-

lic groundwater supplies and sources of contamination and contaminants found in groundwater (i.e., in general). The information on groundwater can be related in part to public supplies, rural, and individual supplies, however, it was not possible to distinguish between these uses and information contained and provided in 305(B) reports.

Sources of Contamination in Groundwater (General)

The 305(B) reports general groundwater quality information compiled from 37 states and showed a total of 74 sources of groundwater pollution or contamination. These groundwater contamination sources included the following:

(1) *Septic tanks*

(2) *Municipal landfills*

(3) *On-site industrial landfills*

(4) *Other landfills*

(5) *Surface impoundments (excluding oil and gas brine pits)*

(6) *Oil and gas brine pits*

(7) *Oil exploration and production activities*

(8) *Underground storage tanks (UST)*

(9) *Underground storage tanks (gas stations)*

(10) *Chemical plants*

(11) *Aboveground storage tanks*

(12) *Injection wells—Classes I–IV*

(13) *Injection wells, including UIC Class V*

(14) *Abandoned hazardous waste sites*

(15) *Regulated hazardous waste sites*

(16) *Storage, handling, and transportation of hazardous substances*

(17) *Saltwater intrusion*

(18) *Land application treatment*

(19) *Agricultural activities*

(20) *Agricultural—abandoned wells, drainage wells, sinkholes,*

(21) *Road salting*

(22) *Road salt storage*

(23) *Chemical leaks and spills at industrial and commercial facilities*

(24) *Agricultural chemicals—pesticides and fertilizers*

(25) *Urban use of fertilizers and pesticides*

(26) *Mining activities*

(27) *Chronic spills and leaks*

(28) *Injection wells*

(29) *Pipeline leaks*

(30) *Naturally occurring radioactive elements*

(31) *Naturally occurring metals*

(32) *Unpermitted disposal*

(33) *Water supply wells*

(34) *Animal feedlots*

(35) *Urban and industrial site runoff*

(36) *Monitoring wells*

(37) *Abandoned wells*

(38) *Septic, sewage, and wastewater treatment sludge*

(39) *Animal waste disposal*

(40) *Industrial and municipal wastewater*

(41) *Unknown sources* (U.S. Department of the Interior, 1971)

Out of the 35 states that provided information on groundwater contaminants without specific reference to public groundwater supplies, a total of 146 organic chemicals were identified. The 10 most frequently cited organic constituents out of the 146 listed included the following:

(1) *Pesticides* (29 states)

(2) *Petroleum products* (25 states)

(3) *VOCs* (22 states)

(4) *SOCs* (15 states)

(5) *Gasoline* (10 states)

(6) *Benzene* (nine states)

(7) *Toluene* (nine states)

(8) *Trichloroethylene* (nine states)

(9) *Agricultural chemicals* (eight states)

(10) *1,1,1-trichoroethane* (eight states)

Of the 30 inorganic chemicals identified in groundwater, the nine most common were the following:

(1) *Nitrates* (29 states)

(2) *Metals* (25 states)

(3) *Brine/salinity* (17 states)

(4) *Arsenic* (14 states)

(5) *Cyanide* (nine states)

(6) *Fluorides* (nine states)

(7) *Iron* (eight states)

(8) *Manganese* (eight states)

(9) *Sulfates* (eight states)

Other constituents found in the groundwater include the following:

(1) *Radioactive material* (10 states)

(2) *Bacteria* (nine states)

Discussion/Limitations/Conclusions

Several observations are appropriate in conjunction with this composite analysis of state 305(B) reports (Canter and

Maness, 1995). One, is that these reports, while required as part of the Clean Water Act of 1987 [URL Ref. No. 287], are not in any common format, thus, there is great diversity in the attention given to groundwater in general. Also, just because a given state has not been identified herein as having groundwater contamination problems or concerns, it does not necessarily mean that the state does not exhibit such problems. A diversity of terminology exists, and terms such as *landfills, sanitary landfills, industrial landfills,* and *industrial waste disposal sites* were often utilized, and the presumption is that these terms could well represent one generic category of groundwater contamination sources.

The composite analysis of information from 42 state 305(B) reports has indicated widespread groundwater contamination problems of a diverse nature. Numerous public groundwater supplies are being affected by groundwater contamination, and this is of great concern from a national perspective in that groundwater contamination problems are often difficult and costly to remediate (Vandermeulen and Hrudey, 1987). However, this information could be utilized to identify key nationwide sources of contamination and specific contaminants and to develop programs to prevent groundwater pollution from such sources/contaminants where possible.

Professionals involved in the planning and implementation of state groundwater quality management programs could benefit by review of the information from 305(B) reports and comparisons between individual states. This information can be utilized to identify key contamination sources and contaminants and then to develop positive programs that protect groundwater resources to preclude or stabilize contamination.

Programs for Corrective Action of Groundwater Contamination Problems

Few comprehensive corrective action programs to solve groundwater contamination problems in the United States have been undertaken relative to the number of sites identified as requiring action. Although federally funded corrective actions authorized by the Comprehensive Environmental Response, Compensation, and Liability Act (CERCLA) [URL Ref. No. 264] address sources and contaminants, the actions have been limited to hazardous waste disposal sites. Corrective actions, in many instances, have not involved cleanup of contaminated groundwater.

Additionally, state corrective action programs, in many instances, are still at early stages of development, although presently, a great number of programs relate to accidental spill situations and detection of leaks from underground storage tanks (USTs) [URL Ref. No. 251]. Other state programs seem to be designed either to retain (e.g., in landfills) [URL Ref. No. 220] or discharge (e.g., via municipal sewerage [URL Ref. No. 222], injection wells [URL Ref. No. 123], etc.) contaminants into surface or subsurface areas. In my opinion, however, in the majority of cases, many actions still result from politically motivated complaints, rather than systematic efforts to identify, evaluate, and control known groundwater contaminated sites.

Consequently, few corrective action programs are addressed in federal, state, and local programs to prevent groundwater contamination when one considers the full magnitude of the problem. State and federal programs still focus on sources associated with toxic materials [URL Ref. No. 207, 256, 304], and implementation and enforcement of many corrective programs are still in infant stages.

Some approaches used to prevent groundwater contamination by some states have included provisions for evaluating design, operational aspects, siting, uses, and closing of point-source contamination areas. In most instances, remedies in this category have been mandatory. Additional approaches to prevent groundwater contamination include implementation of alternatives to a specific groundwater contaminating activity (i.e., make process or product changes that reduce waste volume, initiate waste recycling and recovery, and pretreatment on the site).

Although focusing on a specific source is an approach to groundwater contamination, unfortunately, other types of approaches have not been extensively applied to groundwater contamination problems. For example, few efforts have been initiated to control cause and effect activities located in recharge areas. Apparently to some, if approaches are not source-specific, they seem to carry no merit, even though the federal government provides support for protection of selected recharge areas through the Sole Source Aquifer Program and Wellhead Protection provisions (i.e., Safe Drinking Water Act and amendments) [URL Ref. No. 258].

Another strategy that should be implemented to prevent groundwater contamination involves placing restrictions on the manufacturing, generation, distribution, and use of specific contaminating substances. For example, pesticides [URL Ref. No. 239] may be introduced from a nonpoint source such as a specific land application, from a storage tank, from landfills, and from residential disposal, although the Toxic Substances Control Act (TSCA) [URL Ref. No. 295] and the Federal Insecticide, Fungicide, and Rodenticide Act (FIFRA) [URL Ref. No. 296] authorize regulation of potential groundwater contaminants. In this instance, application of programs addressed to correct or prevent groundwater contamination are limited.

Focus of Programs for Prevention of Groundwater Contamination
[URL Ref. No. 105, 108, 119–121, 127, 141–142]

The effectiveness of federal, state, and local programs to protect groundwater from being contaminated has not only

had a limited focus, but state-of-the-art technical factors to solve problems have not always been applied. For example, adequate *hydrogeologic investigations* [URL Ref. No. 187] are needed to detect existing problems, evaluate the performance of corrective actions, monitor the effectiveness of preventive activities, and assure that adequate quality control/assurance concerns are in place. The methods and technologies for obtaining adequate hydrogeologic information are available, although a certain amount of *uncertainty* [URL Ref. No. 261] about groundwater contamination investigations will always exist because of difficulties in dealing with indirect observations. The reliability of hydrogeologic investigations also depends on the degree of skilled personnel that must tailor their study to the site-specific nature of a specific groundwater contamination analysis(es) problem.

It must be emphasized that water quality data is also difficult to analyze and interpret, especially if trace levels or mixtures of contaminants are present or if changes in chemical and biological observations occur frequently.

In summary, the major constraints that limit corrective action on solving groundwater contamination problems include uncertainty about some techniques that can be used to improve groundwater quality, the high costs of implementation, the need to design measures appropriate for site-specific areas, and the lack of adequate baseline data. An exact definition and evaluation of the nature of contaminants for a specific site are other constraints.

Prevention efforts generally are slowed because unresolved questions about the technical adequacy of certain available methods and an incomplete understanding about the complexities of groundwater contamination exist. In some instances (i.e., or in all instances), political indecisiveness or derivitives thereof, can also be blamed for the problem.

U.S. National Policy Implications

U.S. national policy options presently relate to development and implementation of federal, state, and local programs to protect groundwater. The federal framework has the potential to protect the nation's groundwater from additional contamination, but its realization is dependent on broadening the coverage of authorized programs. Implementation at all levels requires that activities be coordinated among and within federal and state agencies. Political judgments concerning the role of the federal government and the importance of the states making positive progress in their abilities to detect, correct, and/or prevent groundwater contamination are critical. A positive role with local concerns is essential.

The development of national policies related to the protection of groundwater from contamination must include recognition of site-specific problems for given areas. National policy must be flexible and able to accommodate various groundwater contamination problems encountered by varying site conditions. Federal and state government should then be instrumental in providing funding, technical assistance, demonstration projects, and research and development and in maintaining the integrity of the work, with local government being part of this process. Current federal laws [URL Ref. No. 68–69] and programs [URL Ref. No. 8, 19, 45] have assisted states in abating or recognizing their groundwater contamination problems, however, the level of federal support to the states is still not adequate.

Groundwater Technical Assistance to States and Local Governments

Technical assistance to the states by the federal government [URL Ref. No. 8, 19, 45] must involve training programs, guidelines, document distribution, and efficient state-of-the-art information exchange. This is because at the state or local level, qualified personnel generally are limited, although in some areas, they surpass the state or federal level efforts. Federal, state, and local integrated funding for training and education is important and necessary to achieve an increase in the technical capabilities of the nation to deal with groundwater contamination problems. The U.S. Geological Survey [URL Ref. No. 8] Cooperative Program to state and municipalities and other related technical assistance programs along with U.S. EPA [URL Ref. No. 19] efforts must always be part of the solution. Certification programs by the federal government, states, universities, municipalities, private professional contractors, or professional societies must also be part of a professional curriculum and continue to ensure that personnel in the field possess minimum technical and, in some instances, certified or registered qualifications.

From a national perspective, the goal of these guidelines to protect groundwater should then be to ensure that a minimum set of considerations are used to enhance groundwater quality. These guidelines or considerations would also be a way of providing information required by states or municipalities in evaluating their groundwater contamination problems. General guidelines could be developed by federal/state government for assisting and working with local authorities in setting priorities for allocating their resources, specifically, the federal government could provide assistance to states and local authorities in the following areas (Office of Technology Assessment, 1984) [URL Ref. No. 178]:

(1) *With respect to detection:*
- guidelines to assist in conducting reliable hydrogeologic investigations [URL Ref. No. 187], including monitoring of the flow system, sampling and analysis, and data interpretation
- guidelines for addressing contaminants for which there are no federal standards

- guidelines in setting priorities and determining which contaminant sources to monitor and inventory

(2) *With respect to correction::*
- guidelines in selecting and implementing corrective action when necessary
- guidelines for cleanup standards determined on a site-specific basis

(3) *With respect to prevention:*
- guidelines for preventing contamination from contaminating sources, presentation of alternatives for reducing the wastes generated by a source, and provision of information on waste recycling (i.e., as part of preventing contamination from sources)
- guidelines for considering protection of aquifer recharge areas, wellhead protection, and proper land use, and for establishing a balance that enhances water quality

There are also several other ways that the federal government in association with local authorities can facilitate information exchange among the states and municipalities. It could provide information about the different state approaches to protection of water areas and would assist them or be facilitators in implementing programs and learning from their successes and failures (i.e., lessons learned and root cause analysis).

Research and Development Needs [URL Ref. No. 4, 14, 19, 25, 47–48, 52, 72, 83, 97, 99–100, 103 {14}, 130–131, 145 {1}, 154–155, 171, 188, 220, 237–238, 244, 335 {4–5, 21}, 336 {7, 9, 14–15, 57, 60, 73, 75, 83, 96}, 337 {12, 111}, 338 {4, 11, 16, 19, 24, 32}, 339 {1, 30}]]

Many research and development activities provide pertinent information that would support states and local authorities in their efforts to prevent groundwater contamination. The key elements of research activities include the following:

(1) *Detection:*
- research on toxicology and adverse health effects of contaminants that occur in groundwater, with emphasis on synergistic effects of mixtures
- research on development of water quality standards for substances known to occur in groundwater that are not now covered and could be applied to drinking water supply and groundwater quality programs
- research on environmental and economic impacts of contamination
- research on development of reliable techniques for conducting hydrogeologic investigations

(2) *Correction:*
- research on the behavior of contaminants in groundwater, specifically, chemical and biological transformations of organic chemicals

- research on development of techniques for treating water contaminated from multiple sources

(3) *Prevention*:
- research on mechanisms for preventing contamination, including ways to reduce generation of items and disposal volumes

Establishment of Groundwater Protection Programs at the Local Level

As has been the premise in this book, good quality water is essential for health, aesthetic, and economic reasons, and comprehensive groundwater monitoring databases (i.e., including quality and quantity parameters within an aquifer) are therefore important. Databases [URL Ref. No. 128, 130, 150, 335–336, 339 {3}] are valuable, not only for preventing direct contamination of underground aquifers, but also in applying corrective measures when problems have been identified. The methods for establishing systematic groundwater monitoring and surveillance network plans and programs for any underground water sources can differ from one place to another. However, some ideas and recommendations are common to each other, and ways to evaluate the many complexities that arise when trying to prevent, control, or *stabilize* groundwater contamination are now presented.

Protocol for Protection, Control, or Stabilization Programs to Prevent Groundwater Contamination

Establishment of a groundwater contamination evaluation program (i.e., versus a groundwater protection plan) within an aquifer is essential to determine if significant groundwater contamination has, or has not occurred, or will occur within a given area (Rail, 1985a, 1985b, 1986, 1989, 1992). Completely thought-out plans and programs for an aquifer or a series of adjacent aquifers are necessary and should, as a minimum, include baseline information involving water quality (biological, chemical, radiological, etc.) and water quantity (water flow, water level, recharge rates, legal, etc.) [URL Ref. No. 1–339]. Monitoring, surveillance, protection, control, or stabilization programs can then be used to protect and prolong the useful life of domestic groundwater areas or solve problems if (as) they are detected.

To establish a proper and essential monitoring and protection program, the following information needs to be addressed on a *long-term basis*. These initial plans and programs should include information pertinent to water quality and quantity concerns of any given area and should encompass, as a minimum, the following basic steps (Rail, 1985a, 1989):

(1) *Select and assemble participants*
(2) *Establish purpose and a development plan*
(3) *Technically evaluate the aquifer(s)*
- Identify contamination potential
- Use applicable available technologies
- Determine present and future demands on the aquifer(s)
(4) *Select computer models for data summarization*
(5) *Evaluate water law and the responsibilities of municipalities and government*
(6) *Prepare and implement the plans and programs and modify as necessary*

These steps are not listed in chronological order and can be addressed simultaneously depending on the geographical area being evaluated and its water concerns, etc.

Recommended Steps

Step 1: Select and Assemble Participants

Implementation of a groundwater contamination protection plan should depend on a high degree of involvement of those who will benefit from the resource. This is also the idea proposed in local ordinances concerning groundwater in southeastern Minnesota (Gass, 1985; Hale et al., 1965). It is the beneficiaries (i.e., the local owners of the resource) whose future depends on continued availability of good quality groundwater for its intended use, whether it be used for agriculture, domestic, municipal, or industrial needs.

Step 1 can possibly be accomplished by the formation of a regional Aquifer Water Steering Group made up of members from the local or water basin area in question (e.g., farmers, developers, old folks who knew what it used to be like, local, state, and federal agencies, etc.). The main function of the regional group or representatives from each area would be to provide input on developing groundwater protection, control, and stabilization activities for the area. The steering group membership should also include water quality and quantity professionals. In New Mexico, for example, active members on the steering group could come from adjacent Indian reservations [URL Ref. No. 339 {41}], the U.S. Geological Survey [URL Ref. No. 8], the state Environmental Improvement Agency [URL Ref. No. 333], the state Scientific Laboratory System [URL Ref. No. 339 {42}], the state Engineer's Office [URL Ref. No. 339 {42}], the state Soil and Water Conservation Division [URL Ref. No. 339 {42}], the Council of Governments [URL Ref. No. 339 {42}], the New Mexico Conservation Division [URL Ref. No. 339 {42}], the New Mexico Game and Fish Department [URL Ref. No. 339 {42}], the New Mexico Bureau of Mines [URL Ref. No. 339 {40}], the Interstate Stream Commission

[URL Ref. No. 339 {42}], the U.S. Forest Service [URL Ref. No. 339 {20}], the Corps of Engineers, the city of Albuquerque Public Work Department [URL Ref. No. 327], Water Supply and Liquid Waste Divisions, Department of Energy [URL Ref. No. 312], local National Laboratories [URL Ref. No. 73, 97], the county Environmental Health Department [URL Ref. No. 89], state universities [URL Ref. No. 83], military bases [URL Ref. No. 326], and other related entities concerned with water. Membership in the group could also include officials or nonofficials from adjacent cities, towns, and communities and counties that have interests in and share the same aquifer(s).

Step 2: Establish Purpose and a Development Plan

A professional staff should be retained to perform the investigations, if members of the steering group do not provide this function. The professional staff that is retained could consist of outside consultants who would have data available and accessible to them by the various entities that are participating in the steering group. However, whether a professional staff is contracted, hired directly, or consists of active members from the steering group (whose agencies and membership might be willing to allow in-kind services), eventually, they must also be supported by some means. Financial assistance for professional staff might have to be sought from state legislatures [URL Ref. No. 339 {42}], city or county governments [URL Ref. No. 89, 327], and/or federal grant funding sources. The steering group could also investigate other possible funding sources and make recommendations for implementation.

Predictable and actual sources of revenue or in-kind time-sharing allocations from the different members and agencies of the steering group that can be used to set up and maintain the eventual monitoring network must be identified early in its formulation. In-kind time-sharing and the responsibility for developing certain phases of a monitoring network and protection strategy should not be made by members or agencies of the group, unless it is felt they can actively participate in the complete development and implementation process of the proposed groundwater monitoring and protection plan and program.

Step 3: Technically Evaluate the Aquifer(s)

One of the first objectives of any proposed groundwater protection/control/stabilization effort should be to identify the boundaries of the aquifer(s) in question. A critical hydrologic analysis is essential if internal factors that cause significant changes, not necessarily detrimental, in the aquifer(s) are to be identified and described. Compilation of historical water quality data should be assembled from

available sources (Rail, 1986, 1989). In New Mexico, for example, water-related information can be summarized from a review of federal, state, and municipal records. Other states could see or obtain information from some of the federal agencies, such as the U.S. Geological Survey [URL Ref. No. 8] or the U.S. EPA [URL Ref. No. 19] representatives in their area. The agencies and municipalities that would provide water-related information would differ in each state but would be similar to the ones in New Mexico.

After groundwater data, as gathered in the preceding step for the aquifer(s), have been compiled and summarized, gaps or absence of information must then be identified. If many gaps exist in the data from specific geographical areas or geological formations (i.e., including the quality and yield from various well depths), then an additional plan for procuring needed information is necessary. The possible selection and construction of new well monitoring sites and the proposed location and identification of present wells that could be incorporated in any monitoring or surveillance network must be included. However, monitoring wells should not automatically be drilled, unless they are needed or will provide important relevant data.

After existing wells from the aquifer have been selected for monitoring purposes and new wells have been drilled, then water quality or quantity information should be gathered systematically (e.g., monthly, bimonthly, seasonal, yearly, etc.). The time intervals chosen should vary with the tests being conducted, such as qualitative, Safe Drinking Water Act [URL Ref. No. 258] parameters, Priority Pollutants (U.S. EPA, 1979), quantitative water levels, water movement, etc. The schedule should include intervals that are sensitive enough to detect significant and statistical changes within the water measurements taken (Nie, 1983 [URL Ref. No. 339 {31}]; Steel and Torrie, 1960 [URL Ref. No. 339 {31}]; SAS Institute, Inc., 1979 [URL Ref. No. 339 {31}])

Water supply wells used for monitoring purposes, especially those belonging to municipalities, should (i.e., if costs are not too high) constructionally modify their systems, if possible, to provide accurate and reliable water samples and evaluations. Municipal water supply wells that are in the process of being constructed or are yet to be drilled should be completed in such a way that they meet the necessary specifications for their use in an ongoing systematic groundwater monitoring and surveillance program.

Identify Contamination Potential

When naturally occurring sources of water quality degradation (e.g., high nitrates [URL Ref. No. 201] or chlorides [URL Ref. No. 198]) within a aquifer(s) have been located and verified, past, present, and projected human activities should be described, especially, activities that could continue to be detrimental to water supply areas. State, regional, or local comprehensive plans should also be reviewed. Historical maps and aerial photographs should be examined, if available for the area. Potential pollution or contamination threats to groundwater quality, such as leakage from underground fuel tanks [URL Ref. No. 251], location of abandoned wells [URL Ref. No. 253] from domestic systems, abandoned wells from oil and natural gas field exploration activities [URL Ref. No. 237], large pits [URL Ref. No. 231] with high infiltration potential, injection wells [URL Ref. No. 123, 133], ponds, lagoons [URL Ref. No. 231], landfills [URL Ref. No. 220], feedlots [URL Ref. No. 238, 240], septic tanks and leachfields [URL Ref. No. 217–218], disposal of septage [URL Ref. No. 218], industrial wastes [URL Ref. No. 227], and other potential sources in the aquifer should be located, mapped, and verified. Stormwater runoff drainage [URL Ref. No. 249] and ponding areas resulting from urban and industrial sources should be mapped and evaluated along with other potential recharge areas to the aquifer(s).

After a complete listing of potential and actual contamination sources to the aquifer(s) has been compiled, reviewed, and summarized, a Ranking System (RS) must be developed. This ranking system can then be used to provide information concerning where the greatest potential for and what type of contamination occurs in the aquifer(s).

Future development and expansion of municipalities into critical recharge zones can be planned in such a way that the impact can be minimized through proper zoning and enforcement of requirements. Establishment of watershed [URL Ref. No. 128] protection areas, incorporation of open space, preservation of natural habitat, and/or use of water detention facilities (i.e., if acceptable to Water Law [URL Ref. No. 325] and Water Rights practices in the state) are some examples of how some potential water problems (or problem areas) can be addressed.

Use Applicable Available Technologies

There are several technical tools that can assist in groundwater evaluations. Some of these include the following:

- well probes that can be used for determining accurate depths to the water table
- well drilling tools (and techniques) that allow collection of water samples from different zones within a particular aquifer while an initial test hole is under construction
- remote-well sensing equipment that is designed to give continuous readouts, recordings of static and pumping water levels, and other information from wells

Determine Present and Future Demands on the Aquifer(s)

It is critical that any groundwater protection/control/stabilization plan should ensure that the present use of the water does not endanger its use in the future. If water becomes scarce in the future, the costs of importing water into certain areas can be very prohibitive. It is, therefore, necessary to determine how much water is being withdrawn, how much is being recharged, and the effects that current and projected pumping activities have on the hydrologic balance of the aquifer(s).

Records on past water quantity use, present use, projected population and industrial growth, and estimates of per capita water use can help to predict present and projected demands on the aquifer(s). This basic information can then provide knowledge concerning how much water can be withdrawn safely from the aquifer without upsetting its hydrologic balance (e.g., Safe Yield Rate). If more water is being withdrawn than is being replenished and a constant drop in the water table is occurring, then the aquifer is being mined.

Step 4: Select Computer Models for Data Summarization

A computerized database information system should be included in any monitoring and protection plan, along with computers to facilitate analyses and summarize any data generated from the previously mentioned steps (1–3). The database system should also, as a minimum, be able to contain, analyze, and compare information concerning the following:

- hydrologic parameters of the aquifer
- delineation of recharge areas and zones
- historical, current, and projected water levels and water use
- known water quality parameters (biological, chemical, and radiological, etc.)

After groundwater monitoring related data [URL Ref. No. 48–49, 109, 112, 128, 252, 337 {32–33}, 338 {19, 34}] is entered in a computerized data management system, the system should then be used to store incoming information.

Use of Computer Models [URL Ref. No. 12–13, 15, 48, 50, 85 {18}, 141, 252, 254–255, 262, 336 {62, 69–70, 72, 74, 82, 89–90}, 337 {122}, 338 {2}]

Computer models can simulate characteristics of an aquifer and are able to predict how certain human activities will impact it, such as the drilling of an excessive number of high volume water wells in a vulnerable location and what a monitoring and protection program tries to prevent (Tung and Koltermann, 1985; Shirley, 1982; Olsthoorn, 1985; Monogham and Larson, 1985; Icenhour et al., 1995). Some groundwater computer models are also able to translate mathematical results [URL Ref. No. 339 {31}] and interpret them back to physical conditions existing within an underground system. Any model, however, is meant to provide a predictive capacity that can be used to project water quality and quantity demands within an area. In this regard, many models are available, including some that can show the needed proper spacing between wells and provide information concerning where wells should be constructed (Monogham and Larson, 1985; Tung and Koltermann, 1985; Icenhour et al., 1995) [URL Ref. No. 19]. Nevertheless, computer systems and software [URL Ref. No. 12–13, 15, 48, 50, 85 {18}, 141, 252, 254–255, 262, 336 {62, 69–70, 72, 74, 82, 89–90}, 337 {122}, 338 {2}] are now readily available.

Programs such as SAS (Statistical Analyses System) [URL Ref. No. 339 {31}] and other databases [URL Ref. No. 128, 130, 150, 335–336, 339 {30}] and statistical packages can also be modified to store groundwater monitoring data and summarize and conduct in-depth statistical analyses of the input parameters.

ing and protection programs, including coordinating and being responsible for the entire system. State and federal efforts with groundwater protection/control/stabilization, in this instance, could still continue in the consultant capacity. However, in certain areas, if local beneficiaries are unwilling or unable to implement their own groundwater programs, then state and federal governments should.

The federal U.S. EPA strategy [URL Ref. No. 19] should always be to strengthen state groundwater programs, including encouraging states to make use of existing grant programs to develop groundwater protection programs and strategies (U.S. EPA, 1984a, 1984b, 1984c, 1984d, 1986a, 1986b, 1986c, 1987a, 1987b, 1989a, 1989b, 1996a, 1996b, 1996c). The agency must continue to encourage states to prepare or enhance their groundwater program development plans. Some states, such as New Mexico, accomplished this prior to the new U.S. EPA strategy and have already prepared a program plan for the statewide monitoring and surveillance of groundwater quality. The water quality monitoring plan for New Mexico includes sections on the conceptual framework for the statewide monitoring of groundwater quality, identification of problem areas, priorities for data gathering, monitoring methods, and recommendations regarding development of a statewide groundwater monitoring system. Implementation of some of these phases for some areas of the state, however, is still lacking.

Step 5: Evaluate Water Law [URL Ref. No. 325] and the Responsibilities of Municipalities and Government

Water law [URL Ref. No. 325] maintains the rights of owners to utilize water that exists on or beneath their property (Trelease, 1974). The law relates to all individuals, including municipalities that use the same aquifer(s). In New Mexico [URL Ref. No. 339 {42}], for example, the law provides that the surface and underground waters of the state belong to the public and are subject to appropriation for beneficial use. Such use is the basis, the measure, and the limit to the right to use water, with priority in time given to the better right. The underlying principle for water law in New Mexico is known as the Appropriative Doctrine, with water rights in the state being administered in accordance with provisions of the Constitution, the statutes, the terms of interstate water compacts, international treaties, and rules and regulations of the state engineer.

Local governments in most states, besides being regulated by their state water laws, also have the authority and responsibility to regulate urban expansion and provide public services, however, they might also want to seek direct involvement in all aspects of a resource such as water. They might want to get involved directly in groundwater monitor-

Step 6: Prepare and Implement Plans and Programs

Steps (1–5) above should be considered and evaluated in developing any groundwater protection/control/stabilization strategies. The steps presented are straightforward, and undoubtedly, however, as each component is designed and completed, unexpected situations specific to the particular aquifer(s) being evaluated will require additional work, time, and attention.

Development of such programs and plans is always an evolving process. And, if plans and programs are successful, they will be major positive contributors to the longevity of communities because of better understanding of the aquifer(s). An adequate knowledge base of the aquifer(s) is important and necessary in terms of preventing degradation of water quality or implementing steps to remove contaminants from the aquifer(s), should they be discovered.

Effective protection of groundwater against undue contamination/pollution requires well-integrated monitoring/surveillance and protection plans and programs that are implemented at the local, state, and federal levels. A *prevention-based approach* along with practical and reasonable groundwater cleanup is essential in terms of future policy decisions, if future problems are to be neutralized.

Overview of Groundwater Contamination—Summarization

Contamination of groundwater by organic and inorganic chemicals, pesticides, herbicides, radionuclides, microorganisms, and other contaminants, has occurred in every state and is being detected with increasing frequency. Detailed qualitative and quantitative estimates of the extent and effects of groundwater contamination, specifically, radioactive concerns, will never be available. The time, costs, and technical requirements to develop estimates on a nationwide basis are prohibitive and information necessary for predicting future contamination problems (i.e., future uses of groundwater, potential sources, and types of contaminants) is not known with certainty.

Contaminants observed in groundwater, particularly organic chemicals and radionuclides, are known to be associated with adverse health, social, environmental, and economic impacts, although only a small portion of the nation's total groundwater resource is thought to be significantly contaminated on the overall scale, the potential effects of this contamination warrant national attention. Public health concerns [URL Ref. No. 111, 181, 200] arise because some groundwater contaminants are linked to cancers, liver and kidney damage, and damage to the central nervous system. Uncertainties [URL Ref. No. 261] about human health impacts [URL Ref. No. 262] will continue to persist because these types of impacts are difficult to evaluate scientifically, and perhaps this will always be the case. The health issues and the latest analysis of risk assessment (e.g., hazard analysis) [URL Ref. No. 235, 271, 274] related to groundwater contamination also become more complex, because some impacts are not observable or occur long after exposures [URL Ref. No. 272] have taken place.

In addition, environmental impacts [URL Ref. No. 308] involving groundwater contamination concerns are not limited to soil water movement, but must include air and surface water areas because of the complex interrelationships water plays within ecosystems [URL Ref. No. 182, 260] (e.g., groundwater is essential to streams, rivers, vegetation, fish,

wildlife, etc.). The economic costs of detecting, correcting, preventing, or stabilizing groundwater contamination, therefore, can be high. Corrective and preventive actions will involve millions of dollars or more and billions of dollars for Department of Energy [URL Ref. No. 45, 61, 88] facilities. However, practical solutions based on specific time frames with proper goals and objectives to prevent or control groundwater contamination must prevail and continue. Decreases in agricultural and industrial productivity, lowered property values, costs for repair or replacement of damaged equipment, and increased costs of developing usable alternative water supplies for domestic purposes are only some of the consequences.

Although current information about groundwater contamination problems in the nation does not always describe actual situations, the information that has been gathered concerning the problems reflects the way in which investigations have been conducted, what contaminants have been searched, where they have been looked for, and where they have been detected. Because the majority of substances described as contaminants in groundwater [URL Ref. No. 1–339] are necessary for society, including a possible continued military defense initiative [URL Ref. No. 45, 259, 265], more widespread detection of contamination will continue to occur. Additional detection of contaminated groundwater areas will be associated with increased efforts to monitor known problems, locate undetected problems, and evaluate potential problems. Unfortunately, the costs and technical uncertainties [URL Ref. No. 261] associated with detection, correction, or stabilization activities related to groundwater contamination effectively preclude the investigation and correction of known and/or suspected contamination problems. Consequently, an adequate groundwater protection/control/stabilization program is central and essential to any long-term approach to groundwater quality protection. Choices involving detection, correction, prevention, and stabilization (i.e., given limited funds and technical assistance) will always depend on policy decisions regarding the extent groundwater resources are and should be safeguarded, but the effort needs to be continually present at the private and public sectors.

TWO

Ecotoxicological Risk Assessment (Risk Assessment Strategies) and Groundwater Contamination [URL Ref. No. 64, 84, 133, 140, 201, 235, 260–261, 336 {12}, 337 {39, 46–47, 79–80}, 338 {39}, 339 {3}]

Ecotoxicology (Risk Assessment and Groundwater Interactions)

Risk assessment [URL Ref. No. 235] is being seen by some policy makers and legislators as a magic bullet, the application of which will immediately clarify and rationalize environmental regulations (Eklund, 1996; Yosie, 1987; Maxwell et al., 1999a, 1999b; Hartly, 1999) [URL Ref. No. 21, 68–69]. As known, a number of inherent technical and scientific uncertainties [URL Ref. No. 261] underlie any risk assessment, and they include the following (Haas, 1996):

(1) What is the distribution of exposures and their duration?

(2) What is the intrinsic sensitivity among the populations exposed, and how does one translate animal effects data into estimates of consequence to humans—the mouse to man problem?

(3) How does one extrapolate from high-dose (required by limited resources for testing) to low-dose effects?

(4) How does one quantify the uncertainties in derived quantities given the uncertainties in input quantities?

Risk analysis then by this definition and comment, has been and can be a useful tool in relative ranking, priority setting, allocation of resources, and assessment of research needs to improve understanding of environmental problems (i.e., including groundwater contamination) if it is conducted in a consistent way and if underlying assumptions and uncertainties are fully stated and conveyed (O'Neill et al., 1982). To develop policy (i.e., according to Haas, 1996), however, on the basis of attributing certainty to a single-value estimate of risk, encourages inappropriate and wasteful uses of resources.

Risk inherently involves a geographical component (Lantzy et al., 1998), even at locations in space where receptors (i.e., human or environmental) and hazards come together. Maps, therefore, provide a tool for visually displaying information about the distribution of risk and also help an analyst to understand the nature of a specific situation, then evaluate alternative ways to reduce or manage the risk, and then convey the results of the analysis to the affected stakeholders. The focus of the study by Lantzy et al. (1998) was on the use of maps [URL Ref. No. 140] to address the risk to the local community owing to a plume of contaminated groundwater emanating from a chemical plant. In the study, maps were invaluable in evaluating the existing data, and developing a site conceptual model of the nature and extent of the contamination; determining what additional data needed to be collected; evaluating the effectiveness of remedial options; educating the public to provide them with understandable information to shape their perception of risk; and achieving public acceptance of the plan for remediating [URL Ref. No. 303] the contamination and managing the residual risk.

Quantitative risk assessment is a tool used at hazardous waste sites to evaluate the need for treatment and to determine which control strategies should be implemented (Batchelor, 1997). The approach by Batchelor (1997) is based on a material balance around the disposal zone and the groundwater flowing past it. Examples of applying this framework to pesticides, petroleum, and materials treated by solidification/stabilization are also presented by Richte and Safi (1997) and Hoskins et al. (1997).

The effect of parameter uncertainty and overly conservative measures on risk assessment have also been addressed by Hamed and Bedient (1997). Most of the work conducted was based on the use of the classic Monte Carlo Simulation Method (MCS) as a probabilistic modeling tool [URL Ref. No. 339 {31}], although the MCS lacked computational efficiency when the simulated probability was small, however, the application of the reliability methods to the probabilistic assessment of cancer risk due to groundwater contamination was extended. In a related study by Jacob et al. (1996) involving groundwater contaminated by trichloroethylene (TCE) and risk factors associated with possible carcinogens, similar results were obtained.

A technique for screening pesticides with respect to their overall risk to potential contamination of groundwater was developed for the purpose of selecting priority pesticides for groundwater monitoring programs (Shukla et al., 1996). The risk assessment technique considered leaching potential, extent of usage, and toxicity of pesticides. Incorporation of risk screening techniques into environmental management decision processes was useful for evaluation of the pesticide contamination potential of groundwater.

Historical waste disposal practices pose the greatest threat to groundwater in the United States, and common solvents such as trichloroethylene, tetrachloroethane, benzene, and carbon tetrachloride have been recognized in widespread areas (Datskou and North, 1996). Datskou and North (1996) evaluated baseline plutonium processing waste [URL Ref. No. 284] health risks to the public, health risk reduction to the public as a result of remedial activities, health risk to the workers directly involved in cleaning up the site, and costs associated with each remedial activity. The plume studied contained chemical and radioactive wastes [URL Ref. No. 304] that could pose a health threat to people living in the vicinity of the site.

Other studies involving groundwater contamination and risk factors related to pesticides and groundwater ecosystems (URL Ref. No. 182] have been conducted by Christakos and Hristopulos (1996) and Vandijk and Dehaan (1997). A series of primers on various aspects of risk factors of groundwater contamination and livestock holding pens (Harris, 1997a), wellhead management (Harris, 1997b), hazardous waste (Harris, 1997c), fertilizer storage and handling (Harris, 1997d), petroleum product storage (1997e), milking center wastewater (Harris, 1997f), and livestock manure storage (Harris, 1997g) have been reported.

Ecotoxicology [URL Ref. No. 260], the terminology used for a relatively new science, is concerned with the toxic effects of chemicals and their relationship to living organisms within the area they occupy (i.e., such as a defined community or ecosystem). Ecotoxicology relates to specific toxic properties [URL Ref. No. 256] that interplay with biotic species occupying defined communities. The term includes analyses of transfer pathways [URL Ref. No. 302] for specific chemical agents, including their movement within biological systems (Rail, 1985b, 1989).

The term ecotoxicology was first used by Truhaut (1975) as a natural extension from the science of toxicology (Truhaut, 1977; Moriarity, 1983; Butler, 1978; Ramade, 1979). The science of ecotoxicology by definition, then, was meant to provide information on toxic chemical relationships within defined ecosystems in which organisms live. An ecosystem then, by definition, is meant to include communities of living organisms together with their habitat and the interactions that occur there.

Ecotoxicology as interpreted by Paasivirta (1991), however, aimed to discover the chemicals that pose risks in order to apply preventive measures before any serious damage to natural ecosystems occurs. In another related concern, Shen (1998) applied a number of physical, chemical, and biological tests to predict the exposure-related fate and the toxic effects of chemicals in the environment.

One of the main differences between classical toxicology and ecotoxicology is that ecotoxicology is a four-part subject (Butler, 1978; Paasivirta, 1991). Also, according to the literature (Truhaut, 1975; Butler, 1978; Paasivirta, 1991), any assessment of the ultimate effect of an environmental pollutant must take into account, in a quantitative way, each of the following distinct processes:

(1) A substance is released into the environment—the amounts, forms, and sites of such releases must be known if the subsequent behavior is to be understood.

(2) The substance being transported geographically and into different biota and perhaps being chemically transformed, gives rise to compounds that have quite different environmental behavior patterns and toxic properties—the nature of such processes is unknown for the majority of environmental contaminants, and the dangers arising from their ultimate fate are complex, although the effects of certain chemicals have been well documented in recent years.

(3) The substance must eventually affect certain target organisms (e.g., humans, animals, etc.), and for this to be assessed, the type of exposure that is to be manifested must be examined.

(4) The response of an individual and/or groups of organisms to the specified (or perhaps transformed) contaminant or pollutant over an appropriate time period necessary to do harm must be assessed.

Therefore, for a proper groundwater *ecotoxicological assessment* to be made, the previously listed combination of steps must be evaluated in a quantitative fashion. And, because of the need for quantitative precision, another facet to the previous steps must also occur. This other facet should include recognition that uncertainty [URL Ref. No. 261] and possible error in the current understanding of the other four also occurs.

These summarizations of quantitative data, however, will eventually lead to an ordering of priorities for ongoing action and/or further research. And, it is surmised that problems given the highest priority should have to do with lethal toxics [URL Ref. No. 256] present in certain pathways or those that involve considerable health hazards.

Assessing ecotoxicological danger, environmental inputs, and their effects on the total ecosystem of humans in relationship to groundwater contamination is difficult. However, evaluation of risk factors can be used as a tool if they can be explained as the expected frequency of undesirable effects arising from a specified (e.g., unit) exposure to a pollutant (Butler, 1978). The quantitative relations between exposure to a pollutant and risk of magnitude or undesirable effects in groundwater under specified conditions defined by environmental and target variables, have to be the criteria.

In reference to *groundwater cleanup*, a paradigm *in which we attempt to bring groundwater into compliance with drinking water standards* is being questioned (Bredehoeft, 1996). One alternative or approach is to look for a new way to approach cleanup, specifically via use of the proposed method of *health-risk-based* cleanup (i.e., the U.S. EPA [URL Ref. No. 19] is also pushing this policy). By use of this health-risk-based approach, groundwater cleanup can be designed to reduce to an acceptable level, health risks posed by contaminants as they reach the *accessible environment* or the place where humans (i.e., or other life forms) come in contact with the contaminants. This then would mean that hydrogeologists [URL Ref. No. 13–14] must predict the movement of the contaminants and their concentrations as they are transported by moving groundwater to an accessible environment, which is not an easy task.

Clearly then, if one of the reasons for cleaning up groundwater is to improve human health, then the health-risk-based approach has merit, however, where it is implemented, it has generally resulted in less stringent cleanup (Bredehoeft, 1996). So, what then happens? According to Bredehoeft (1996), while the regulators are moving toward health-risk-based cleanup, there are still constraints that have not been considered. These constraints include issues of property and stoppage of pumping so that contaminants move off-site, adversely impacting property values. The general idea is then that the area that is underlain by contaminated water would be restricted, rather than have drinking water standards [URL Ref. No. 258] imposed where contaminated groundwater is contained.

Reducing Uncertainty in Assessing the Risk of Environmental Contaminants and Relationship to Groundwater Contamination

Reducing uncertainty [URL Ref. No. 261] in assessing the risk of environmental contaminants is important to regulatory agencies at the state and federal levels and to nonregulatory agencies that work for environmental health and safety. Efforts to manage risk are driven by *actual risk,* which can rarely be measured; *calculated risk,* which is based on science but whose inclusion is often restricted through regulatory policy; and *perceived risk* (McKone and Bogen, 1991). Relating this to groundwater contamination is difficult, but it can be done as follows.

Calculated health risk within a population exposed to an environmental contaminant in groundwater is determined by using a release rate or concentration at the source; the expo-

sure function converts the source into the amount contacted by each individual, the organ or tissue dose per unit exposure, and the toxic potency associated with the delivered dose. Although the actual risk can be more complex, it can include temporal and spatial relations and functional dependencies among the source, exposure, dose, and the incidence of detriment. And, in practice, risk must be characterized as a product of four factors that include source term, exposure factors, fraction absorbed, and toxic potency. The publication by McKone and Bogen (1991) illustrates a strategy for evaluating the sources of uncertainty in predictive exposure and health risk assessments. The transformation of uncertainties to groundwater contamination can then be determined.

Two obvious methods are available for reducing uncertainty. These include improving the models and expanding the data. It is possible to use the analytic framework of statistical decision analysis to determine when additional information is beneficial (McKone and Bogen, 1991).

Through case studies, McKone and Bogen (1991) have been working to improve the characterization of uncertainty in human exposure models [URL Ref. No. 273] and the combined uncertainty in source, exposure, and dose-response models. In one of their studies, they described the volatile organic chemical [URL Ref. No. 247, 267], tetrachloroethylene perchloroethylene (PCE), in California water supplies derived from groundwater. Their analysis was divided into five steps that included the following:

(1) Consideration of the magnitude and variability of PCE concentrations available in large public water supplies in California

(2) Characterization of pathway exposure factors (PEFs) for groundwater exposures and estimation of the uncertainty for each PEF

(3) Examination of models that described uptake and metabolism to estimate the relation between exposure and metabolized dose

(4) Consideration of carcinogenic potency of the metabolized PCE dose

(5) Combination of results to estimate the overall magnitude and uncertainty of increased risk to an individual selected at random from the exposed population and to explore the particular contribution to overall uncertainty

Ecotoxicological Testing

Several laws require ecotoxicological testing to predict the hazards that chemicals may pose to the aquatic environment [URL Ref. No. 182], including groundwater, and most hazard evaluations are based on data derived from acute toxicity tests with fish and daphnias (Slough et al., 1986). Some ecotoxicologists have expressed concern about the use of an LC_{50} value [URL Ref. No. 273] to protect ecosystems

(Slough et el., 1986). Such use to some is not scientifically justified and so such information should not be considered a satisfactory basis for proper decision making. Other scientists have conclude, for purposes of environmental protection that include groundwater interactions, that the determination of an acute median lethal concentration (LC_{50}) can be regarded as a waste of time, money, and materials. Their major criticism is related to the margins of uncertainty introduced by the need to extrapolate from *acute* to *chronic* effects and from one level of biological integration to another. In the article by Slough et al. (1986), attention is given to the uncertainties involved in the extrapolation from one species to another, from acute to chronic exposures, and from a single species to the ecosystem level. Relationships to groundwater contamination would then need to be extrapolated.

Statistical measurements [URL Ref. No. 339 {31}] of margins of uncertainty in predicting toxicity from one species to another, from acute to chronic exposures, and from single species to higher levels of biological organization were determined by tests of regression and correlation analyses (Slough et al., 1986). Based on the acute sensitivities of 35 aquatic species (bacteria, algae, protozoa, Coelenterata, Tubellaria, Clitellata, Crustacea, Insecta, Mollusca, Pisces, and Amphibia) to 16 chemical compounds (cadmium, copper, mercury, zinc, arsenic, ammonium sulfamate, sodium chlorate, potassium permanganate, TPSB, chlorendic acid, trichloroethylene, tetrachloroethylene, 2,4-dichloroaniline, 3,4-dichloroaniline, 4-chlorophenol, and 2,4-dichlorophenol), no species was found to be particularly sensitive to all chemicals, and the 95% uncertainty factor (UF) ranged from three to 1,985. Analyses of acute and chronic sensitivities for the same species to 164 chemicals resulted in the acute/chronic relationship log NOEC = − 1.28 = +0.95 log L(E)C50 (r = 0.89) and the UF of 25.6 (i.e., where NOEC is the no observed effect concentration). Comparison of the lowest acute and corresponding ecosystem effect levels for 34 chemicals indicate that the relationship log NOEC(ecosystems) = 10.55 + 0.81 log L(E)C50 (r = 0.77) and the UF is 85.7. As to the predictability of ecosystem effect levels from chronic single species data, the following relationship was expressed: log NOEC(ecosystem) = 0.63 + 0.85 log NOEC (r = 0.85), with a UF of 33.5.

Also, according to Slough et al. (1986), these data indicate that acute testing is not pointless, and it offers a statistical base for the use of acute toxicity information in the hazard assessment of chemicals in the aquatic environment, including groundwater. These data also show that acute toxicity determinations have their merits in that interspecies toxicity prediction is more uncertain than prediction of chronic from acute effect levels in the same species. Further, predictions of ecosystem effect levels from acute tests are unreliable, and so there is no reason to propose expensive and complex tests as additional or alternative research tools for routine *hazard assessments*, including studies in groundwater.

Such statistical data and inference as previously discussed, combined with those on environmental factors like physical-chemical, morphometrical, and biological characteristics of the water of concern (i.e., groundwater included), as well as the toxicant budget (e.g., input, partition, output), would yield a matrix that may give a sound estimate of the (site-specific) ecological impact of a given substance when applied to mathematically derived *risk models*. However, risk models are always in a state of development, but they are still urgently needed for further evaluation, for calibration verification, and for standardization of models, especially those that relate to groundwater contamination.

Process of Human Health Risk Assessment [URL Ref. No. 262, 272] *and Relationship to Groundwater Contamination*

In general, the process of human health risk assessment (HRA) is judged by its ability to predict adverse outcomes of particular environmental contaminants or exposures for individual humans (Burger, 1994; National Research Council, 1983). Likewise, environmental scientists often examine the adverse outcomes of chemical or physical hazards on individual species. However, this ecotoxicological approach to ecological risk assessment (ERA) fails to encompass the potential range of adverse outcomes to animal, plant, or human populations, communities, and ecosystems, including groundwater interactions. Moreover, whereas the success of HRA can be evaluated by examining the health of individual humans, the success of ERA cannot because populations of species are the important unit ecologically, rather than individuals; the overall structure and complexity of the system is more important than the structure or organization within one species (i.e., humans in the case of HRA); and the overall functioning of the system is more important than only the functioning of one species.

The risk assessment paradigm then, that includes hazard identification [URL Ref. No. 271], dose-response analysis [URL Ref. No. 273], exposure assessment [URL Ref. No. 272], risk characterization [URL Ref. No. 235], and the relationship to groundwater contamination, should have a parallel phase or discipline of research [i.e., evaluation of predicted and actual outcomes (Burger, 1994)]. This phase, termed predictive accuracy, is particularly critical for ecological risk assessment or the relationship to groundwater contamination because actual or potential outcomes may occur long after an initial perturbation, and some contamination problems are not discussed until many years after the initial incident.

Burger (1994) proposed that a fifth step, *predictive accuracy* (i.e., which determines the relationship between the predicted outcome and the actual outcome) be added to risk assessment strategies. If this is done, then an overall probability can be determined that reflects the reality of risk from actual hazardous events, including groundwater contamination, rather than merely the probability of a future event or occurrence. Evaluation of data of this type that involves predictive accuracy would also be useful to groundwater managers and environmental regulators in making future decisions, recognizing what hazards to tolerate and the magnitude of risk that is acceptable [i.e., based on past risk assessments of similar hazards (Burger, 1994)].

Risk [URL Ref. No. 235], also, in a general sense, when considered in terms of environmental risk assessment, recognizes that no known general methodology is available for conducting an evaluation that includes humans and nonhumans, ecological receptors, or groundwater contamination concerns (Cornaby et al., 1982). The terminology in the literature is vague and suggests that views and knowledge of environmental risk assessment continue to evolve. Therefore, assessing risks that can occur during aquifer restoration is not easy. This is because selecting an effective remedial technique involves the balancing of the need to contain contaminants within acceptable levels against the costs associated with specific cleanup measures.

Canter and Knox (1986) describe two generally accepted approaches to risk assessment and the relationship to contamination. The first approach includes utilization of criteria or standards for a pollutant. The second approach analyzed the effectiveness of various alternatives and compared their resultant concentrations with a given standard. However, no matter which approach is used for risk assessment and relationship to groundwater contamination, a criterion, standard, or acceptable level is involved and necessary, and for the most part, data of this type are sparse or nonexistent.

Additionally, to complete an assessment of risk and relate it to groundwater contamination, one must deal with the question of exposure (Walker, 1985; Farrara et al., 1984), and the exposure concentration of a particular substance is the concentration to which humans, fish, and other organisms at some level in the food chain, or even the environment as a whole, are exposed to at a particular time. And, according to Haque et al. (1980), three basic elements are required to estimate exposure concentration. These include the following:

(1) Information on the source of the toxicant, including such items as production rate and release rate to the environment

(2) Characteristics of the toxicant that describe its ability to travel and react in the natural environment

(3) Data that can be used to estimate the population at risk, including occupational characteristics, medical surveillance data, and socioeconomic use habits

According to Farrara et al. (1984), the first two elements previously listed [i.e., (1) and (2)] can be incorporated into mathematical models for estimating the *transport and fate* [URL Ref. No. 302] of the toxicant and, therefore, the con-

centrations that will appear in different sectors of the environment (i.e., the outcome can be a time-history of toxicant concentrations at various locations in the environment), including aqueous areas. Fararra et al. (1984) also mention that the last step in exposure assessment [see previous item (3)] [URL Ref. No. 272] coincides with the first step [see previous item (1)] in hazard assessment [URL Ref. No. 271]. Consequently, the degree of hazard then deals not only with toxicity but also with the degree of exposure (i.e., highly toxic substances with no exposure are not hazardous, whereas mildly toxic substances with high exposure could be very hazardous) can have a direct effect on the overall toxicity of groundwater.

In a related study (Schuller et al., 1991), the objective of their risk assessment (RA) was to evaluate the potential risks to human health for the no-action alternative at a U.S. EPA Superfund site. The RA was to evaluate the potential risks posed by compounds detected at the site under present conditions and hypothetical future-use conditions at the receptor points. A hypothetical future-use scenario examined in detail potential changes resulting from an instantaneous and complete liner failure at the site, with subsequent downgradient use of the contaminated groundwater as a potable water supply. The RA addressed potential risks to public health posed by this future-use scenario, but a groundwater modeling effort was necessary to estimate future concentrations of compounds. The hypothetical, future-use scenario established by Schuller et al. (1991) and used in assessing the no-action alternative included the following:

(a) Instantaneous, 100 percent failure of the pits' liners, which is unrealistic since at best, 25 percent could fail in a year, based on installation over a four-year period

(b) Use of contaminated groundwater as a potable water supply for the residential areas located directly downgradient of the site

Available site data were also used to determine potential exposure concentrations, and a two-dimensional, transient, instantaneous-release, multipoint-source model for the landfill was used. This model calculated groundwater flow velocities and estimated mechanical dispersion and linear partitioning to predict potential groundwater quality at the exposed population after a hypothetical liner failure. The exposure routes for the different populations in the scenario consisted of *inhalation* of and *dermal* contact with compounds detected in the modeled groundwater during showering and ingestion of these compounds.

Schuller et al. (1991) also demonstrated that the potential risks from exposure to carcinogens (e.g., organic compounds [URL Ref. No. 210]: ethylbenzene, styrene, toluene, benzene, 1,2-dichlorobenzene, phenol, cresols, and several pthalate esters), *noncarcinogens* (e.g., iron [URL Ref. No. 195], calcium [URL Ref. No. 192], and magnesium) [URL Ref. No. 193], and extraction procedure (EP) toxicity metals analysis indicated that the wastes had individual metal concentrations below the Resource Conservation Recovery Act (RCRA) [URL Ref. No. 280] limits for classification as hazardous waste. The carcinogens and noncarcinogens were compared to current exposure point concentrations with applicable, or relevant and appropriate, requirements (ARARs); calculated subchronic intakes with acceptable subchronic intakes for noncarcinogens; calculated chronic intakes of noncarcinogens with reference doses; calculated risks with range for potential carcinogens; and calculation of upper-bound (worst-case) chronic hazard index and carcinogenic risks with guidelines. Additionally, the predicted groundwater concentrations were significantly lower than the respective drinking water ARARs (i.e., the hazard indexes for subchronic, weighted-chronic, and upper-bound cases were calculated as 0.199, 0.154, and 1.22, respectively).

In *summary,* the magnitude of the public health impact was evaluated [i.e., a risk assessment for this Superfund NPL Site (Schuller et al., 1991)], assuming that no remediation had or would occur. The purpose of the groundwater modeling effort then was to predict future groundwater concentrations after liner failure would occur at a residential well that used the groundwater, and therefore, a two-dimensional, transient, instantaneous-release, multipoint-source model estimated the groundwater concentration at the residential well for 120 years after the release. The modeled groundwater concentrations that were simulated did not exceed standards, and the predicted worst-case exposure concentrations were unlikely to be reached, and the human population ultimately, might not be exposed.

Acute or *chronic* dangers from toxic chemicals in groundwater [URL Ref. No. 256] must be considered with acute toxicity problems because they present a more easily evaluated problem in terms of the risk associated with contact and exposure. Chronic toxicity problems on the opposite extreme involve more complex evaluations, and, therefore, to assess the effect of low-level exposure to humans, one must rely on established theories describing relationships between exposure and response mechanisms. One must extrapolate the data in essentially two ways, with the first recognizing that there is a threshold level below which no response is observed and the second that a response occurs regardless of how small the exposure level is (i.e., zero response exists at zero exposure). Since zero exposure in today's technological society is not practical or economically feasible, it then becomes difficult to define a safe level, although the acceptance of a threshold level depends on whether an individual accepts or rejects the threshold theory. How risk assessment evaluations can be conducted for aquifer restoration use follows.

Risk Assessment [URL Ref. No. 235] and Aquifer Restoration [URL Ref. No. 263]

Much of the work addressing groundwater contamination problems and solutions has been conducted in response to the Comprehensive Environmental Response, Compensation, and Liability Act, PL 95-5110 (known as CERCLA [URL Ref. No. 264] or Superfund). Superfund sites require the development of a remedial action master plan (RAMP). The purpose of the RAMP is to identify the type, scope, sequence, and schedule of remedial projects that may be appropriate in meeting an identified need. The RAMP initially was designed as an approach for developing an optimal solution for meeting a given need. The RAMP analysis involves consideration of human health under various conditions of exposure, including a description of *uncertainties* [URL Ref. No. 261] involved with environmental impacts, costs, and risks.

The main problem faced by risk assessment techniques is that a large portion of the needed information, such as risk pathways or acceptable concentrations, is for all practical purposes, unknown or changes through time. However, by knowing the limitations, extrapolations or estimations can still be conducted.

Contamination from toxic chemicals and their waste also generates concern because it affects human health, the ecology of an area, and nonliving systems such as buildings, soil, water resources, and air quality (Goldblum et al., 1992). Management of these hazardous chemicals and disposal techniques and areas includes assessment of the risks of exposure and regulations to control these risks. For example, further analysis into risk assessment extrapolations for a drum storage area at a Department of Defense (DOD) [URL Ref. No. 265] installation rendering airlift support for airborne forces resulted in elimination of an unnecessary and costly remedial action.

Quality data are necessary to assess the risk associated with toxic chemicals situated and disposed of at a hazardous waste site, and it is also required to set priorities for cleanups at such sites (Goldblum et al., 1992). This risk assessment process estimates total carcinogenic and noncarcinogenic risk at each site with the total risk being the sum of the individual risk components associated with the intake of each toxic chemical along each exposure pathway (i.e., inhalation, ingestion, and dermal contact). However, if the risk assessment outcome is borderline, then the individual risks may have to be further broken down into separate target organ systems within an overall living organism such as the human body. Thus, the risk assessment process is very complex and is very important, and the interpretation given to quality data is fundamental to making decisions on remediations.

The critical review of a contractor's risk assessment for a drum storage area at a military installation, as previously mentioned (Goldblum et al., 1992), prevented a costly and unnecessary remedial action and, consequently, taxpayer money was saved. Goldblum et al. (1992) showed how faulty mathematical calculations and the questionable inclusion of a chemical of concern led to the apparent need for cleanup. When the risk was recalculated using valid data (i.e., correcting the faulty mathematical calculations and using appropriate chemicals of concern), a cleanup was not required. The calculated risk reduction that resulted from using the corrected average TCE concentration then allowed the Air Force to place this site in the *No Further Action* (NFA) status. The Air Force had been considering an expensive pump and treat remedial action [URL Ref. No. 336 {34}, 337 {27}] that would have been of little value at this particular site. Consequently, when performing a risk assessment and relating it to groundwater, particular attention must be paid to the media of the pathways analyzed, methods of analysis, fate and transport [URL Ref. No. 302] of the environmental contaminants, and mathematical operations [URL Ref. No. 339 {31}] to insure a valid portrayal of risk.

Risk-Based Management of Hazardous Waste and Groundwater

In 1992, the U.S. Environmental Protection Agency (EPA) [URL Ref. No. 19] proposed risk-based management of hazardous waste [URL Ref. No. 266] (Chiang et al., 1995). A major component of the proposed rule included the determination of non-site-specific screening concentration levels from waste leachate and a rule that groundwater at a downgradient exposure point must not exceed initial screening levels, or more stringent requirements would apply. The screening concentration level was determined with verified models and equations that simulated the transport and attenuation of chemicals as they traveled from the source area to an exposure point.

Chiang et al. (1995) also focused on the development of non-site-specific screening levels for the leachate-groundwater transport pathway as shown in the EPA's proposed rule (*Federal Register*, 1992) [URL Ref. No. 68–69] regulating the identification and listing of hazardous waste. Dilution attenuation factors (DAFs) were also determined by Chiang et al. (1995) based on output from verified models and equations that simulated the transport and attenuation of chemicals as they traveled from a source area to a water well.

In 1992, the EPA proposed (*Federal Register*, 1992) [URL Ref. No. 68–69] two sets of risk-based levels with one set not requiring conditions that demand management of the waste (*tier 1*) and the second set requiring subsequent management of the waste in a specified manner (*tier 2*). The U.S. EPA analysis then showed that a conservative value for the universal DAF for *tier 1* should be at least 150, when compared to the EPA proposed DAF of 10. The DAF of 150 was developed based on the mismanagement scenario for landfills with natural soil covers and no clay liners under the steady-state condition. Under transient conditions, however, the universal DAF for *tier 1* was 11,596 after 70 years of landfill operation, which is the commonly accepted exposure period for calculating a risk factor. A conservative

value for the universal DAF for *tier 2* should be at least 320, which was also developed based on steady-state condition. EPA proposed a DAF of 100 for the *tier 2* scenario. Under transient conditions, the universal DAF for *tier 2* was 3.23×10^{11} after 70 years of landfill operation. Therefore, the DAF of 320 is considered conservative.

When considering volatile compounds (e.g., ETEX) [URL Ref. No. 247, 267] with their adsorption characteristics, the 85th percentile DAF was calculated to be 1,447 after 1,000 years of landfill operation, and when a conservative biodegradation rate of 0.1 percent/day^{-1} (i.e., the lowest reported rate in the literature is 0.6 percent/day^{-1}) was considered, DAFs increased by more than one order of magnitude to 16.38 and 6,689 for mismanagement and contingent management scenarios, respectively, as compared with DAFs of 150 and 320 for their respective scenarios when biodegradation was not considered.

Concern over the potential adverse health effects of chemically contaminated groundwater has existed for many years (Germolec et al., 1989), and, in general, studies concerning chemically contaminated groundwater have focused on retrospective epidemiological studies related to cancer risk. In the study conducted by Germolec et al. (1989), the immune function in female B6C3F$_1$ mice exposed to a chemical mixture in drinking water for either 14 or 90 days was monitored. The mixture consisted of 25 common groundwater contaminants frequently found near toxic waste dumps, as determined by EPA surveys (Yang et al., 1989).

None of the animals studied developed overt signs of toxicity such as body or liver weight changes, and mice exposed to the highest dose of this mixture for 14 or 90 days showed immune function changes that could be related to rapidly proliferating cells, including suppression of hematopoietic stem cells and of antigen-induced antibody-forming cells. Some of these responses (e.g., granulocyte-macrophage colony formations) were also suppressed at lower concentrations of the chemical mixture. There were no effects on T cell function or T and B cell numbers in any of the treatment groups. Altered resistance to challenge with an infectious agent also occurred in mice given the highest concentration (i.e., related to threshold limits and overwhelming of the defense mechanism and high doses), which correlated with the immune function changes. Paired-water studies indicated that the immune effects were related to chemical exposure and not to decreased water intake. These results in the broadest sense suggest that *the long-term exposure to contaminated groundwater may represent a risk to the immune system in humans.*

The previously mentioned study by Germolec et al. (1989) combined with current knowledge about the pathogenesis of disease resulting from immunodeficiency and the potential for large-scale human exposure to groundwater contaminants indicate a need for greater awareness among clinicians and epidemiologists as to the subtle effects that may occur with groundwater contamination. In addition, examination of targets such as the immune system should be conducted more routinely (i.e., studies concerned with high dose versus low dose). Data by investigators, however, should be interpreted with caution since the chemical mixture studied, while approximating concentrations that can be found in environmental samples, was prepared in the laboratory and was designed to mimic a worst-case scenario. Thus, given the heterogeneity that would occur when testing the immune system of humans and the limitations in assays that are normally performed in the clinic, it may be difficult to detect subtle (e.g., chemical) immune changes in humans, as occurred here in mice, unless special studies (e.g., stem cell function) are conducted with a large sample population.

In another study, Yang and Rauckman (1987) cooperated with the Agency for Toxic Substances and Disease Registry, the National Toxicology Program [URL Ref. No. 268–269], and the Public Health Service [URL Ref. No. 200] in activities related to the Comprehensive Environmental Response, Compensation, and Liability Act (Superfund Act) [URL Ref. No. 264] by conducting toxicology studies on chemicals detected in high priority hazardous waste sites and for which adequate toxicological data were not yet available. As part of this effort, a project on the toxicology of chemical mixtures of groundwater contaminants was initiated, and the first study centered on the health effects of groundwater contaminants, including 19 organic and six inorganic chemicals selected from more than 1,000 known groundwater contaminants that were given in drinking water to Fishcher 344 rats and B6C3F$_1$ mice for three or six months. Controls and five dose levels, based on average concentrations of individual component chemicals, or 0.1-, 10-, 100-, or 1,000-fold (i.e., baseline level), were adopted. The toxicology endpoints included mortality, clinical signs, water and food consumption, body and organ weights, clinical pathology analytes (e.g., hematology, clinical chemistry, and urinalysis), gross and histopathology, neurobehavioral test, sperm morphology and vaginal cytology evaluations (SMVCE), and cytogenetics. The study by Yang and Rauckman (1987) summarized the rationale behind their experiments and included the factors that needed to be considered when designing studies such as those that involve complex chemical mixtures.

Also, as discussed by Yang and Rauckman (1987), their objective was to investigate the possible adverse health effects in rodents following subchronic ingestion of a prototype mixture of groundwater contaminants. Consequently, any adverse health effects would be directly related to any treatment-related toxic responses observed in the proposed subchronic toxicity study. For example, significant body weight depression without obvious differences in food consumption would suggest a growth retardation effect, additionally, marked elevation of alanine aminotransferase might suggest liver damage, and distinct neurobehavioral signs followed by Wallerian degeneration in the nervous

system would suggest neuropathy. Therefore, when a chemical mixture has more than two compounds, it became difficult to speculate on the experimental outcome. Consequently, with a mixture of 25 chemicals, it then becomes impossible to predict target organs and toxic endpoints routinely employed under the NTP Task I in the 13-week subchronic toxicity study [see the NTP General Statement of Work (Yang and Rauckman, 1987)]. Two extreme outcomes of the study warrant further discussion. The first outcome was that no observable effects were reported by the investigators or seen after subchronic exposure of the animals to the 25-chemical mixture at the proposed dose levels. The second outcome reported that severe toxicity resulted at even some of the lower doses. In either case, *such information was useful in the risk assessment of contaminated groundwater and stimulated further research into the mechanisms of toxicity induced by chemical mixtures.*

In another related study (Heindel et al., 1995) concerning groundwater contaminants, the potential reproductive toxicity of a mixture of 25 chemicals (MIX) formulated to represent contaminated groundwater supplies near hazardous waste dumps was evaluated in CD-1 Swiss mice and Sprague-Dawley rats using the reproductive assessment by continuous breeding protocol. Male and female mice and rats were exposed to MIX (acetone, aroclor 1260, arsenic trioxide, benzene, cadmium acetate, carbon tetrachloride, chloroform, chlorobenzene, chromium chloride, 1,1-dichloroethane, 1,1-dichloroethylene, 1,2-t-dichloroethylene, di(2-ethylhexyl)phthalate, ethylbenzene, lead acetate, mercuric chloride, methylene chloride, nickel acetate, phenol, tetrachloroethylene, toluene, 1,1,1-trichloroethane, trichloroethylene, and xylene) in the drinking water at concentrations of 1, 5, and 10 percent of a technically achievable stock solution.

Results of the previously presented study showed that for mice, body weight and feed consumption were not affected by MIX, but water consumption was decreased for concentrations of 5 and 10 percent in both groups. A cocktail of 25 chemicals commonly found in contaminated groundwater at or near hazardous waste sites was also administered in drinking water at doses that resulted in severely decreased water consumption in mice and rats. And, despite the presence of many known reproductive toxicants including cadmium, mercury, lead, chloroform, di(2-ethylhexyl)phthalate, and methylene chloride, only minimal reproductive effects were observed in F$_0$ and F$_1$ mice and rats. However, some specific reproductive effects noted during the rat and mice studies deserve mention. In the F$_0$ generation of both studies, isolated effects of MIX were statistically significant. In the rat study, the number of live male pups and live pup weight were reduced in the F$_0$ litters of the high-dose group. This effect was due to a significant decrease in live male rat pups per litter, which occurred only in the first and second litters. The authors (Heindel et al., 1995) believed that this change lacked biological significance and, similarly, the weight of live pups was statistically significant only for litters one and five in the high-dose group, with each effect being only an 8 percent reduction and the overall effect being only 6 percent. Neither of these effects were seen in the F_1 generation rats, again suggesting they were not chemically related.

Pesticides and Fertilizer Toxicity and Groundwater Contamination

Pesticides [URL Ref. No. 239] and fertilizers, as used in modern agriculture, contributed to the overall low-level contamination of groundwater sources (Heindel et al., 1994). In order to determine the potential of pesticide and fertilizer mixtures to produce reproductive or developmental toxicity at concentrations up to 100 times the medium level found in groundwater, Heindel et al. (1994) prepared and studied two mixtures of pesticides and a fertilizer (ammonium nitrate). One mixture contained aldicarb, atazine, dibromochloropropane, 1,2-dichloropropane, ethylene dibromide, and simazine plus ammonium nitrate that was considered to be representative of groundwater contamination in California. The other, containing alachlor, atrazine, cyanazine, metolachlor, metribuzin, and ammonium nitrates, simulated groundwater contamination in Iowa. Unlike conventional toxicology studies, the purpose of this study was to evaluate the health effects of realistic human concentrations and administration of these pesticide/fertilizer mixtures at levels up to 100-fold greater than the median concentrations in groundwater supplies in California or Iowa but at levels that did not cause any detectable reproductive problems in mice or developmental toxicity problems in rats.

Pesticide and fertilizer contamination of groundwater (i.e., at relatively low concentrations) is widespread, particularly in areas where intensive farming takes place or where hazardous waste disposal sites are located, and because approximately one-half of the U.S. population is dependent on groundwater, it is important to identify any health effects of long-term low-level intake of such chemical mixtures. The study reported by Heindel et al. (1994) was designed to address the potential reproductive and developmental toxicity of mixtures of pesticides [URL Ref. No. 239] and fertilizers formulated to represent median or higher concentrations of pesticides in groundwater in California or Iowa. The study was conducted with recognition that the pattern of rodent fluid consumption differs markedly from that of humans, and it may influence the pharmacokinetic profile of the components of the mixture.

Nitrates as Fertilizers and Toxicity in Groundwater

Heindel et al. (1994) included nitrates [URL Ref. No. 201] in both mixtures at environmentally relevant levels in their studies. Rats in the developmental study consumed 130–140 mg/kg/day of ammonium nitrate via drinking water. Exposure of F_0 and F_1 generation mice or pregnant rats to a mixture of pesticides and ammonium nitrate at levels up to 100-fold greater than the median concentration in groundwater in California or Iowa did not cause detectable reproductive toxicity, developmental toxicity, or other adverse effects.

Hepatic and renal effects of repeated exposure to a mixture of 25 chemicals frequently found in groundwater near hazardous waste disposal sites and the effect of such exposure on carbon tetrachloride (CCl_4) toxicity were examined by Simmons et al. (1994). Slight but statistically significant alterations, of uncertain biological significance, resulted from the water treatments: 10 percent MIX increased alanine aminotransferase, urea nitrogen (BUN), and BUN/creatinine ratio; restricted water increased 5-nucleotidase and decreased alkaline phosphatase. Relative kidney weight was increased by 10 percent MIX and restricted water. CCl_4 resulted in significant dosage-dependent hepatotoxicity in all three water treatment groups but had little or no effect on renal indicators of toxicity. Relative to AD Lib Water, significantly greater hepatotoxicity occurred in 10 percent MIX and restricted water rats. The response to CCl_4 in the restricted water rats was similar to that of 10 percent MIX rats, indicating that a substantial portion of the effect of 10 percent MIX on CCl_4 hepatotoxicity is due to decreased water and feed intake.

In rats exposed to 10 percent MIX alone, hepatic lesions were not observed, and the incidence and severity of renal nephropathy were not altered from control levels. This was consistent with the lack of hepatic and renal histopathological alterations in male B6C3F1 mice exposed for 90 days to 1 percent, 5 percent, and 10 percent MIX. Exposure to 10 percent MIX alone resulted in a slight but statistically significant increase in serum ALT levels. The significance of this increasing the absence of either microscopically evident hepatic damage or elevations in serum or other enzymes reflective of hepatic parenchymal damage is unknown. Similarly, the biological significance, if any, of the slight but significant increases in serum BUN and BUN/CREAT in rats exposed to 10 percent MIX is unknown.

For complex mixtures such as the ones previously presented, a variety of interactions may be occurring among various components, including less than additive, and greater than additive interactions, with each contributing to an observed effect. Chemicals known to interact upon simultaneously or temporally separated exposures were present in the 25-chemical mixture, for example, prior exposure to acetone increased the hepatic toxicity of $CHCl_3$ as well as the hepatotoxicity of CCl_4.

In related study, Hong et al. (1991) assessed the potential health effects of chemically contaminated groundwater, and a toxicological program was initiated on a mixture of 25 frequently detected groundwater contaminants derived from hazardous waste disposal sites. As part of this study, myelotoxicity [URL Ref. No. 270] studies were conducted. Bone marrow parameters were examined in mice exposed to 0, 1, 5, or 10 percent of chemical mixture stock solution for 108 days, and they showed suppressed marrow granulocyte macrophage progenitors (CFU-GM), however, this suppression disappeared in 10 weeks following the cessation of treatment. The possible toxicological interaction of groundwater contaminants and radiation on hematopoiesis was investigated by using the number of bone marrow CFU-GM as an index. When mice were exposed to 200 rads whole body irradiation at two and nine weeks during this 10-week recovery period, the combined treatment (i.e., chemical mixture followed by irradiation) group showed a significantly slower recovery of bone marrow progenitors as compared with the control group (i.e., radiation without prior chemical mixture treatment).

This study by Hong et al. (1991) also showed that even 10 weeks after the cessation of chemical mixture treatment when all hematological parameters were normal, a residual effect of the chemical mixture may still be demonstrated as lower progenitor cell numbers following irradiation. Thus, residual damage of hematopoiesis in mice exposed to groundwater contaminants for 108 days rendered the mice more sensitive to subsequent irradiation-induced injury. Cyclotoxic damage to bone marrow cells was also related to conditions such as pancytopenia or anemia, and genotoxic damage can be correlated with tumor induction.

Additionally, in the study by Hong et al. (1991), several chemicals that were present in the test chemical mixture of groundwater contaminants were reported to be myelotoxic at relatively high concentrations, and include arochlor 1260, benzene, phenol, toluene, xylene, and heavy metals. Given the low exposure levels in this study, however, it was not certain that any single chemical by itself could account for the myelotoxicity observed. Also, since systematic investigation of all combinations of the 25 components of this mixture was initially impossible, studies on submixtures of those components that are known myelotoxic agents were still fruitful.

Yang et al. (1989), also as part of an effort to evaluate the toxicology of groundwater contaminants, reported the formulation and analytical chemistry of mixtures. Their chemically achievable mixture and ppm is as follows: acetone 530, arochlor (1260) 0.1, arsenic (III) 90, benzene 125, cadmium (II) 510, carbon tetrachloride 4, chlorobenzene 1, chloroform 79, chromium (III) 360, DEHP 0.15, 1,1-dichloroethane 14, 1,2-dichloroethane 400, 1,1 Dichloroethylene 5, 1,2-trans-dichloroethylene 25, ethylbenzene 3, lead (II) 700, mercury (II), methylene chloride 375, nickel (II) 68, phenol 290, tetrachloroethylene 34, toluene 70, 1,1,1-trichloroethane 65, and xylene 16.

Many problems were anticipated by the authors in a study of this type, including limitation of solubility, chemical inter-

actions, and extreme volatility in the aqueous solution of 25 chemicals. The final technically achievable stock solution was prepared based on EPA survey concentrations of these chemicals in groundwater around hazardous waste disposal sites, their toxicity information, and solubility of the individual compounds in the matrix of the aqueous solution of these 25 chemicals. Analyses of all 25 chemicals in the drinking water mixture required six different chromatographic and spectroscopic methods and some loss of organic volatiles during mixing of the substocks and during the first 24 hours following preparation occurred. Solutions held under simulated animal cage conditions for 96 hours showed losses of the organic volatiles [URL Ref. No. 247, 267], the majority of which occurred within the first 24 hours. This study also showed that it is possible to conduct animal experiments on an aqueous mixture containing 25 groundwater contaminants and that a reasonable estimate of intake of individual chemicals can be achieved provided that dosing solutions are prepared fresh at frequent intervals (e.g., 48 to 72 hours).

Two findings from another related study by Hong et al. (1991) warrant special attention. These include, first, even at the medium dose level (5 percent chemical mixture stock), a reduction of 24 percent of progenitor cells was observed without any significant changes in body and organ weights, bone marrow cellularity, histopathological and hematological parameters after 108 days of treatment, and second, when mice pretreated with a mixture of chemicals were subjected to subsequent whole body irradiation during the recovery period, there was a significant dose-related CFU-GM difference between the vehicle control and the chemically treated mice. In *summary*, the previously listed study indicates that female mice treated *with a simulated chemical mixture of groundwater contaminants showed significant suppression in granulocyte-macrophage progenitor cells that suggested that environmental pollutants, acting in concert at a relatively high level, may cause residual marrow damage in mice with effective compensation to maintain normal circulating leukocyte or erythrocyte levels. Thus, long-term exposures to highly contaminated (i.e., causing health problems) groundwater represent potential long-term risks to the hematopoietic system.*

Other models developed from studies on contaminated groundwater around hazardous waste sites have been used to investigate the effects of 25 chemical mixtures on spermatogenesis in B6C3F1 mice (Chapin et al., 1989; Germolec et al., 1989; Yang et al., 1989) for chemical mixture formulation and contents. In these studies, the animals consumed three different concentrations of a mixture for 90 days, after which time they were euthanatized. Although there was a concentration-related decrease in the amount of fluid consumed at the higher two concentrations, there were no differences in body weight among the groups. Similarly, there was no effect of mixture consumption upon the histology of liver, kidney, testis, epididymis, or seminal vesicles or upon the absolute organ weights of these organs. The study by Chapin et al. (1989) then showed that at exposure levels that decreased fluid intake and increased adjusted kidney weight, there were no effects of this mixture on gametogenesis in male mice. Also, the exposure of mice to this mixture of 25 chemicals in the drinking water did not alter the microscopic structures of the liver, kidney, testis, epididymis, prostate, or seminal vesicles, and permatid production by the testis was unaltered, providing quantitative support for the qualitatively negative histopathology findings.

Additionally, body weights in the treated groups did not differ from controls at any time point measured and absolute kidney weights appeared to be increased, though not significantly, in the two highest dose groups. However, when expressed per body weight, these increases became significant. The lack of significant association between fluid volume and kidney weight suggested that these weight changes were a result of exposure to the 25-chemical mixture and not to dehydration. In general, phenol and all six inorganics were consistently recovered quantitatively with the volatile organics varying in their recovery rates and also displaying significant analytical differences.

Risk Factors and Radium

Several states in the U.S. have significant levels of ^{226}radium and ^{228}radium in groundwater used for public water supplies (Hallenbeck, 1989). The cancer risk posed by this exposure can be assessed by utilizing data developed from the long-term follow-up of a particular occupationally exposed group, the radium dial workers, although the derivation of the EPA risk factors is somewhat obscure according to Hallenbeck (1989). Hence, a single risk factor was developed for assessing the risk due to exposure to ^{226}Ra and ^{228}Ra in drinking water. The value of this risk factor assessed was 2×10^{-7} (upper 99 percent limit $= 3.1 \times 10^{-7}$) excess cases of cancer per person exposed per year per pCi/liter of exposure. This risk factor has then been used to develop equations for the calculation of excess lifetime risk and number of cases generated per year of exposure.

The risk factors developed by Hallenbeck (1989) were based on unusually complete human exposure-response data derived from the long-term observation of radium dial workers, although there were, however, two sources of unquantifiable uncertainty in both analyses. First, there was the need to extrapolate the existing dial worker data downward over 2–3 orders of magnitude of systemic intake, and second, the health effects (bone sarcoma and head carcinoma) observed in the female dial workers reflected long-term observation (about 45 years) following a relatively short period of radium intake (2.8 years) at a certain age at first exposure (20.3 years).

Overview of Risk Assessment [URL Ref. No. 235]

Perspectives on risk assessment include many aspects, and with groundwater, concerns range from its contamination to its effects on human health. Approaches dealing with risk assessment and groundwater cover consideration of selected chemical properties of materials to calculation of numerical indices and presentation of information on the probabilities and likely consequences of catastrophic events. For example, commonly used risk assessment guidance is not adequate when applied to solvent-contaminated soils in arid environments (Korte et al., 1992). The equations that are recommended for calculating show that such soils will affect groundwater assuming that liquid phase leaching controls contaminant migration. Also, if vapor phase migration is to be considered, diffusion is assumed to be the dominant process. Although, in contrast, a field study performed at an industrial site in Southern California, as shown in Korte et al. (1992), demonstrated that leaching could not account for the transport of contaminants to the water table, and the recent technical literature suggested that gravity-induced vapor migration may be the principal mechanism for vapor phase migration.

An *overview* of *the risk assessment process* pertains to various aspects of groundwater contamination and remediation [URL Ref. No. 303] (Kaplan and McTernan, 1993). And, pollution control programs generally do not focus on the concept of a residual risk or failure, although assessments and risk management must be incorporated by groundwater professionals into the design, construction, and operation of processes associated with groundwater pollution evaluation and control. Consequently, risk assessment has assumed major status in the regulation and management of groundwater resources.

Spurred by federal legislation and by requirements for liability insurance, risk assessment has moved from a conceptual process to an implemented tool in many instances. Statistical formulations [URL Ref. No. 339 {31}], for example, have been incorporated into models that use computers to estimate contaminant source terms, to calculate the extents of solute migration, to estimate populations at risk and impact thresholds, and finally, to estimate risk to exposed human or animal populations of concern. Additionally, the somewhat optimistic view held by many concerning the usefulness of these techniques must be balanced by a view that the use of probability theory can generate a false sense of precision, and the practicing professional must ultimately face questions such as, How sure is sure enough? How comprehensive is comprehensive enough? And, how clean is clean enough?

Large uncertainty still remains about the efficacy of various risk assessment techniques under the many different site-specific conditions that exist at most sites conducive to groundwater contamination problems. Part of the problem stems from a lack of appropriate data, part from lack of quantitative methods and models, part from a hesitancy to spend the funds necessary to develop needed methodologies, and part from regulatory safety margins set with little recognition of the desires of the costs to society.

The general principles of health risk assessment and management of toxic substances [URL Ref. No. 256, 295] are understood, but application of these principles is sometimes difficult because hazardous waste management and risk assessment processes are complex, and optimal decisions require multidisciplinary approaches, which are not always achieved. Inter- and intra-agency communication are very important and, if possible, risk assessment decisions should be made with pertinent facts related to the following:

(1) *Hazard identification* [URL Ref. No. 271]—the qualitative evaluation of the adverse health effects of a substance(s) in animals or in humans

(2) *Exposure assessment* [URL Ref. No. 272]—the evaluation of the types (routes and media), magnitude, time, and duration of actual or anticipated exposures and of doses, when known, and when appropriate, the number of persons who are likely to be exposed

(3) *Dose-response assessment* [URL Ref. No. 273]—the process of estimating the relation between the dose of a substance(s) and the incidence of an adverse health effect situation

(4) *Risk-characterization*—the process of estimating the probable incidence of an adverse health effect

Risk assessments may still be performed in response to short-term (acute) exposures from toxic substances, long-term (chronic) exposures in which no immediate threat to the public health is apparent, or to combinations of these. Risk assessments, in some instances, may require quantitative mathematical models [URL Ref. No. 339 {31}] to estimate exposure of the probability of an event or may even require immediate action to manage potentially adverse human health and environmental effects.

The following technical considerations, skills, and work tasks/activities then would be necessary when evaluating risks associated with toxic chemicals. The interplay and relationships with groundwater are evident.

(1) *Work knowledge and skills:* toxicology, environmental toxicology, ecotoxicology, chemistry (analytical and organic), biochemistry, engineering, environmental, safety, quality assurance, epidemiology, statistics, biometry, biostatics, medicine, mathematical modeling, dispersions, predictive capacity, hydrogeology, industrial hygiene, environmental health, risk assessment, ecology, biology, microbiology, cartography, physical geography, computer use, data management, explosives, sociology, physics, communications, and other related disciplines

(2) *Work tasks and activities:* environmental surveys, epidemiological studies, analysis and collection of data, geological, hydrogeological, and meteorological investigations, sampling and analysis of groundwater, air, soil, etc.; integration and blending of data; detection and reporting of problems, cooperation with other agencies; dispersion modeling, communication, risk assessment, inventory of materials located in area, monitoring programs (air, water, and other), survey of human population and other biota in area, assessment of potential risks from exposure, engineering surveys, and assessment of potential release of toxics

The reader of this book should recognize and remember that groundwater contamination restoration [URL Ref. No. 303] is not a distinct isolated hydrogeologic problem as it relates to risk assessment and evaluation. And that, the solution to groundwater risk assessment and evaluation as it relates to groundwater contamination risk analysis(es) and reclamation/restoration problems will always require involvement from many entities and disciplines, including some managerial personnel, technical personnel, and remedial-related personnel from the construction industry. Institutional personnel from different levels of government, such as, individuals from international concerns, also need to participate. Multidisciplinary approaches involving teams and individuals from each of the previously mentioned categories are also needed. Hydrogeologists, environmental health scientists, epidemiologists, environmental engineers, scientists, ecotoxicologists, chemists, geologists, risk analysts, and other staff personnel must work together toward common goals and objectives.

Hazard Ranking Information (HRI) and Site Rating Methodology [URL Ref. No. 274] in Relation to Risk Assessment and Groundwater Contamination

The U.S. Environmental Protection Agency [URL Ref. No. 19] developed the Hazard Ranking System (HRS) as a method for ranking toxic material facilities for remedial action according to risks to health and the environment (Canter and Knox, 1986). HRS was meant to be a scoring system designed to address problems resulting from movement of toxic substances through many sources, including groundwater. It reflected the potential for harm to humans or the environment as a result of migration of a toxic substance away from a central facility via different routes (e.g., surface water, air, groundwater, etc.). The listing of factors considered and involved in transport included the following:

• migration, containment routes, characteristics of waste, hazardous waste quantities, targets; fire and explosion, containment, waste characteristics, quantity, targets; direct contact, accessibility containment

- depth to aquifer, net precipitation, permeability, physical state, toxicity, waste quantity, distance to nearest well (downgradient), population served by groundwater, evidence of ignitability or explosivity, reactivity, distance to human population (sensitive environments and habitat), land use, and accessibility

Additionally, the HRS was designed not to result in quantitative estimates of the probability of harm from a facility, but to provide a rank-order mechanism in terms of a potential hazard or hazards. The HRS was also designed to evaluate: the physical containment of toxic substances, routes by which releases would occur, characteristics of harmful substances, and potential targets.

The SRM (Site Rating Methodology) process has been used to prioritize Superfund [URL Ref. No. 264] sites for the U.S. EPA in terms of remedial actions that are necessary. Basically, the SRM includes the following:

(1) A system for rating the general hazard potential of a site [Rating Factor System (RFS)]
(2) A system for modifying the general rating based on site-specific problems [Additional Points System (APS)]
(3) A system for interpreting the ratings in meaningful terms (Scoring)

The RFS is used to establish an initial rating of a site based on a set of 31 factors (Kufs et al., 1980; Canter and Knox, 1986) that which include the following: population within 1,000 ft, distance to nearest drinking water well, distance to nearest offsite building, land use/zoning, critical environments, evidence of contamination, level of contamination, type of contamination, distance to nearest surface water, depth to groundwater, net precipitation, soil permeability, bedrock permeability, depth to bedrock, toxicity, radioactivity, persistence, ignitability, reactivity, corrosiveness, solubility, volatility, physical state, site security, hazardous waste quantity, total waste quantity, waste incompatibility, use of liners, use of leachate collection systems, use of gas collection systems, and use and condition of containers.

For each of these previously listed 31 factors, there is a four-level rating scale that has been developed (i.e., 0, 1, 2, 3: with 0 indicating no potential hazard to 3 indicating a high hazard). The scales are defined so that they could be summarized on the basis of information obtained from published materials, public records, or other sources of reliable and verified information. And, once a site has been rated using the SRM, the scores are then interpreted with the following concerns in mind:

(1) Is collection of additional background information still necessary?
(2) Have complete and adequate surveys of sites been conducted?
(3) Are complete investigations of sites current and reliable?

(4) Has implementation of remedial action been started?
(5) Have enforcement cases been prepared in situations where this is necessary?

A Case History Hazard Ranking System (HRS) Example Taken from the U.S. Department of Energy Hanford Site

[URL Ref. No. 72, 132, 164–165, 274, 279, 281, 292]

More than 1,500 waste-disposal sites have been identified at the U.S. Department of Energy (DOE) Hanford Site (Sherwood et al., 1990), and at the request of the U.S. Environmental Protection Agency, these sites were aggregated into four administrative areas for listing on the National Priority List. Within these four established aggregate areas, 646 inactive sites were selected for further evaluation using the Hazard Ranking System (HRS). Evaluation of inactive waste sites by the HRS process provided valuable insights into designing a focused radiological and hazardous substance monitoring network that was expanded to address not only radioactive constituents but also hazardous chemicals. Since the HRS scoring process also considered the likelihood of groundwater contamination from past disposal practices at inactive waste sites, the network at Hanford was essentially designed to monitor groundwater at the facilities identified for ^{129}I, ^{99}Tc, ^{90}Sr, uranium [URL Ref. No. 283], chromium, carbon tetrachloride, and cyanide.

Additionally, in 1985, the DOE published DOE Order 5480.14 [URL Ref. No. 275], a directive to organize an inactive-waste-site evaluation program that was to parallel the Comprehensive Environmental Response, Compensation, and Liability Act (CERCLA) [URL Ref. No. 264] of 1980. In this case, the CERCLA Act was used by the U.S. EPA to regulate nongovernment inactive waste sites (U.S. EPA, 1988). And, in order to identify contaminants, a focus on characteristics of the HRS groundwater route that applied to the groundwater monitoring network design was established. Results were then used to aid future characterization, assessment, and remediation of inactive Hanford sites.

The HRS Groundwater Route

The breakdown evaluation of relative hazards from the groundwater pathway at Hanford included five components: observed release, route characteristics, containment, waste characteristics, and target populations. However, only four components were considered in a groundwater route evaluation, with either observed release or route characteristics being used with the other four components.

Observed releases were also known or circumstantial evidence for contaminant release into groundwater were also presented for the Hanford facility (Sherwood et al., 1990). If no releases were observed, route characteristics (i.e., including depth to aquifer), net precipitation, unsaturated permeability, and physical state of the waste were used to evaluate potential releases to groundwater with the final component including the distance to potentially exposed target populations and their groundwater uses. After appropriate values were then assigned to each property, a groundwater route score was calculated and ranking was evaluated.

Application to Network Design

Information on observed releases and waste characteristics at Hanford were used to select additional monitoring locations and analyses, with the observed constituents being used to identify inactive waste sites likely to have contaminated groundwater. This was done, however, before direct evidence of observed releases at the 646 inactive Hanford waste sites was available. This was because monitoring wells did not initially exist near all facilities, and groundwater plumes emanating from operating areas overlapped, inhibiting initial source identification. Consequently, to identify suspected releases of contaminated water to the aquifer, observed releases needed to be based on volume of liquid disposed at each site, physical size of the facility, and depth to groundwater.

The disposal facility area at Hanford, multiplied by the distance to groundwater, also provided a total soil-column volume between the waste site bottom and water table. And, consequently, sites with observed releases were then considered obvious locations for groundwater contamination, thus, adjacent wells were added to the sampling and monitoring network. Waste data were reviewed to identify radioactive and chemical substances present in each inactive waste site that had an observed release.

A broad spectrum of radioactive and chemical substances [URL Ref. No. 232] potentially presented in groundwater at Hanford were also identified from inactive-waste-site inventories and their knowledge of contaminant mobility. Additionally, radionuclide inventories and radionuclide mobility were used to augment the list of radioactive constituents.

Tritium [URL Ref. No. 293], *gross alpha, gross beta,* and *gamma* scans were historically used to monitor radionuclide contamination in Hanford groundwater. Strontium-90 and ^{129}I were analyzed in order to assess their offsite migration route. Thus, radionuclide-specific analyses for ^{14}C, ^{63}Ni, and ^{99}Tc were added, and ^{90}Sr and ^{129}I analyses were expanded near observed release sites. If ^{14}C, ^{90}Sr, ^{99}Tc, and ^{129}I were present in liquid streams, their high mobility would then result in their release to Hanford groundwater.

A different approach at Hanford was undertaken for evaluation of the presence of hazardous chemicals. Nitrate ion [URL Ref. No. 201] was the contaminant most often investigated in past groundwater monitoring efforts. Expansion of

hazardous chemical monitoring included establishing background or naturally occurring concentrations of certain constituents and identifying anthropogenic substances from past liquid discharges. Since January 1, 1987, 226 of 484 wells previously monitored for radiological constituents have been analyzed for a broad spectrum of radioactive, inorganic, and organic constituents, including select radionuclides, cations, anions, trace metals, volatile organics, and cyanide. And, the results of the expanded monitoring network identified several new contaminants, including ^{99}Tc, carbon tetrachloride, and cyanide, thus establishing a link between past disposal practices and existing contaminant plumes. A more direct link was established between disposal activities and the presence of ^{129}I, ^{90}Sr, uranium [URL Ref. No. 283], and chromium in Hanford groundwater

Technetium-99 was observed in groundwater from wells across the site with concentrations greater than the maximum concentration limit (MCL) of 900 pCi/l (U.S. EPA, 1976a) detected. Maximum ^{99}Tc concentrations, 29,100 pCi/l, were also detected north of the 200-East area. Carbon tetrachloride exceeded the 5-ppb MCL in 48 wells, and maximum concentrations, 5,550 ppb, were reported from wells near the Plutonium [URL Ref. No. 284] Finishing Plant (PFP).

Cyanide was also detected in isolated locations within the 200-East and 200-West areas, and north of the 200-East area, and concentrations of ^{129}I exceeded the 1-pCi/l MCL (U.S. EPA, 1976a) in a widely dispersed area between the 200-West and 200-East areas at Hanford and the Columbia River. Maximum ^{129}I concentrations, 87.8 ppb, were also detected near the 200-West area, and concentrations of ^{90}Sr greater than the 8-pCi/MCL were reported from throughout the Hanford site.

The expanded groundwater monitoring program at Hanford, in essence, identified and reported contaminants in groundwater on a site-wide basis. In many cases, the data linked groundwater contamination with known sources, and each known source then became the focus of a Remedial Investigation/Feasibility study under CERCLA [URL Ref. No. 264] or a Facility Investigation/Corrective Measures Study under RCRA [URL Ref. No. 280]. Information obtained through the expanded groundwater monitoring program in turn provided the technical basis to conduct waste-site investigations at inactive waste sites at Hanford.

General Summary of Risk Assessment
[URL Ref. No. 235] Strategies
and Groundwater Contamination

Risk assessment strategies should focus on groundwater contamination prevention activities that additionally include risk management concepts. For example, preventive activities that involve conducting underground storage tank [URL

Ref. No. 85 {20}, 251] and underground pipeline evaluations prior to installation contribute to the ultimate goal of protecting groundwater supplies. Since preventive programs involve many disciplines, individuals working in this area must be willing to participate in a broad integrated approach to hazardous waste regulations and programs [URL Ref. No. 68–69], hazardous waste assessment methods and techniques, survey techniques, sampling techniques and strategies, enforcement procedures/legal support, environmental toxicology (ecotoxicological concepts and principles) [URL Ref. No. 260], epidemiological investigations, site assessment and remediation [URL Ref. No. 303] knowledge, personal protection and safety/environmental health interactions [URL Ref. No. 166], and other related hydrological disciplines/concerns [URL Ref. No. 13–14, 187]. Individuals should also have knowledge concerning how to interpret risk assessment [URL Ref. No. 140, 235, 260–262, 336 {12}, 338 {39}], what information still needs to be generated, and the relationship of groundwater to risk assessment. These specifically should include the following:

(1) The physical and chemical characteristics of the waste in the regulated unit, including its potential for migration

(2) The hydrogeological characteristics of the facility and surrounding land

(3) The quantity of groundwater and the direction of groundwater flow

(4) The proximity and withdrawal rates of groundwater users

(5) The current and future uses of groundwater in the area

(6) The existing quality of groundwater, including other sources of contamination and their cumulative impact on the groundwater quality

(7) The potential for health risks caused by human exposure to waste constituents

(8) The potential damage to wildlife, crops, vegetation, and physical structures caused by exposure to waste constituents

(9) The persistence and permanence of the potential adverse effects

Nonradioactive Hazardous Waste and Groundwater Contamination Interactions [URL Ref. No. 144, 266, 279, 336 {68}, 337 {111}, 338 {17}]

Use of Selected U.S. Department of Energy Facilities as Case Models

Baseline Information on U.S. Department of Energy (DOE) [URL Ref. No. 88]

In order to comprehend this chapter relating to the U.S. Department of Energy (DOE), some basic fundamental information on the DOE is initially presented so that a logical systematic flow of information related to groundwater contamination occurs. How the DOE interrelates to hazardous waste, radioactivity, and restoration/remediation groundwater contamination concerns is presented in this chapter and in Chapter 4. After the U.S. Department of Energy information is presented, the Los Alamos National Laboratory and other DOE subcontractor nuclear facilities and national laboratories, are used as main case history models in which information is presented in more depth on how these particular DOE facilities are participating and approaching groundwater contamination problems, in terms of the past, present, and the future. Internet WWW references that do not always support or agree with the U.S. DOE are also presented [URL Ref. No. 339 {43–101}] in this chapter.[1] For purposes of this chapter, I have adapted information from U.S. Department of Energy publications (1995a–1995l), the Federal Register, [URL Ref. No. 21, 68–69], and U.S. DOE Baseline Environmental Management Reports [URL Ref. No. 88].

General Summary of the U.S. Department of Energy (DOE)

During World War II and the Cold War, the United States developed a vast network of industrial facilities for the research, production, and testing of nuclear weapons, known as the nuclear weapons complex (U.S. DOE, 1995c, 1995d,

[1] URL Ref. No. 339 {43–101} provides information from the U.S. DOE complex concerning various whistleblower complaints and actions.

1998; Anonymous, 1965, 1996; MacDonald, 1999; Atomic Energy Commission, 1973). It included thousands of large industrial structures such as nuclear reactors, chemical processing buildings, metal machining plants, and maintenance facilities. During the last 50 years, this enterprise manufactured tens of thousands of nuclear warheads and detonated more than a thousand. The U.S. Department of Energy [URL Ref. No. 61, 64], the name of the present federal agency responsible for managing the nuclear weapons complex, still manages more than 120 million square feet of buildings and 2.3 million acres of land (i.e., an area larger than Delaware, Rhode Island, and the District of Columbia combined).

In addition to creating an arsenal of nuclear weapons, the DOE complex left an unprecedented environmental legacy, and because of the priority on weapons production, the treatment and storage of radioactive [URL Ref. No. 304] and chemical waste [URL Ref. No. 256] was handled in such a way that led to contamination of soil, surface water, and *groundwater,* including an enormous backlog of waste and dangerous materials.

The cost of dealing with these problems precipitated by the U.S. nuclear complex can be considered a Cold War Mortgage (U.S. DOE, 1995c, 1995d). Paying the mortgage will take decades and substantial resources comparable to the level of effort initially expended for nuclear weapons production and research activities (Long, 1995). This could even be delayed if something such as a World War III were to break out.

The DOE Environmental Management Program (U.S. DOE, 1995c, 1995d) [URL Ref. No. 61]

The Office of Environmental Management at DOE [URL Ref. No. 61] was created in 1989 to help address the environmental legacy of nuclear weapons production and other sources such as nuclear research programs. Activities that encompass the Environmental Management Program include environmental restoration, waste management, nuclear material and facility stabilization, technology development, and landlord functions (e.g., fire-fighting response, road maintenance, and utilities).

All of these activities are simplified as cleanup, but it is clear they involve a lot more than that (Higley and Geiger, 1990; Shanklin et al., 1995). Although most Environmental Management program work involves dealing with the legacy of contamination and the backlog of accumulated wastes, a significant amount of work also involves handling newly generated waste from various research programs that involve high and significant groundwater contamination problems.

The U.S. DOE is also responsible for storing, treating, and disposing of an extraordinary array of wastes and spent nuclear fuels. These wastes include a variety of physical forms (e.g., solids, liquids, and sludges), chemical types (e.g., solvents, metals, and salts), and sources (e.g., high-level waste [URL Ref. No. 305] from reprocessing, spent nuclear fuel from production reactors, and naval reactors), transuranic waste [URL Ref. No. 289] from plutonium [URL Ref. No. 284] operations, and low-level waste [URL Ref. No. 290] (i.e., which includes everything else that is radioactive waste) [URL Ref. No. 304].

Most of the wastes included in the Baseline DOE 1995 Report (U.S. DOE, 1995c, 1995d) [URL Ref. No. 88] were generated during the production of nuclear weapons during the Cold War. In the future, the DOE expects that the quantities of waste from these sources will decrease as pollution prevention efforts become more effective and nuclear weapons production activity decreases and that a new source of waste will become increasingly important, such as secondary waste generated as a result of environmental restoration and nuclear material and facility stabilization. Consequently, with the end of the Cold War, production of most nuclear weapons materials has been indefinitely slowed down or halted and many DOE facilities were now not needed as per their previous missions.

Hazardous Waste at DOE Facilities
[URL Ref. No. 279]

The U.S. Department of Energy (DOE) generates and manages large volumes of *hazardous* [URL Ref. No. 279] and radioactive wastes [URL Ref. No. 304] as part of their routine operation, and the activities related to groundwater contamination are ongoing and dynamic with the U.S. DOE [URL Ref. No. 45] and the U.S. Environmental Protection Agency (EPA) [URL Ref. No. 19] signing a Memorandum of Understanding agreement in February of 1984 that stipulated that DOE facilities that manage *hazardous waste* or *hazardous components* must comply with all EPA regulations [URL Ref. No. 21, 68–69] governing the generation and management of these wastes (U.S. DOE, 1990a). This requirement was formalized on April 13, 1994, when the U.S. District Court of Tennessee ruled in LEAF v. Hodel (586

F. Supp. 1163) that RCRA [URL Ref. No. 280] requirements are not inconsistent with the Atomic Energy Act (AEA). Additionally, this ruling was reinforced in May of 1987, when DOE issued an interpretive rule clarifying that RCRA [URL No. 280] applied the hazardous component of by-product material (52 FR 15937) [URL No. 67], while the radioactive component was still to be regulated under the AEA [URL Ref. No. 330] (10 CFR 962) [URL No. 67].

RCRA [URL Ref. No. 280] Subtitle C regulations, in Chapter 40 of the Code of Federal Regulations, Parts 260–272 [URL No. 69 {10}], therefore, set cradle-to-grave standards for the generation, transport, treatment, storage, and disposal of hazardous wastes. And, included among the many facility design and operating standards that must be specified in RCRA [URL No. 280], permits *became strict closure and post-closure care requirements for hazardous waste management units* [URL Ref. No. 300]. The RCRA [URL No. 280] Subtitle C closure and post-closure care requirement then stipulated that DOE Operations staff, DOE facility staff, and facility contractor staff at many DOE facilities located throughout the U.S., were responsible for oversight and compliance.

Federal Regulations [URL Ref. No. 21, 68–69] and Hazardous Waste Units [URL Ref. No. 300] at DOE facilities governing the closure of hazardous units on DOE facilities also centered around RCRA [URL Ref. No. 280] requirements that included permitting programs, facility design and operating standards, and proper requirements for closure and post-closure of hazardous waste management units (i.e., which include landfills, waste piles, container storage areas, incinerators, underground injection wells, surface impoundments, land treatment units, tanks, miscellaneous units, thermal treatment units, and chemical, physical, and biological treatment).

Hazardous wastes, under RCRA [URL No. 280], in turn, are then defined as a subset of solid wastes, therefore, unless a waste is first a solid waste, it cannot be considered a hazardous waste. Solid wastes are explained in 40 CFR 262.1 [URL Ref. No. 69 {10}] as any material that is disposed of, burned or incinerated, recycled, or considered inherently waste-like, regardless of whether it is a solid, semi-solid, or liquid, and the terms recycled and inherently waste-like, also mean regardless of whether it is of solid, semi-solid, or liquid material content (52 *FR* 11147) [URL No. 67].

Hazardous waste by RCRA [URL Ref. No. 280] definition (40 CFR 261.3) [URL No. 69 {10}] includes answering the following questions to determine if a solid waste qualifies:

(1) Does the waste exhibit any one of the four *characteristics* of a hazardous waste identified in 40 CFR Part 261 Subpart C [URL Ref. No. 69 {10}]?

(2) Has the waste been listed as a hazardous waste in 40 CFR Part 261 Subpart C [URL Ref. No. 69 {10}]?

(3) Is the waste a *mixture* containing a listed hazardous waste and a nonhazardous waste?

(4) Is the waste *derived from* the treatment of a listed hazardous waste?

(5) Is the waste not *excluded* from regulation as a hazardous waste?

(6) Is the waste a hazardous waste *contained in* an environmental medium such as *groundwater* or soil?

If the answer is *yes* to any one of the preceding questions, *the solid waste then can be considered a hazardous waste.*

On March 29, 1990, the EPA adopted a rule that replaced the Extraction Procedure (EP) with the TCLP and expanded the list of toxic constituents from 14 to 39. This new rule (51 *FR* 21648) [URL No. 67] had a major impact on whether a solid waste is designated as a hazardous waste. This rule also significantly increased the amount of wastes DOE facilities must consider as hazardous (e.g., radioactive mixed wastes) [URL Ref. No. 291].

A number of common waste streams were excluded from regulation under RCRA's [URL Ref. No. 280] hazardous waste program. These wastes included household wastes [URL Ref. No. 334], municipal resource recovery wastes, agricultural wastes, and mining overburden returned to the mine site. Oil and gas wastes [URL Ref. No. 237, 254], mining wastes [URL Ref. No. 230, 309], and cement kiln dust were also exempted from Subtitle C regulation. Used solids that were generated by small quantity generators or used oil that exhibits a hazardous characteristic but is recycled in some way other than being burned for energy recovery were also exempt. Used oil burned for energy recovery was also exempt from Subtitle C but was regulated under 40 CFR 266 [URL No. 69 {10}]. In addition to hazardous wastes, RCRA identified a large number of hazardous constituents. These constituents pose a threat to human health and the environment and form the basis for *listing* a solid waste as hazardous. These constituents also play an important role in RCRA [URL Ref. No. 280] closure regulations, although by themselves, their presence in a waste stream does not automatically define the waste stream as hazardous under RCRA.

General Information Related to Hazardous Waste Generation at U.S. DOE Facilities

The creation of each gram of plutonium [URL Ref. No. 284], reactor fuel element, and container of enriched uranium [URL Ref. No. 283], remains with us today (Anonymous, 1965). Not only do all the wastes remain, they pose a variety of hazards, and many are so toxic that they must be isolated for hundreds of centuries. Consequently, they need special treatment before they can be permanently disposed of. This is because waste has been the most abundant product of every step in the weapons production process: uranium [URL Ref. No. 283] mining and milling, uranium enrichment, handling spent fuel, spent fuel reprocessing, and plutonium [URL Ref. No. 284] production and plutonium

parts manufacture also, are still with us. We might have only changed the location of storage.

Waste Tanks [URL Ref. No. 281]

Waste tanks such as the ones at the Savannah River Site, West Valley, and Idaho all have serious problems that deserve attention, but the leaking tanks at the Hanford site are infamous and are summarized next.

The tank farms at the Hanford Site [URL Ref. No. 281] hold 61 million gallons of liquids and sludges that include radioactive waste and spent fuel from nine weapons production reactors mixed with assorted hazardous chemicals, including nitrates [URL Ref. No. 201] and nitrites, chromium, mercury, and cyanide. Even as early as the 1990s, 24 of the tanks were considered to be in some danger of exploding. The 177 tanks hold wastes that were initially neutralized and then concentrated and mixed to reduce them in volume before storage. Because the wastes were mixed, each tank has different contents and ultimately presents different problems and remediation concerns.

Sixty-seven of 148 single-wall tanks that were built between 1943 and 1964 are known to have leaked or are suspected of leaking their toxic contents into the ground. And the most famous of Hanford's potentially explosive tanks, the 101-SY, slowly builds up hydrogen gas, which it periodically vented. However, in July, 1993, a pump that slowly mixed and circulated the liquid wastes was installed in this burping tank that prevented gases like hydrogen from accumulating to dangerous levels in the thicker sludges at the bottom of the tank.

Presently, Hanford's tanks [URL Ref. No. 281] must be properly managed until the wastes can be moved to a deep geologic repository [URL Ref. No. 315]. But, that transfer will not be easy because hydraulic sluicing may not be effective or useful on hardened sludges, and it may not be environmentally sound. The strategy that needs to be used must have proven separation technologies, and advanced methods are needed to remove a greater proportion of radionuclides or decrease the actual amount of high-level wastes [URL Ref. No. 305].

Environmental Cleanup at U.S. DOE Facilities

The Galvin Report [URL Ref. No. 282] strongly criticizes the DOE's cleanup efforts, particularly management's perceived failure to provide leadership (Campbell and T.Z.C., 1995). Some particularly angry words were also reserved for Congress, Superfund legislation, and excessive oversight by the DOE. The report suggests that too many regulations, combined with litigation, force contractors to perform ineffectual make-work cleanup before a good technical solution can be developed. The task force proposed establishing an Environmental Advisory Board to help coordinate basic research, applied research, and field engineering and remediation efforts. The U.S. DOE National Laboratories could help by identifying technical barriers to cleanup and pointing out problem areas within various DOE complexes.

Sources of Pollutants Related to Groundwater Contamination at U.S. DOE Facilities

Production of nuclear weapons in the United States requires the use of a vast array of facilities—mines, laboratories, nuclear reactors, chemical plants, machine shops, and test sites—and at all sites where these activities take place, some environmental contamination in the form of groundwater pollution occurs. In some instances, the groundwater contamination is contained and poses no immediate risk to people and the environment. In other areas, however, the contamination is extensive enough to have polluted not only the surrounding soils but also vast areas. Since most waste generated by the DOE is radioactive, it therefore cannot be eliminated—it can only be contained in a safe manner while its radioactivity diminishes through time.

At the core of the weapons-manufacturing process is the production of three materials—highly enriched uranium [URL Ref. No. 283], plutonium [URL Ref. No. 284], and tritium [URL Ref. No. 293]. The production of these nuclear materials required the most complicated facilities in the weapons complex and was responsible for most of the environmental legacy of the Cold War. In addition, some major waste problems have been created that in turn can have a severe detrimental effect on groundwater contamination. A listing of these wastes is presented next.

Uranium Mining and Milling Wastes
[URL Ref. No. 310]

The United States mined about 60 million tons of ore to produce uranium for nuclear weapons production. Mining and milling produced large volumes of a sand-like by-product called mill tailings, which contain toxic heavy metals [URL Ref. No. 317] and radioactive radium and thorium [URL Ref. No. 339 {35}]. Although there is a large volume of this material, it represents only a small fraction of the total radioactivity managed by the Environmental Management program [URL Ref. No. 61].

Uranium Enrichment Wastes

To make highly enriched uranium, enrichment plants removed and separated ^{235}uranium from ^{238}uranium. Enrichment plant operations produced large volumes of enriched uranium and environmental contamination with radioactive materials, solvents, polychlorinated (PCBs), heavy metals, and other toxic substances.

Fuel and Target Fabrication Wastes

The conversion of uranium hexafluoride gas into metal is included, and the main types of environmental legacies from these operations are unintended releases of uranium dust, landfills contaminated with chemicals, and contaminated facilities.

Reactor Irradiation Wastes

Uranium targets were irradiated in production reactors to produce plutonium. Their main environmental legacy is highly radioactive spent fuel and contaminated facilities.

Chemical Separation Wastes

The chemical separation of fission products from uranium and plutonium generated more than 100 million gallons of highly radioactive *and hazardous chemical waste, some of which was discharged directly into the ground.* Waste from reprocessing contains the vast majority of the total radioactivity managed by the Environmental Management program, much of it emitted from long-lived radioactive elements *that could pose hazards for tens of thousands of years.* Chemical separations also left a legacy of contaminated facilities.

Fabrications of Weapons Components Wastes

Plutonium [URL Ref. No. 284] was machined into warhead components. The weapons laboratories also used plutonium to make and test prototype designs for weapons. Waste from this process is mostly plutonium-contaminated (transuranic) waste.

Weapons Assembly and Maintenance Wastes

Factories contributed nonnuclear components for the final assembly of nuclear weapons. The environmental legacy includes soil contaminated with high-explosive waste, fuel and oil leaks, and solvents.

Research, Development, and Testing Wastes

More than 1,000 nuclear devices were exploded in atmospheric, underwater, and underground tests. The environmental legacy includes hundreds of highly radioactive underground craters and soils and debris contaminated with highly explosive materials and other chemicals.

Other Related Source Wastes

Although the environmental costs of nuclear weapons production are substantial, the Environmental Management program at DOE addresses a legacy of waste from nonweapons production as well as wastes generated by ongoing activities. The nonweapons legacy wastes are those wastes associated with cleaning up waste generated from past activities, such as energy research, basic science, and the Three Mile Island nuclear plant accident. For example, Brookhaven National Laboratory's [URL Ref. No. 45, 64] environmental restoration activities were focused on remediation of contamination of soil, surface water, and possibly *groundwater*, resulting from research and development work by the U.S. Army and the DOE since 1947. At the Princeton Plasma Physics Laboratory, New Jersey, which carried out nuclear fusion research and development for the DOE for more than 40 years, contamination sources included former wastewater treatment plant facilities, a cooling tower and its adjacent soils, the chromate reduction pits, and *a hazardous waste accumulation area.*

Environmental Remediation/Restoration at DOE Facilities and Relationship to Groundwater Contamination
[URL Ref. No. 285]

Environmental restoration within the DOE [URL Ref. No. 61] involved 10,500 potential release sites, which were grouped into 614 subprojects, or operable units (U.S. DOE, 1995c, 1995d). The subprojects, costs, and activities from these sites, formed the basis for tracking the costs of projects as presented in Volume II of the DOE publications (U.S. DOE, 1995c, 1995d).

To establish the case for environmental restoration, the DOE depended on ongoing baseline efforts [URL Ref. No. 88] where all sites in the complex have or are completing baseline estimates for all potential release sites and all surplus contaminated facilities that have been stabilized. Information gathered in the baseline studies included an extensive set of site-specific assumptions about the nature and extent of contamination, ultimate land use, and remedial strategies. And, once a level of contamination was established or assumed, remedial actions were divided into *two categories that included those directed at containing contaminants to prevent them from migrating from the source and those directed at eliminating the contamination.*

Groundwater Contamination at U.S. DOE Sites

The groundwater has been contaminated at most major DOE sites, and the principal contaminants are volatile organic compounds [URL Ref. No. 247, 267], heavy metals, and radionuclides. And, because current technologies are in-

effective in most instances, DOE report estimates do not assume that all groundwater will be remediated to drinking water standards. Instead, the U.S. DOE report estimate reflects a spectrum of measures aimed mainly at preventing further contaminant migration and protecting off-site populations (U.S. DOE, 1995c, 1995d). The measures used include the following [URL Ref. No. 88]:

- *Source of elimination:* Most sites eliminate the source of groundwater contamination by removing the contaminant or capping the contaminated area to prevent further leaching. Generally, the baseline report [URL Ref. No. 88] estimate includes the cost of source elimination at all sites.

- *Containment:* Some DOE sites are planning to contain contaminant migration in groundwater by using slurry walls or barriers or by innovating pumping actions. Where containment is the most cost-effective option, the baseline report [URL Ref. No. 88] estimate reflects it.

- *Natural attenuation:* The concentration of some naturally occurring contaminants (e.g., uranium) in groundwater will return to natural levels before the contaminants can reach any off-site users. And, certain short-lived radionuclides (e.g., tritium) will decay to safe levels before they reach off-site receptors. Where natural attenuation is the assumed strategy, the baseline report estimate includes costs for monitoring but not for remediation.

- *Pumping and treating:* Costs for this remedial action are included in the report estimate for a few sites, mainly those where remediation has already started (e.g., Kansas City Plant, Savannah River Site). Because this costly method can take many years, and its efficacy has not been established, it is not the dominant strategy reflected in the report estimate.

Groundwater contamination and prevention or stabilization of such, is a technical challenge for the program [URL Ref. No. 61], and remediation of currently contaminated groundwater accounts for less than 5 percent of the estimated life-cycle cost of environmental restoration. The DOE, however, also plans to monitor and contain contamination to the extent possible. All primary sources of contaminants that can migrate to groundwater are assumed by the DOE to be addressed in their baseline program [URL Ref. No. 88] and are within the scope of this estimate. The DOE also hopes that in the future they will address these problems more effectively through their technology development efforts that they have defined as involving activities for waste management that include treatment, storage and handling, and disposal of waste. These activities are discussed next (U.S. DOE, 1995c, 1995d).

Treatment

Four treatment projects are planned or are in progress by the U.S. DOE for *high-level wastes*[1] stored in tanks at Hanford, the Savannah River site, the Idaho Chemical Processing Plant, and the West Valley Demonstration Project in New York. Secondary wastes (e.g., *low-level* and *low-level mixed wastes*) from these treatment facilities are assumed to already have been processed and prepared for disposal in approved facilities.

Treatment facilities for low-level mixed waste are being planned through the consultant process under the Federal Facility Compliance Act that includes treatment at 34 generator sites. The treatment facilities include new ones planned for large sites that use a variety of technologies. Unless otherwise specified by an individual site, facilities are assumed to require 10 years for research, development, design, permitting, construction, and start-up activities. Treatment for *low-level* and *transuranic waste* consists of characterization, packaging, and, if necessary, processing to meet criteria for disposal. These functions will be performed at the generating sites.

Additionally, according to U.S. DOE (1995c, 1995d) [URL Ref. No. 88] reports, all sites manage hazardous and sanitary wastes within their waste management programs, and efforts are made to prevent groundwater contamination.

Storage and Handling

High-level waste [URL Ref. No. 305] is stored in tanks or bins at Hanford, Savannah River, the Idaho Chemical Processing Plant, and West Valley. The DOE Case cost estimate includes complete life cycle for the tank farms, storage facilities, and transfer facilities, and costs include facility upgrades and decommissioning once the mission is complete. Estimates for proper disposal also include costs for storing canisters of vitrified high-level waste pending disposal in a permanent repository.

Spent nuclear fuel is currently stored at 13 sites within the complex, with approximately 99 percent stored at the Hanford Site [URL Ref. No. 292], the Idaho National Engineering Laboratory, the Savannah River Site, and the Oak Ridge Reservation. The Base Case cost estimates include life cycle cost of storing spent nuclear fuel prior to disposal at a national geologic repository, which is assumed to be available by 2016. The Base Case does not include costs for the remaining 1 percent of spent fuel stored at various locations throughout the complex.

Additionally, transuranic wastes [URL Ref. No. 289] are stored at 10 installations, primarily at the Idaho National Engineering Laboratory, Hanford, Savannah River, and Los

Alamos National Laboratory and that are now headed to the Waste Isolation Pilot Plant (WIPP) [URL Ref. No. 286] which is located in southern New Mexico.

Disposal

High-level waste and spent nuclear fuel, according to the U.S. DOE, will eventually be disposed of in a deep geologic repository developed and operated by the Department's Office of Civilian Radioactive Waste Management. And, it is assumed that DOE wastes would be accepted by a permanent repository beginning in 2016. Transuranic waste is stored in a retrievable manner pending the opening and use of a geological repository such as WIPP [URL Ref. No. 286]. Low-level waste [URL Ref. No. 290] also continues to be disposed of at the existing six disposal sites, which include Hanford, Savannah River, the Oak Ridge Reservation, Los Alamos National Laboratory, Idaho National Engineering Laboratory, and the Nevada Test Site. Engineered disposal vaults are assumed to be constructed at Savannah River and Oak Ridge, and shallow land disposal was assumed to continue at the remaining sites. Low-level mixed waste, after treatment to meet regulatory standards, is also assumed by the U.S. DOE to be disposed of in approved disposal facilities. The Base Case assumes that disposal facilities for low-level mixed waste will be provided at the six sites previously listed.

Regulatory Requirements for Groundwater Monitoring Networks at U.S. DOE Hazardous Waste Sites [URL Ref. No. 109]

In the absence of an explicit national legislative mandate to protect groundwater quality and because there is no coordination between federal and state agencies in some instances, those responsible for hazardous waste management and cleanup must utilize a number of statutes and regulations as guidance for detecting, correcting, and preventing groundwater contamination (Keller, 1990; Lesage and Jackson, 1992). Many federal/state environmental pollution control statutes/regulations that relate to prevention of groundwater contamination are relevant to such programs and include the Resource Conservation and Recovery Act (RCRA, 1984) [URL Ref. No. 280]; the Comprehensive Environmental Response, Compensation, and Liability Act (CERCLA, 1986) [URL Ref. No. 264]; the Safe Drinking Water Act (SDWA, 1986) [URL Ref. No. 258]; the Clean Water Act (CWA, 1987) [URL Ref. No. 287]; the Low-Level Radioactive Waste Policy Act (LLRWPA, 1980) [URL Ref. No. 290]; and the Nuclear Waste Policy Act (NWPA, 1982) [URL Ref. No. 278].

The RCRA [URL Ref. No. 280] establishes monitoring requirements for hazardous and solid waste facilities that

---

[1] Low-level, low-level mixed wastes, transuranic wastes, and high-level wastes are defined in Chapter 4.

might leach contaminants into groundwater, and under RCRA, operators of such facilities must implement programs to determine the site's impact on groundwater quality. RCRA also contains guidelines for establishing groundwater monitoring systems, applying protection standards, and determining points of compliance.

CERCLA [URL Ref. No. 264] requires groundwater monitoring in connection with cleanup activities, and monitoring begins when the site is characterized as to type, rate, and extent of contamination. Monitoring then continues through various planning stages to provide information necessary to design site-specific cleanup plans, and after cleanup, groundwater monitoring then must continue to determine if it was a success.

The SDWA [URL Ref. No. 258] protects drinking water primarily by setting standards, and the standards are often used to ensure that groundwater protection is appropriately considered at active and inactive hazardous waste sites. Strategies, policies, and guidelines such as those developed under the wellhead protection and the sole-source aquifer programs of SDWA can be used to develop comprehensive groundwater cleanup and protection programs.

The CWA [URL Ref. No. 287] is also the U.S. Environmental Protection Agency's (EPA) mechanism for helping states develop and implement groundwater protection strategies, and in addition to containing surface water standards, the CWA contains guidelines for controlling and monitoring discharges to surface waters from a single point. Amendments to the CWA, however, have authorized EPA to establish programs for managing discharges in surface waters from multiple sources.

The LLRWPA implements regulations that apply to release of radionuclides into groundwater, and they are of particular interest to managers of facilities handling radioactive or mixed radioactive wastes. Nuclear Waste Policy Act (NWPA) requirements apply to release of high-level or transuranic radionuclides into groundwater and in addition to regulatory requirements, operators must understand the technical requirements associated with their specific hazardous waste site.

Migration of Hazardous Waste Constituents to an Aquifer

To evaluate potential migration of hazardous waste constituents from a given site to an aquifer, the following must be determined:

- water balance, precipitation, evapotranspiration, runoff, and infiltration
- unsaturated zone characteristics, geologic materials, physical properties, and depth to groundwater
- saturated zone characteristics, geologic materials, physical properties, quantity and chemical quality, rate of groundwater flow, and groundwater discharge points

- proximity of facility to supply wells or to surface water, proximity of users and rate at which groundwater is withdrawn
- volume, physical, and chemical characteristics of waste, including potential for migration and behavior and persistence of contaminants in the groundwater environment
- hydrogeological characteristics of the site and the surrounding land
- interaction among groundwater systems and between groundwater and surface water systems

Additionally, the following must also be considered:

(1) Current and future groundwater or surface water uses, including any established water quality standards

(2) Existing quality of ground or surface water, including other contaminant sources and their cumulative impact on water quality

(3) Potential for health risk

(4) Potential for damage to ecological systems, including wildlife, crops, vegetables, and physical structures

(5) Persistence and permanence of potential adverse effects

The SDWA and RCRA regulations [URL. No. 280] can be used to ensure that groundwater protection related to hazardous waste and aquifers is appropriately considered at active and inactive hazardous waste sites. This is because CERCLA relies on applicable or relevant and appropriate requirements as standards, and although CWA standards are not particularly useful for groundwater monitoring, they will be useful for monitoring ground and surface water connections. Also, where EPA and state regulatory agencies have not clearly defined the applicability of groundwater classification and protection standards to active and inactive hazardous waste sites, operators of such sites should consider groundwater schemes in developing protection and monitoring programs. Additionally, as EPA and the states continue to develop and modify groundwater protection strategies, operators of hazardous waste sites should track regulatory changes and integrate them into their technical program related to migration of hazardous waste constituents into aquifers.

U.S. Department of Energy Case History Studies and Evaluations Related to Groundwater Contamination

Los Alamos National Laboratory
[URL Ref. No. 97]

Ever since the laboratory was established, many of its operations have required the use of hazardous chemicals and radioactive materials like plutonium [URL Ref. No. 284] and uranium [URL Ref. No. 283]. The continued use of these materials through time has resulted in the contamination of facilities and, in some instances, the surrounding environment (Los Alamos National Laboratory, 1992). In fact, a major source of environmental pollution was the actual management of waste, which was initially discharged into water, air, or buried in land disposal areas in accordance with standards in effect at that time. In addition to hazardous chemicals and radioactive materials, the contaminants of concern included explosive residues and asbestos. Asbestos [URL Ref. No. 339 {36}], while no longer used, is still generated as a waste during facility modifications and during decommissioning and decontamination of facilities.

The environmental restoration program at Los Alamos has identified approximately 2,100 potential release sites for wastes, with many primary mechanisms for the release of contaminants being runoff of surface water that carries contaminated sediments in addition to the erosion of soil to exposed buried contaminants. At Los Alamos, the main pathways by which released contaminants can reach people living beyond the boundaries of the laboratory site include infiltration into alluvial aquifers and airborne dispersion and settlement of particulate matter. The potential release sites slated for further action at Los Alamos have been combined into six field units for continued investigation/remediation, and a description of the six Field Units follows, including related comments on *potential groundwater contamination* concerns as necessary.

Field Unit 1

Field Unit 1 consists of 664 potential release sites. It includes all of the Los Alamos County sites that are on land no longer owned by DOE, and it also includes sites at the old plutonium processing facility (Technical Area 21, TA-21). This field unit contains several of the Laboratory's old material-disposal areas as well as the Los Alamos Municipal Sanitary Landfill. The primary potential contaminants of concern in *Field Unit 1* that relate to groundwater contamination include radionuclides [URL Ref. No. 232], volatile organic compounds [URL Ref. No. 247], and inorganic compounds, including heavy metals.

Field Unit 2

Field Unit 2 consists of 301 potential release sites associated with 14 technical areas. This unit includes active and inactive firing sites, a facility for research on nuclear criticality, a 0.5 mile-long linear proton accelerator, and associated experimental research areas. The primary contaminants of concern that can contaminate groundwater include radionuclides, high explosives, organics, and heavy metals.

Field Unit 3

Field Unit 3 consists of 555 potential release sites associated with 10 technical areas. It includes old sites, where high explosives were developed and processed, initiators for nuclear weapons were tested, and reactor components were developed. The primary contaminants of concern and possible groundwater contaminants include radionuclides, high explosives, volatile and semivolatile organics, polychlorinated biphenyls (PCBs), asbestos, pesticides, and herbicides.

Much of the contamination in this field unit resulted from operations established during World War II to develop, fabricate, and test explosive components for nuclear weapons. Various other facilities included areas for photofission experiments, a mortar impact area, an air-gun firing range and other gun-firing sites, a burning ground, laboratories, storage buildings, sumps, and material disposal areas. In many of the experiments, beryllium-containing weapons were tested, and in some experiments, uranium components were used. A high-pressure tritium [URL Ref. No. 293] facility was also in operation until 1990.

Field Unit 4

Field Unit 4 consists of 260 potential release sites and 19 canyons on the Pajarito Plateau, a reactor site, and various heavily industrialized sites. The primary contaminants of concern that could contaminate groundwater include radionuclides, high explosives, volatile and semivolatile organic compounds, and inorganics, including heavy metals. Most of the contamination that has been detected has resulted from various operations dating from as early as 1944, and most contamination has been associated with such facilities as surface impoundments and disposal areas, experimental reactors, wastewater treatment and septic systems [URL Ref. No. 217–218], aboveground and underground storage tanks [URL Ref. No. 251], sanitary and industrial waste effluent lines, PCB transformers, firing sites, incinerators, chemical processing, and shops for machining radioactive materials.

The Pajarito Plateau is a system of finger-like mesas extending from the Jemez Mountains, with canyons between each mesa. Contamination may have occurred in 19 canyons from various Laboratory operations on the mesas and within the canyons. Many of the canyons extend beyond the current boundaries of the Laboratory and eventually drain into the Rio Grande [URL Ref. No. 55, 339 {12–13}] in New Mexico. The environmental restoration activity will investigate any potential off-site contamination from potential release sites that discharge into the canyons.

Radioactive contaminants (e.g., tritium, cesium 137, and strontium 90) have been detected in alluvial groundwater downgradient of two sites located in one of the main canyons within the Laboratory's boundaries. One of the sites houses the Omega West Reactor. This reactor, no longer operational, was an 8-megawatt water-cooled reactor fueled with enriched uranium; it was used for basic research in nuclear physics. The other site was used in developing weapons-boosting systems and conducting long-term studies on weapon subsystems.

Field Unit 5

Field Unit 5 consists of 312 potential release sites associated with several areas used for explosives development, primary waste management facilities, and one off-site area located on land owned by the U.S. Forest Service [URL Ref. No. 339 {20}] and leased by DOE. Many of the Laboratory's material disposal areas are also located within this field unit. The primary contaminants of concern and those that could contaminate groundwater include radionuclides, high explosives, volatile organic compounds, and metals.

Much of the contamination in this field unit area has resulted from high-explosives research and development and from testing at aboveground firing sites. Other contributors to contamination were research into various methods for assembling fissionable material to produce nuclear bombs and the testing, development, and production of bomb detonators.

This unit contains all of the Laboratory's retired and operating waste management facilities other than the very early landfills, which are part of Field Unit 1. One of the retired facilities established in 1948, consists of several pits and shafts that contain a very diverse mixture of contaminants, including low-level [URL Ref. No. 290], transuranic [URL Ref. No. 289], hazardous [URL Ref. No. 229], and mixed waste [URL Ref. No. 291]. Another landfill was established in 1974 to replace this historical site and continues to operate today. The Laboratory's radioactive landfill is also part of this field unit. Another buried material disposal area was used in the early 1960s and currently contains large amounts of various waste materials, including plutonium and lead. This unit also contains the Laboratory's liquid-waste treatment plant, built in 1963. The plant receives liquid waste from across the Laboratory, treats it to remove target contaminants, and monitors and then releases the treated liquid effluent.

Field Unit 6

Field Unit 6 covers activities related to decommissioning facilities that are no longer needed, and when it is determined that a contaminated facility is no longer needed for its original purpose, the decommissioning program decontaminates the facility but does not demolish it if it can be used for another purpose. If the building cannot be used for another purpose, it is demolished.

The decommissioning projects include buildings from the former plutonium-processing facility (Rail, 1992) that was used from the late 1940s to the early 1970s, a phase separator pit used from the mid-1960s through the early 1990s, a former tritium facility used from the mid-1950s through the late 1980s, many abandoned buildings contaminated with high explosives and used from the 1950s to the 1980s, and the Omega West Reactor (discussed under Field Unit 4) that was used from the mid-1950s to the early 1990s. Potential contaminants that can enter the groundwater are the same as those presented in the discussions of *Field Units 1–5.*

U.S. Department of Energy Hanford Site
[URL Ref. No. 292]

More than 1,500 waste disposal sites have been identified at the U.S. Department of Energy (DOE) Hanford Site (Sherwood et al., 1990). At the request of the U.S. Environmental Protection Agency (EPA), these sites were aggregated into four administrative areas for listing on the National Priority List. Within the four aggregate areas, 646 inactive sites were selected for further evaluation using the Hazard Ranking System (HRS).[2] Evaluation of inactive waste sites by HRS provided valuable insight to design a focused radiological and hazardous substance monitoring network. Hanford site-wide groundwater monitoring was expanded to address not only radioactive constituents but also hazardous chemicals. The network designed to monitor groundwater at those facilities identified ^{129}I, ^{99}Tc, ^{90}Sr, uranium, chromium, carbon tetrachloride, and cyanide.

Primary pathways of concern from inactive waste sites at Hanford are through ground and surface waters. The dominant contaminant transport pathway is from inactive waste sites through unsaturated sediments to groundwater and through groundwater directly to potentially exposed populations or from groundwater via surface water to potentially exposed populations.

To identify contaminants at Hanford, the focus was established on characteristics of the HRS groundwater route that applied to the groundwater monitoring network design. Results were meant to aid future characterization, assessment, and remediation of inactive Hanford sites.

A broad spectrum of radioactive and chemical substances potentially present in groundwater were also identified from inactive-waste-site inventories and knowledge of contami-

---

[2] In 1985, the DOE published DOE Order 5480.14 [URL Ref. No. 275] to organize an inactive-waste-site evaluation program paralleling the Comprehensive Environmental Response, Compensation, and Liability Act (CERCLA) [URL Ref. No. 264] of 1980. The Act was used by the EPA to regulate nongovernment inactive waste sites. Both programs used the EPA Hazard Ranking System (HRS) (U.S. EPA, 1988) to evaluate relative hazards from inactive hazardous waste sites along five exposure routes: groundwater, surface water, air, direct contact, and fire and explosion.

nant mobility. Radionuclide inventories and radionuclide mobility were used to augment the list of radioactive constituents. Constituents such as tritium [URL Ref. No. 293], gross alpha, gross beta, and gamma scans were historically used to monitor radionuclide contamination in Hanford groundwater. Strontium-90 and [129]I were also analyzed but only to assess their off-site migration, not to identify their sources. Thus, radionuclide-specific analyses for ^{14}C, [63]Ni, and ^{99}Tc were added, and ^{90}Sr and [129]I analyses were expanded near observed release sites. If ^{14}C, ^{90}Sr, ^{99}Tc, and [129]I were present in liquid streams, their high mobility would result in their release to Hanford groundwater.

A different approach was undertaken for identification of hazardous chemicals at Hanford, with nitrate ion [URL Ref. No. 201] being the contaminant most often investigated in past groundwater monitoring efforts. Some chromium analyses were also performed in the 100 and 300 areas, but few other chemicals were routinely analyzed. Expansion of hazardous chemical monitoring included establishing background or naturally occurring concentrations of certain constituents and identifying anthropogenic substances from past liquid discharges. Since January 1, 1987, 226 of 484 wells previously monitored for radiological constituents were analyzed for a broad spectrum of radioactive, inorganic, and organic constituents, including select radionuclides, cations, anions, trace metals [URL Ref. No. 207], volatile organics [URL Ref. No. 247, 267], and cyanide.

The results of the expanded monitoring network at Hanford identified several new contaminants, including ^{99}Tc, carbon tetrachloride, and cyanide, thus establishing a link between past disposal practices and existing contaminant plumes. A more direct link was also established between disposal activities and the presence of [129]I, ^{90}Sr, uranium, and chromium.

[99]Technetium was detected in wells across the site with concentrations greater than the maximum concentration limit (MCL) of 900 pCi/l in the 100-H, 200-East, 200-West, and 600 areas (U.S. EPA, 1976c). Maximum ^{99}Tc concentrations, 29, 100 pCi/l, were detected north of the 200-East area. Carbon tetrachloride had also been detected beneath much of the 200-West area and the concentrations exceeded the 5-ppb MCL in 48 wells, with the maximum concentration being 5550 ppb, near the plutonium plant.

Concentrations of [129]I exceeded the 1-pCi/l MCL (U.S. EPA, 1976c) in a widely dispersed area between the 200-West and 200-East areas and the Columbia River, and maximum [129]I concentration, 87.8 ppb, was detected near the 200-West area. Concentrations of ^{90}Sr greater than the 8-pCi/l MCL (U.S. EPA, 1976c) were detected throughout the site with most values slightly above the MCL. Peak ^{90}Sr concentrations that far exceeded the MCL occurred in the 100-N area and in isolated locations within the 200-East area where maximum concentrations were 10,400 and 6,270 pCi/l, respectively.

Uranium [URL Ref. No. 283] concentrations in groundwater have been monitored for many years throughout the site, although because uranium is a primary product of Hanford operations, its presence is expected. However, maximum uranium concentrations, 11,500 pCi/l, were detected in the 200-West area near the uranium purification plant, and elevated concentrations were also reported near the uranium fuel fabrication waste sites in the 100-H and 300 areas.

The groundwater monitoring program at Hanford has also identified contaminants in groundwater on a site-wide basis, and in many cases, this monitoring program will provide the technical basis to design additional waste-site investigations for inactive waste sites at Hanford.

The Fernald [URL Ref. No. 294] *Groundwater Concerns*

The Fernald Environmental Management Project (FEMP) in Ohio, is a 1,050-acre U.S. Department of Energy (DOE) facility located 18 miles northwest of downtown Cincinnati near the farming community of Fernald (Nelson and Janke, 1995). While in active operation from 1952 until 1989, the Feed Material Production Center (FMPC), as it was, produced highly purified uranium metal [URL Ref. No. 283] for ultimate use in the manufacture of nuclear weapons. In 1986, the U.S. Environmental Protection Agency (EPA) and the DOE entered into a Federal Facility Compliance Agreement covering environmental impacts associated with the FMPC and in response to the FFCA, a site-wide Remedial Investigation/Feasibility Study (RI/FS) was initiated pursuant to the Comprehensive Environmental Response, Compensation, and Liability Act (CERCLA) [URL Ref. No. 264] as amended by the Superfund Amendment and Reauthorization Act. Production was permanently suspended at the facility in 1989, and the focus has since shifted to environmental restoration and waste management activities [URL Ref. No 276, 285].

The original RI/FS work plan identified five units that were related to potential groundwater contamination concerns, and these were categorized as follows:

(1) *Operable Unit 1*—waste pit area
(2) *Operable Unit 2*—other solid waste units
(3) *Operable Unit 3*—former production area
(4) *Operable Unit 4*—silos 1 through 4
(5) *Operable Unit 5*—environmental medial

On June 29, 1990, a consent agreement (under Sec. 120 and 106[a] of CERCLA) between the DOE and EPA became effective, and the purpose of this agreement was to achieve consistency between the operable units and ensure commitments to the RI/FS program without altering the underlying objectives. The consent agreement was amended the following year to revise the schedules for completing the remediation of the five operable units and to direct operable unit integration to ensure compliance with the residual risk requirements of the National Hazardous Substances and Oil Contingency Plan.

In accordance with provisions of the ACA, a methodology was prepared for performing risk assessments and establishing risk-based remedial action goals at the FEMP. This Risk Assessment Work Plan Addendum (RAWPA) presents this methodology and was prepared to fulfill the requirements of the ACA.

Problems at the Department of Energy's Hanford Nuclear Reservation

Problems at the Department of Energy's Hanford Nuclear Reservation included the condition of high-level nuclear waste storage tanks on the site being poor and deteriorating (Illman, 1993; Valenti, 1993). The initial report outlining the problem was prepared by the Red Team, a group of technical experts drawn from national labs and consulting firms that conducted an independent technical review of the Hanford tank farm operation [URL Ref. No. 6, 97]. The team's findings were also echoed in the conclusions of a Washington State Department of Ecology report (Illman, 1993) that was highly critical of the Westinghouse Hanford Company management of the site for DOE. That report charged that the company had failed to install and maintain monitoring equipment in the most dangerous radioactive and mixed-waste tanks and did not have sufficient emergency equipment in place to respond to tank waste spills, leaks, and explosions. The team also reported that 3,000 pieces of equipment were out of service, and the list included pumps, compressors, gauges, and ventilation systems. The repair of older, failed equipment often exceeds nine months. Much of the equipment was apparently not designed for calibration and was producing output of indeterminate quality. Also, according to the report, a critical factor was that insufficient analytical capability at Hanford did not exist and became a bottleneck because the first step in any program to improve monitoring and safety was to be able to characterize the waste tanks properly.

The Savannah River Site (SRS)

The Savannah River Site (SRS) has reached the 2-billion-gallon treatment milestone in its program to clean up contaminated groundwater (Anonymous, 1995), and the SRS groundwater remediation program, which removed industrial solvents from groundwater, was the largest such program within the Department of Energy complex and was considered one of the largest groundwater cleanup programs in the nation.

The 2 billion gallons of water SRS has treated to date would fill 20,000 Olympic-sized swimming pools. The SRS is a 310-square-mile nuclear materials production facility on the western border of South Carolina and is owned by the Department of Energy and operated by the Westinghouse Savannah River Co.

The remediation program at SRS was initiated in 1983 with a pilot groundwater extraction and treatment system using an air stripper. A full-scale air stripper was installed in 1985, and in 1992, an additional air stripper was installed. The treatment program has removed more than 315,000 pounds of chlorinated solvents such as trichloroethylene (TCE) and tetrachloroethylene (PCE), which had been used since the 1950s for cleaning and degreasing operations at SRS.

Twelve groundwater recovery wells pump approximately 550 gallons of groundwater per minute to two stainless steel stripper towers, and as the water cascades downward through the column, pumped air is forced upward from the bottom of the column. Then, when the water mixes with the air, the contaminants move from a liquid phase to a vapor phase where they are stripped out and eventually dissipate into the surrounding air, where sunlight eventually breaks them down. The water enters the stripper with PCE and TCE concentrations as high as 50,000 ppb. The water is treated to well below drinking water standards (i.e., actually below the level of detection, which is approximately 1 ppb). The water is then returned to the environment via a nearby stream.

The groundwater remediation program is augmented by vacuum extraction technology that uses horizontal and vertical wells to remove organic contaminants from the soil in the zone above the groundwater and a project for *in situ* cleanup of groundwater and soils in the area will move from demonstration to operation phase during the next few years. This technology, which involves *in situ* bioremediation, injects nutrients such as methane plus phosphate [URL Ref. No. 202] and nitrous oxide that stimulates the growth of naturally occurring microbes [URL Ref. No. 213–215] that eventually break down the contaminants.

Trace Element Distribution in Various Phases of Aquatic Systems of the Savannah River Plant

Elevated concentrations of potentially toxic metals in water and bottom sediments are commonly associated with many industrial processes and human activities (Sandhu, 1991), and metals that enter the aquatic systems partition into different components of solid and dissolved phases. The bioavailability and toxicity of these metals are not solely a function of their total concentrations but relate to partitioning between solid and solution phases. Even in the solution phase, the relative toxicity of these metals is often related to their chemical forms, such as inorganically complexed ions, exchangeable ions, and organically complexed ions. The free ionic form of metals in general is relatively more toxic than the complexed form as it tends to interact more readily than other forms that may also be present in solution.

In a study conducted by Sandhu (1991) at the Savannah River Plant (SRP), the distribution of Cd, Cu, Fe, Mn, Ni, and Zn species was estimated for the dissolved solid phases of thermally impacted and nonimpacted aquatic SRP systems. The major fractions of Cd, Cu, Ni, and Zn were present as dissolved ions, while most of the Fe was present in the solid phase. Dissolved species of Cu, Fe, Mn, and Ni were insensitive to natural Ca and alkalinity gradients across SRP aquatic systems, whereas the dissolved species of Cd and Zn and solid phase exchangeable Zn responded to this gradient. Solid-phase Cd was primarily observed in the exchangeable and carbonate phase, although Zn and Ni did not display a clear distribution pattern between various components of the solid phase. The increase in the percentage of dissolved Cd can be accounted for by source water chemistry and thermal conditions associated with cooling water activities.

Applicability of Land Disposal Restrictions to RCRA and CERCLA Groundwater Treatment Reinjections [URL Ref. No. 123] at U.S. DOE Facilities

On December 15, 1989, the Environmental Guidance Division of DOE (EH 231) issued a memorandum to all Program and Operations Offices entitled, "Fact Sheets: Natural Resource Trusteeship Under CERCLA [URL Ref. No. 264] and Management of Contaminated Ground Water as Hazardous Waste" (U.S. DOE, 1990d). The fact sheet on groundwater as hazardous waste described pertinent definitions and facts about the U.S. Environmental Protection Agency's (EPA) approach to managing contaminated groundwater under the Resource Conservation and Recovery Act (RCRA) [URL Ref. No. 280]. It also alerted readers that EPA's Office of Solid Waste and Emergency Response (OSWER) [URL Ref. No. 48, 98, 311, 328, 337 {73}] planned to issue an interpretive memorandum that would describe whether Land Disposal Restrictions (LDR) apply to groundwater that is reinjected during environmental restoration pump and treat operations [URL Ref. No. 336 {34}, 337 {27}]. Additionally, the RCRA LDR may be applicable or relevant and provide appropriate requirements for certain response actions taken under CERCLA.

The LDR interpretation for reinjected groundwater [URL Ref. No. 123] has been announced by EPA and briefly states that the EPA has determined that under certain circumstances, the LDR does not apply to reinjections of groundwater during pump and treat operations. The EPA's LDR interpretation for reinjected groundwater was consistent with EPA reports.

Need for Interpretation

In its management review of the Superfund Program, EPA identified the misapplications of RCRA LDR, which are proscriptive regulations designed to prevent contamination before it happens, as contributing to the inefficient implementation of CERCLA. Additionally, there is recognition on the part of EPA that the problem of cleaning up large-scale contamination (i.e., under CERCLA) is quite different from the problem of how hazardous wastes should be properly managed by an ongoing operation (the focus of RCRA).

In general, RCRA LDR prohibit the land disposal of restricted wastes (i.e., after the effective date of the restriction), unless such waste meets promulgated treatment standards based on best demonstrated available technology (BDAT) identified by EPA for that particular type of waste. Requiring compliance with the fundamentally preventative provisions of RCRA could place unnecessary constraints on CERCLA response actions, although the BDAT regimes can be difficult to apply at CERCLA sites because the wastes that are encountered are usually a mixture of different types of restricted wastes, nonrestricted hazardous substances, and debris. Each of the restricted wastes in a CERCLA mixture may require a different BDAT treatment, and since restricted wastes subject to LDR may be mixed with other restricted wastes, it can be difficult to determine the appropriate BDAT(s) for all of the restricted wastes within a CERCLA mixture. These previously listed difficulties can be expected to be magnified at DOE environmental restoration sites [URL Ref. No. 276, 285], because there can be additional technical problems associated with DOE hazardous and radioactive wastes.

Because injection of groundwater containing restricted wastes constitutes land disposal under RCRA section 3004(k) [URL Ref. No. 280], the question of whether the LDR are applicable to reinjected groundwater during CERCLA pump and treat operations has been raised. If LDR are applicable, groundwater containing restricted wastes would require treatment to attain standards based on BDAT prior to each reinjection, and since pump and treat remedies may have to operate many years, the cleanup action could become overly burdensome, technically impractical, and/or prohibitively expensive.

Basis for the LDR Interpretation

RCRA Section 3020(b) [URL Ref. No. 280] prohibits the injection of hazardous waste into or above an underground source of drinking water, with the following exception: the prohibition does not apply to the injection of contaminated groundwater into the aquifer from which it was withdrawn. If the injection is a CERCLA response action (or a RCRA corrective action), the groundwater has been treated to substantially reduce hazardous constituents, and the action will protect human health and the environment. The EPA interpretation that LDR are not applicable to reinjection of treated groundwater during RCRA corrective and CERCLA response actions is based on the traditional principle that the more specific of two overlapping statutory provisions should control. In this case, the language of the LDR, which refers generally to the land disposal of wastes, was found to be less specific than another RCRA provision [Section 3020(b) that directly focuses on the injection of treated contaminated groundwater into Class IV injection wells (40 CFR 146.5)] [URL No. 69 {10}].

In determining whether RCRA LDR may be relevant and appropriate (i.e., for CERCLA response actions), EPA indicates that the requirements must address problems or situations similar to the circumstances of the response action contemplated and be well-suited to the (CERCLA) site. Comparing the CERCLA response objectives with the purpose and objective of the LDR requirement is the key to EPA's interpretation of the potential relevance and appropriateness of the LDR to pump and treat operations conducted under a CERCLA response action. Treating and reinjecting groundwater into Class IV injection wells is ultimately performed to restore the groundwater (aquifer) to drinking water quality. EPA believes that standards that have been specifically developed to establish drinking water quality levels, such as Maximum Contaminant Levels (MCLs), are well-suited to the accomplishment of the CERCLA response action (i.e., pump and treat) objective. Thus, the EPA interpretation provides that where drinking water standards are available (e.g., MCLs), those standards, and not the standards set by the LDR, will generally be the relevant and appropriate requirements to use in setting treatment standards for CERCLA response actions involving the cleanup of drinking water aquifers.

Necessary Conditions

In order to reinject treated groundwater during pump and treat operations at the DOE's environmental restoration sites without triggering the RCRA LDR, three conditions must be met:

- the reinjection must be part of a CERCLA section 104 or 106 response action or be a RCRA corrective action
- the contaminated groundwater must be treated to substantially reduce hazardous constituents prior to such injection
- the response action or corrective action must be sufficient to protect human health and the environment upon completion

While the language of RCRA section 3020(b) is straightforward in its application to RCRA corrective actions, it is not explicit with respect to CERCLA response actions conducted at federal facilities, although, the federal government is directed by Section 120 of the Superfund Amendments and Reauthorization Act (SARA) to comply with CERCLA to the same extent as private parties, so that a question arises as to which statutory authority federal facilities employ for CERCLA response actions. DOE employs CERCLA section 120, and, consequently, DOE's CERCLA response actions satisfy the first condition.

The second requirement of RCRA section 3020(b) is that the reinjection must be treated to substantially reduce hazardous constituents prior to such injection, and there is no quantitative guidance available at this time that will provide environmental restoration managers with the knowledge that they need to meet this requirement. EPA suggests, however, that the steps necessary to substantially reduce hazardous constituents during a RCRA corrected action or CERCLA response action should be decided on a case-by-case basis. The U.S. DOE (ESH-231) will disseminate EPA guidance to all Field Organizations once it becomes available, and as a final condition, the corrective action or response action must be sufficient to protect human health and the environment upon completion.

Groundwater Contamination As Defined by Federal, State, and Local Statutes

There is a seemingly endless number of toxic or hazardous substances that can become groundwater contaminates as defined in volumes of federal, state, and local statutes (Missimer, 1992; Barber, 1992) [URL Ref. No. 68–69]. These statutes, with the exception of the Surface Mining Control and Reclamation Act (SMCRA) [URL Ref. No. 297], are generally administered and enforced by EPA at some level. The major federal statutes that relate to EPA include the Comprehensive Environmental Response, Compensation, and Liability Act (CERCLA or Superfund) [URL Ref. No. 264], the Resource Conservation and Recovery Act (RCRA) [URL Ref. No. 280], the Safe Drinking Water Act (SDWA) [URL Ref. No. 258], the Clean Water CWA (CWA) [URL Ref. No. 287], the Toxic Substances Control Act (TSCA) [URL Ref. No. 295], and the Federal Insecticide, Fungicide and Rodenticide Act (FIFRA) [URL Ref. No. 296].

Cercla [URL Ref. No. 264]

CERCLA (Superfund) gives EPA the extremely broad authority to act against spills or leaks of pollutants involving hazardous substances, and because the word release, is so broad, the applicability of CERCLA cannot be ignored in the groundwater context. The Superfund itself, is actually a multibillion dollar trust fund created by CERCLA, which allows EPA to conduct its own cleanup at affected sites, however, this program is unevenly enforced from industry and insurance perspectives. Specifically, if EPA can find potential responsible parties for creating hazardous waste, it can administratively enforce the cleanup or can choose to sue potentially responsible parties for reimbursement for costs associated with cleanup. This is true notwithstanding the fact that many other equal parties may be responsible for the pollution. Moreover, not only does the waste have to be cleaned up, cleanup has to be followed according to the very stringent procedures set forth in the National Contingency Plan (NCP), which sets forth specific bureaucratic guidelines regarding how cleanup activities must be implemented.

RCRA [URL Ref. No. 280]

In the groundwater context, RCRA protects groundwater by regulating disposal of wastes with a specific focus on hazardous waste. RCRA and the regulations implementing the statute have, by design, created what have come to be known as cradle-to-grave handling procedures, governing the affected wastes from the point of their manufacture until they are disposed of in an approved landfill or other disposal site. RCRA primarily establishes uniform federal standards (i.e., states can't implement weaker ones) to be followed by the manufacturers and haulers of hazardous waste as by the facilities such as landfills that dispose of or otherwise treat the hazardous waste transferred to them.

SDWA [URL Ref. No. 258]

The SDWA was enacted by Congress to make sure that the drinking water supplies of the country are maintained at safe levels. In the groundwater context, EPA is required by the SDWA to test for, maintain, and enforce maximum exposure levels of health-threatening contaminants in the drinking water provided to citizens through the public water systems of the country. From the waste-generating standpoint, EPA is required by the SDWA to protect underground sources of drinking water from hazardous waste disposal primarily by the injection method. Moreover, from the real estate planning perspective, the impact of certain programs under the Safe Drinking Water Act can be significant. The wellhead program [URL Ref. No. 106, 108], for example, directs state enforcement agencies to protect the areas surrounding water wells and the water source itself from contaminants entering the groundwater from the surrounding area. Likewise, designation by EPA of an aquifer [URL Ref. No. 116] as a sole source aquifer results in federal agencies being allowed to refuse loans or financial assistance to projects that can contaminate the aquifer.

CWA [URL Ref. No. 287]

The focus of the Clean Water Act is to regulate the pollution of surface waters in the United States, specifically lakes, rivers, and streams. This regulation is accomplished through the National Pollutant Discharge Elimination System, or NPDES, [URL Ref. No. 313] which was also designed to prohibit the discharge of pollutants into water except in accordance with a permit issued by EPA. Overseen and regulated by the EPA, the NPDES permit application process is very complicated and includes very specific standards that the applicant-generator must follow in order to protect surface water quality in the United States. Although groundwater would appear to have little to do with surface water, environmental groups typically use the CWA to get at generators of groundwater contamination, given the physical link that occurs between groundwater and surface water in the United States.

TSCA and FIFRA [URL Ref. No. 295, 296]

The Toxic Substances Control Act and the Federal Insecticide, Fungicide and Rodenticide Act are closely related in that they require manufacturers and formulators to register their chemicals (in the context of TSCA) or pesticides [URL Ref. No. 239] (in the context of FIFRA) so that EPA can impose any appropriate restrictions on their use. EPA then determines whether the application mixture can leach into local groundwater sources and whether the leachate has a measurable effect on human health.

Surface Mining and Control and Reclamation Act [URL Ref. No. 297]

This Act regulates surface coal mining activity to prohibit the runoff from aboveground mines from contaminating local groundwater sources. The Department of the Interior [URL Ref. No. 339 {38}] or an authorized state agency is responsible for implementing the program by requiring operation of the surface mine so as to protect the groundwater from contaminated area drainages.

Groundwater Contamination and Analysis at Other Non-DOE Hazardous Waste Sites

Groundwater contamination at various hazardous waste sites has proved to be a challenge to hydrogeologists, chemists, and other professionals because of the complexity and diversity of wastes and of the sites where these wastes are eventually deposited and collect. For example, Lesage and Jackson (1992) gathered information regarding the investigation of the fate of toxic chemicals emanating from hazardous waste sites. Lesage and Jackson (1992) presented their discussion in four sections that included analytical methodologies, monitoring strategies, site investigations, and geochemical investigations. Information that was presented included the behavior of toxic chemicals, the methods used for analysis and their validity, and the interpretation of data. They also asked authors from diverse backgrounds to describe and interpret hazardous waste site investigations from their own perspective, which led to them presenting information on known types of organic contaminant plumes and monitoring strategies used to analyze and evaluate these areas.

FOUR

Radioactivity, Including Occurrence/Fate/Transport and Remediation/Restoration Groundwater with Case History Example from U.S. DOE Facilities [URL Ref. No. 232, 233, 301, 302, 303]

General

Information concerning the U.S. Department of Energy (DOE) *hazardous waste/groundwater contamination* concerns and interactions (U.S. DOE, 1990b) within the National Laboratory complexes have been presented in *Chapter* 3. A discussion on *radioactivity* [URL Ref. No. 304] *and remediation/restoration* [URL Ref. No. 285, 303] concerns at these DOE facilities is now presented in this *chapter,* and at the end is presented a general summary on information concerning subsurface *occurrence/fate/transport* of groundwater (U.S. DOE, 1993a) [URL Ref. No. 302]. The U.S. DOE facilities (i.e., with emphasis on Los Alamos National Laboratory) [URL Ref. No. 97, 101, 288], again as in *Chapter* 3, are used as case history models to be able to present a discussion that can be extrapolated to non-DOE areas.

As has been discussed, subsurface contamination of groundwater is a complex environmental problem (Institution of Civil Engineers, 1990). The movement of chemical, biological, and radiological constituents in the subterranean environment always involves multiple phases of interactions with a myriad of potential reactions in an inherently nonhomogeneous, anisotropic porous media (Knox et al., 1993; Gray, 1990). Also, because the ultimate fate of constituents introduced to the subsurface is so difficult to predict in most instances and altering the subsurface to control the fate of these constituents in order to effect remediation [URL Ref. No. 285, 303] is a continuous dynamic formidable challenge, and research is always essential (International Atomic Energy Agency, 1959). Consequently, in order to predict or control the fate of any substance in the subsurface environment, one must possess an understanding of the basic processes that influence the movement, transport, and fate [URL Ref. No. 302] of the substance.

U.S. Department of Energy (DOE)
[URL Ref. No. 45, 61, 64]

Since the end of the Cold War, the news about the Department of Energy (DOE) has been grim: thousands of acres poisoned with radioactive [URL Ref. No. 304] and toxic wastes [URL Ref. No. 256, 317], massive cleanup cost overruns due to contractor waste and fraud, and a nuclear waste vitrification plant that has yet to be built and that may not solve any problems when it is (Austin, 1994). But after five years and after putting $12 billion into the world's largest environmental cleanup, slowly, the DOE is developing technology that may help restore its battered land (Gray, 1990).

When it came to developing remediation technology, however, the DOE had little choice. In its 1989 five-year plan, the DOE promised Congress it would bring all facilities into environmental compliance with applicable federal, state, and local laws and regulations by 2019, at a cost of about $100 billion. But, in February 1991, a report to Congress from the Congressional Office of Technology Assessment [URL Ref. No. 178] (U.S. DOE, 1995c, 1995d) explained that the DOE's goal was not based on meaningful estimates of work to be done, the level of cleanup to be completed, or the availability of technologies to achieve certain cleanup levels. The Office of Technology Assessment (OTA) further explained that the capability of existing technologies to clean up or even contain weapons complex contamination is uncertain, and that for some problems, no proven technology exists.

Eventually, the DOE had to agree, and when officials began negotiating cleanup agreements with state and local governments, they discovered that DOE would be held to strict treatment standards. Consequently, in 1990, for example, the department reached an agreement with EPA and the Colorado Department of Health on the parameters of cleanups for 178 sites at the Rocky Flats [URL Ref. No. 179] weapons complex near Denver. As part of the pact, the DOE agreed to treat plutonium-contaminated [URL Ref. No. 284] water to a level of 0.05 picocuries per liter. Since Colorado usually requires treatment only to 0.15 picocuries per liter, the lower amount of the already difficult-to-remediate contamination would be harder to attain.

By August 1991, in outlining its five-year plan for 1993–1997 (U.S. DOE, 1995c, 1995d), the DOE summarized to Congress that technological constraints hindered effective characterization of the subsurface environment and treatment and disposal of DOE-unique process wastes, and that without a new or improved technologies and a well-developed infrastructure, the DOE will not be able to comply with applicable federal, state, and local regulations. So, the DOE upgraded its Office of Technology Development, increasing its budget by more than half from 1991 to 1992, and currently, the office receives almost $400 million annually to oversee and fund agencies to conduct public and private environmental research.

The real challenge, however, for the DOE, has come in changing the focus at the nation's largest research and development institutions, the DOE nuclear weapons laboratories. For example, three labs—Lawrence Livermore National Laboratory, Livermore, CA; Los Alamos National Laboratory, Los Alamos, NM.; and Sandia National Laboratory, Albuquerque, NM [URL Ref. No. 73, 96–97]—which have combined research budgets of $3.4 billion and more than 24,000 researchers, had not helped industry develop much technology, environmental or otherwise. Until the past couple of years, more than two-thirds of the labs' money still went to military missions, despite more than a decade of legislative attempts to encourage technology transfer.

However, in fact, in an update to Congress during 1993, the OTA [URL Ref. No. 178] said that despite earlier disappointments in technology transfer, industry interest in working with these laboratories is now at an all-time high. But, the OTA warns that although there appears to be lively interest on both sides in cooperative research and development, significant problems could dampen the newfound enthusiasm.

Various Types of Radioactive Wastes

High-Level Radioactive Wastes [URL Ref. No. 305], *Transuranic Waste* [URL Ref. No. 289], *Mining and Milling Wastes* [URL Ref. No. 309], *Mixed Waste* [URL Ref. No. 291], *and Low-Level Wastes* [URL Ref. No. 290], *within Department of Energy (U.S. DOE) Facilities*

Background on Radioactive Wastes [URL Ref. No. 232, 304][1]

Radioactive wastes [URL Ref. No. 232] are not a singular material, instead, they are generated in diverse forms that have traditionally been distinguished by their sources, not by their physical characteristics. Radioactive wastes vary greatly in their chemical and radioactive composition and, therefore, in their potential for environmental and public health impacts [URL Ref. No. 200] (Gershey et al., 1990; Mays et al., 1985). A brief discussion is, therefore, presented on the subject and follows. For more detailed basic information on radioactivity or radionuclides in the environment, the following reference is recommended: Gershey et al., 1990.

Definition of Radioactive Wastes

Radioactive wastes [URL Ref. No. 232] are defined by their source, not their characteristics. The three classes, A, B, and C, are classified depending upon their concentration, energy levels, half-life, and the source of the radionuclides present. However, the designation of material as low-level radioactive waste (LLRW) [URL Ref. No. 290] does not necessarily imply low hazards, since the radiation hazard is a function of radionuclide concentration, half-life, emission type, mode of decay, energy, level of protection, and mobility (i.e., through the environment and the body).

Tykva and Sabol (1995) presented information on low-level radioactivity assessment, sources, measurement techniques, instrumentation, and radiological effects that included radionuclides and radiation emitted (i.e., radionuclides and radioactivity and properties of emitted radiation and its interaction with matter), experimental arrangement for low radioactivities (i.e., fundamental conditions for the determination of low-level radioactivity, low-background laboratories, low-level detectors, instrumentation for the processing and evaluation of detector signals, elimination of the extraneous counts, counting statistics and errors, standardization and calibration in low-level radioactivity measurement, and sample treatment), and selected fields of low-level radiation (i.e., starting data, transport of radionuclides in the environment, radiochronology, activation analyses, whole body counting, field and area monitors, and assessment of radon [URL Ref. No. 107, 233] and its decay products).

Types of Radiation [URL Ref. No. 232]

The four basic types of radiation in the nuclear industry are divided into two classes: directly ionizing radiation (alpha particles, beta particles) and indirect ionizing radiation such as gamma or x-rays and neutrons (Los Alamos National Laboratory, 1993).

---

[1] *It is important to understand radioactive wastes before exploring their interactions with groundwater concerns. It is not my intent in this Volume II to discuss the political and managerial implications of the DOE, but simply to let the reader know what has happened, what is happening as of this writing, and what is going on within the DOE system as can best be determined from reviewing available publications and WWW sites as related to groundwater contamination.*

Acute exposure [URL Ref. No. 272] in humans, for example, occurs when a dose of radiation, usually a high-level dose, is received (inhaled, ingested, or otherwise introduced into the body) in a short period of time, typically from seconds to days (Los Alamos National Laboratory, 1993). And since the body cannot repair or replace cells fast enough from an acute, high-level dose, detrimental physical effects occur. Most radioactive effects, however (i.e., once deposited, radioactive decay products in intimate contact with tissue lead to possible DNA damage and/or cell death; the hazard of internally deposited radionuclides is a function not only of radioactive decay but also of metabolism and elimination of the compounds to which they are bound), from acute, high-level exposures will appear within minutes to weeks, depending on the dose. A localized dose of radiation can result in hair loss, skin burns, and temporary or permanent sterility (i.e., an example is the dose received by atomic bomb victims). A whole-body dose decreases blood cell numbers, causes diarrhea, vomiting, fever, disorientation, coma, and eventually death.

Chronic exposure [URL Ref. No. 273] when compared to *acute* exposure, however, occurs when a dose of radiation, usually a low-level dose, is spread over a long period of time, typically from months to years. A chronic, low-level dose is usually less harmful because the body has time to repair or replace damaged cells.

The effects of chronic exposure may not appear until years after the radiation dose is received. Examples of chronic, low-level exposure include the dose received from background radiation (i.e., naturally occurring and human-made sources; the average nationwide radiation dose equivalent to a member of the gene population from naturally occurring and man-made background sources is about 360 mrem per year) and the dose typically received from an occupational exposure. A chronic, low-level dose may slightly increase the risk [URL Ref. No. 235] of developing cancer, and the factors that affect the risk of biological damage, in turn, include the following:

- *total dose*—the greater the dose, the greater the effects
- *dose rate*—the faster the dose is delivered, the less time the cell has to repair the damage
- *type of radiation*—alpha and neutron radiation are more damaging than beta or gamma radiation
- *area of body exposed*—the larger the area of the body exposed, the greater the biological effects (extremities are less sensitive than internal organs)
- *cell sensitivity*—actively dividing cells are most sensitive
- *individual sensitivity*—the developing embryo/fetus is most sensitive, children are more sensitive than adults, the elderly are more sensitive than middle-aged adults

Interactions between radioactive isotopes and a specific environment in question can lead to waterborne and air re-leases. Like certain pesticides [URL Ref. No. 239] and other organic chemicals [URL Ref. No. 210], some radionuclides are concentrated in the food chain, and can lead to high exposures to humans (e.g., ^{90}Sr). Radionuclides that, however, are not bioaccumulated, may instead pass through the soil and contaminate groundwater, or if volatile, accumulate in the atmosphere, and at a later time they can again enter subsurface areas. Similar dynamic processes can occur in water, where the ocean is then considered a large reservoir of radionuclides.

A valuable perspective on LLRW can be gained by analyzing the facilities that generate the wastes. These facilities basically fall into five sources: utilities with nuclear power plants, industrial, institutional (clinical, academic, and biomedical research), military, and nonmilitary sources (Gershey et al., 1990).

Origin and Types of Radioactive Wastes

In the United States, as in other nations involved in nuclear arms manufacturing, the majority of radioactive waste originates as by-products from nuclear weapons production (Los Alamos National Laboratory, 1993). Department of Energy (DOE) defense-related wastes account for more than 70 percent of all radioactive wastes generated in the United States (U.S. DOE, 1984, 1988a) and most of this defense waste has been generated and managed under a shroud of national defense secrecy, without the public scrutiny that marks commercial nuclear activities. However, remedial action programs and environmental restorations are now underway at most of the larger DOE facilities (Gershey et al., 1990).

Thomson (1991) mentions the risk-based priority system that will be used to assist in the process of allocating funds to an estimated 3,000 DOE sites for cleanup activities (i.e., comply with public health and environmental protection laws and regulations, contain contamination at inactive sites, and ensure that the DOE's compliance actions reduce risk to human health and the environment). According to Thomson (1991), assessment and cleanup of inactive DOE sites consists of six elements or phases that include the following:

(1) Preliminary assessment
(2) Inspection
(3) Site characterization
(4) Evaluation of cleanup alternatives
(5) Development and implementation of appropriate remedial actions
(6) Continued compliance through monitoring and inspections

Also, because of the variety of radioactive wastes, it is important to understand the differences between the various kinds of wastes before focusing on them in relationship to groundwater concerns, because many orders of magnitude separate the wastes into different classes. Spent fuel and defense high-level radioactive waste (HLW) [URL Ref. No. 305] have the highest concentration and account for most of the activity associated with nuclear wastes. Low-level radioactive waste (LLRW) [URL Ref. No. 290] and uranium mill tailings [URL Ref. No. 311] have the lowest concentrations but have higher volume.

High-Level Radioactive Wastes [URL Ref. No. 305]

Spent fuel is intact nuclear fuel that has been used in a nuclear reactor. It is highly radioactive and poses serious radiation hazards requiring shielding, containment, remote handling, and initially necessitated underwater storage for cooling. Although spent fuel contains plutonium [URL Ref. No. 284] and enriched uranium [URL Ref. No. 283] in economically recoverable amounts, commercial fuel reprocessing has been discontinued because of government security concerns. The DOE, however, continues to reprocess most of its spent fuel. Reprocessing increases the overall energy efficiency of the nuclear fuel cycle by recycling enriched uranium and recovering plutonium, which can provide a partial fuel substitute for ^{235}U in mixed-oxide and breeder reactors (Gershey et al., 1990).

Negative aspects of reprocessing are the serious potential health consequences from accidental releases of plutonium, which is highly carcinogenic. Plutonium can also end up in unwanted nuclear weapons production and the bomb-grade material could end up in improper hands or organizations or be the source of radioactive or ecological contamination.

High level wastes (HLW) are generated during the reprocessing of spent reactor fuel (Gershey et al., 1990; U.S. DOE, 1984, 1988a; Mays et al., 1985). Most of the inventory of HLW in the United States is related to DOE and defense activities and involves the Savannah River Plant (SR), South Carolina; Idaho National Engineering Laboratory (INEL), Idaho; and Hanford Reservation, Washington [URL Ref. No. 45]. And, additionally, during the spent fuel reprocessing, large volumes of acid and other solvents are used to extract radionuclides chemically from the fuel rod assemblies. These liquids are then treated to precipitate plutonium and uranium and are pumped to storage tanks for additional processing. Disclosure of leaking HLW storage tanks at the Hanford Reservation [URL Ref. No. 292] in eastern Washington helped bring notoriety to the DOE (Gershey et al., 1990). In the tanks, particulates settled out to form sludges and slurries that eventually had to be dewatered. The sludges and precipitates contained high concentrations of radioactive cesium, strontium, plutonium, uranium, and other nuclides, although most of the initial radioactivity in these wastes came from ^{90}Sr, ^{137}Cs, and other fission products that decay within the first few hundred years after disposal (i.e., HLW and spent fuel will retain hazardous levels of uranium,

plutonium, and other actinides for thousands of years). These wastes, therefore, must be physically isolated from the biosphere. Great attention has been focused on methods to contain them, and like spent fuel, HLW is being stored on an interim basis pending development and use of a final repository for these wastes.

As part of the Nuclear Waste Policy Act of 1982, Congress mandated that the DOE select and construct one or more repositories [URL Ref. No. 315] for HLW and spent fuel (Gershey et al., 1990). The research and development of such a repository has a long history, beginning with a National Academy of Sciences [URL Ref. No. 306] recommendation in 1957 that long-term disposal would be best managed by deep geologic burial (National Academy of Sciences, 1957). After many investigations, the DOE chose the crystalline rock formations at the Yucca Mountain [URL Ref. No. 307] site in Nevada, located in southern Nevada approximately 100 miles northwest of Las Vegas, for the nontransuranic waste [URL Ref. No. 289] (e.g., discussion and definition of transuranic waste follows this section). However, the DOE anticipates that many years will still be required to fully characterize that site, and work is still being conducted within areas (Gertz and Cloke, 1993) that involve starter tunnels leading to the Exploratory Studies Facility (ESF), deepening trenches, drilling neutron boreholes (60 to 270 ft deep), drilling a deep dry borehole (UZ-16; 1686), drilling additional geotechnical boreholes near the portal of the north ramp to confirm the stratigraphy and for engineering design, conducting seismic monitoring efforts following the earthquake at Little Skull Mountain, extensively investigating past volcanic activity at Latrrop Wells cinder cone, and digging numerous test pits for acquiring data required for related engineering designs and construction activities. Many permits are required including the need to implement U.S. Regulatory Commission (NRC) [URL Ref. No. 339 {34}] reviews and for acceptance of a quality assurance program. An optimistic date to open the site is still unknown (Gershey et al., 1990; Gertz and Cloke, 1993; Williams, 1995; Rothstein, 1995).

There is still much to do within the weapons complex when it comes to environmental cleanup (Rothstein, 1995). Many sites still need to be characterized (i.e., the contents of three-fourths of the units at weapons production sites that may leak contaminants into the environment remain to be assessed), nuclear materials that are now stored in aging facilities must be stabilized, and a variety of toxic wastes must be stored safely until they can be moved to permanent repositories.

Environmental remediation will need forceful advocates over the 70 years the baseline study estimates it will take to accomplish the task (U.S. DOE, 1995c). As the authors warn, "Estimating the Cold War Mortgage" (U.S. DOE, 1995c), presents only a gross estimate of costs, in part because the Energy Department's environmental tasks are without technical precedent [URL Ref. No. 61]. The department faces some problems for which no solutions are yet available.

"Estimating the Cold War Mortgage" (U.S. DOE, 1995c) [URL Ref. No. 88] estimates that remediation will include disposing, somehow, of 403,000 cubic meters (106 million gallons) of HLW, 2,600 metric tons of spent fuel, 107,000 cubic meters of transuranic wastes [URL Ref. No. 289], 1,800,000 cubic meters of LLRW, and 780,000 cubic meters of mixed waste (chemical and radioactive) [URL Ref. No. 291]. And, there are still the yet unanswered questions about what can be done in the many cases where plant operations have contaminated the soil and groundwater. Where wastewater was dumped on the ground and stored wastes leaked into the earth, volatile organic compounds [URL Ref. No. 247, 267], heavy metals, and radionuclides have spread to surface streams and groundwater. The "Closing the Circle" (U.S. DOE, 1995a) [URL Ref. No. 88] report recommends two approaches that might be called holding actions: trying to eliminate further contamination, which includes repairing still-leaking storage sites; and, in some cases, blocking the migration of contaminated groundwater before it reaches major sources of drinking water.

The cost of remediation in the case of some river systems that also affect groundwater recharge areas—the Columbia River (Hanford Site), the Clinch River (Oak Ridge), and the Savannah River (Savannah River Site)—was omitted from the plan because no effective remediation technique is available. In some cases, remediation efforts themselves could cause unacceptable ecological damage. Some water is being treated at the Savannah River Site, but the treatment is expensive and of unknown efficacy.

Because it is impossible to destroy radionuclides and other contaminants like heavy metals, the Energy Department's "Closing the Circle" (U.S. DOE, 1995a) and "Estimating the Cold War Mortgage" (U.S. DOE, 1995c) reject the greenfields concept (i.e., the idea that all nuclear weapons production sites can or should be returned to their original condition). Instead, the department's cost estimate is based on in-place containment whenever possible. Containment also offers the advantage of producing little or no secondary waste. Nearly every removal technology will produce additional waste during the transportation, storage, treatment, and final disposal stages.

Transuranic Waste [URL Ref. No. 289]

Transuranic (TRU) wastes [URL Ref. No. 289] consist of material contaminated by radionuclides with atomic numbers greater than uranium, such as plutonium, americium, and curium (Gershey et al., 1990). TRU wastes generally contain less activity but are more voluminous than HLW or spent fuel.

TRU wastes result from every industrial process involving transuranic materials but are predominantly by-products from the fabrication of plutonium for nuclear weapons (U.S. DOE, 1988a). In the United States, the DOE is the principal generator of TRU wastes. These wastes pose high health hazards because they tend to be water soluble, respirable (i.e., up to 1 percent by weight, can be less than 10 μm in diameter), and contaminate a variety of physical forms, ranging from unprocessed trash (e.g., absorbant papers, personal protective equipment, plastics, rubber, wood, ion-exchange resins, etc.) to discarded tools and glove boxes. Major producers of TRU waste are the Rocky Flats Arsenal [URL Ref. No. 179], Colorado; Savannah River Plant, South Carolina; Hanford Reservation [URL Ref. No. 292], Washington; and Los Alamos National Laboratory [URL Ref. No. 97], New Mexico. Smaller producers include the Mound Facility, Ohio; Argonne National Laboratory, Illinois; Oak Ridge National Laboratory, Tennessee; and Lawrence Livermore Laboratory [URL Ref. No. 96], California.

Prior to 1970, the TRU category did not exist. TRU wastes were buried at their production sites in open, unlined trenches and then were covered with several meters of earth. At the time of their burial, no plans for the future retrieval of these wastes were made and a decision has not yet been made to systematically exhume the estimated 150,000 m^3 of previously buried TRU wastes. However, in 1970, the U.S. Atomic Energy Commission (AEC) [URL Ref. No. 330], the NRC's predecessor, adopted a policy requiring that wastes contaminated with a concentration greater than 10 nCi/g (370 Bq/g) of alpha particles be packaged, stored, and disposed of separately from other radioactive wastes; this limit was raised to 100 nCi/g (3,700 Bq/g) in 1983. Because of limited storage space at several of the major producing facilities, TRU waste has been shipped to INEL in Idaho Falls, Idaho, since 1970. After its redefinition in 1983, most of the unregulated TRU-containing wastes have been shipped to the Nevada Test Site for disposal by shallow land burial. Approximately 57,000 m^3 of regulated TRU waste (>100 nCi/g or 3,700 Bq/g) is currently stored on a temporary, retrievable basis at INEL.

The Los Alamos National Laboratory (LANL) [URL Ref. No. 97] has been disposing of radioactive wastes since 1944. The LANL Materials Disposal Areas, Areas A, B, C, D, E, T, G, and T, were solid radioactive disposal areas during the earlier years of LANL (Rogers, 1977a, 1977b; Hakonson et al., 1973; Nyhan et al., 1985; Penrose et al., 1990a, 1990b; Christensen et al., 1958). During the period from 1944 to the present, a large volume of radioactive and hazardous waste was buried in shallow trenches and pits at LANL in Los Alamos, New Mexico (Gerty et al., 1989). As part of the DOE Environmental Restoration Program [URL Ref. No. 288], personnel from the Laboratory or their subcontractors are examining several possible methods for locating and managing this waste material.

Radioactive wastes generated by LANL are categorized as routine or nonroutine. Most of the waste is routine, consisting of Laboratory trash (mostly combustible), equipment, chemicals, oil, animal tissue, chemical treatment sludge, cement paste, hot-cell waste, and classified materials (Rogers, 1977a). Nonroutine waste, generated during facility renovation and decommissioning projects, consists of building debris, large equipment items, and soil or rock removed during site cleanup.

The wastes may be contaminated by transuranic radionuclides (^{239}Pu, ^{238}Pu, or ^{241}AM), uranium (enriched, depleted, normal, or ^{238}U), fission products, induced activities, or tritium [URL Ref. No. 293]. Wastes contaminated by fission products, induced activities, and tritium are small in volume, 1–3 percent of the whole, but are high in total curies disposed of by LANL (Rogers, 1977a). More will be said about the various waste disposal areas that were used by LANL in the early years of the Manhattan Project and their relationship to groundwater concerns in a later section (Los Alamos National Laboratory—Case History Study for Radionuclides in Groundwater and Its Groundwater Protection Plan).

Disposal of TRU Wastes

It is anticipated that final disposal of TRU and HLW wastes will occur at the Waste Isolation Pilot Plant (WIPP) [URL Ref. No. 286] constructed near Carlsbad, New Mexico, *or the Yucca Mountain project in Nevada* [URL Ref. No. 307] (Devarakonda and Seiler, 1995). The disposal standards applicable to TRU waste disposal at WIPP [URL Ref. No. 286], as promulgated in December 1993 in the final standards of the Code of Federal Regulations (CFR) Title 40, Part 191 [URL Ref. No. 69 {10}], EPA reevaluation, concluded that geologic repositories [URL Ref. No. 315] are not a form of underground injection.

Again, however, and, as is still occurring presently, the DOE is being reinvented (Rezendes, 1995; U.S. DOE, 1994a, 1994b, 1994c, 1994e; U.S. DOE News, 1994d) *and its overall mission is still not clear.* Rezendes (1995) described that the U.S. General Accounting Office (GAO), in its review of DOE, analyzed management and contracting practices, organizational structure, performance in major mission areas (i.e., such as environmental cleanup and activities of the national laboratories), and as part of its management review, GAO also surveyed former DOE executives and experts on energy policy about how the Department's missions relate to current and future national priorities.

In the environmental area, the DOE faces the daunting task of cleaning up the contamination resulting from half a century of nuclear weapons production. The costs of restoring the nuclear weapons complex to a safe and stable condition are estimated by DOE to be at least $300–500 billion. Developing new technology will help cut costs, as will improved management efficiencies. These measures alone, however, will not allow the DOE to meet its current cleanup commitments under conditions of budget restraint. The DOE now also acknowledges that it will need to change its current process and work toward developing a national *risk-based strategy* that results in a more cost-effective approach to environmental cleanup. Unfortunately, DOE's past history of contamination, along with its long-standing contracting problems, makes it unclear how successful the DOE's new process will be (Rezendes, 1995; Naturman, 1995).

The overall changes that have occurred and are still in progress at the DOE, will certainly, most significantly, relate *to what is done with groundwater concerns at all of the previously discussed DOE areas.* Only time will tell what will happen, but certainly, the problems of the past still need to be corrected or stabilized.

Mining and Mill Tailings Waste
[URL Ref. No. 309–311]

From the early 1940s through the 1960s, much of the uranium ore mined in the United States was processed by private companies under contract to the federal government. The uranium [URL Ref. No. 283] ore was processed for use in national defense research, weapons development, and the developing nuclear energy industry. When the contracts for uranium terminated, the mills shut down, and large uranium tailings piles were left behind.

The process of mining and milling uranium [URL Ref. No. 283] and thorium [URL Ref. No. 339 {35}] ores generated large quantities of rock, sludge, and liquids. These wastes contain daughter nuclides such as radium, polonium, bismuth, and lead (Gershey et al., 1990; U.S. DOE, 1994e), and they are generated during the exploratory and operational phases of mining and consist of large amounts of rock from excavations and liquids from surface drainage, seepage, and *in situ* leaching [URL Ref. No. 246]. During mine operations, liquid and airborne effluents bearing gases and dusts constituted a significant hazard to workers and the public. Once high-grade ore is excavated, a typical milling operation involved the chemical extraction or leaching of radioactive minerals from the ore. Heap piles were built and solutions of acids and solvents were recirculated through the pile until extraction yields of acceptable quality were achieved. Leachates were then shunted to evaporators that furthered the crude product by roasting the concentrate. The concentrate is known as yellowcake, and large volumes of liquid containing acids, their salts, heavy metals, organic solvents, and residual radionuclides are then pumped to tailings impoundments for settling and decantation of the barren liquor. Unless well controlled, these impoundments contaminated local groundwater and surface water by runoff and seepage.

Radon [URL Ref. No. 107, 233] from the decay of ^{226}Ra was considered the most serious potential health hazard, particularly if the tailings were misused as building materials or fill. Windborne dust also posed a significant long-term, off-site hazard. Although most mining and milling activities occurred in sparsely populated areas of the Western states, further processing stages occurred throughout the United States. Mining and mill tailings wastes were poorly managed in the past, therefore, resulting in the need for federally sponsored programs to upgrade earlier disposal sites.

Uranium Mill Tailings Remedial Action (UMTRA) [URL Ref. No. 310]

The UMTRA Project implemented the Uranium Mill Tailings Radiation Control Act (UMTRCA) [URL Ref. No. 311] of 1978 (42 USC 7901 *et seq.*). This act established a program of assessment and remedial action at the uranium mill tailings sites. DOE had been directed to stabilize residual radioactive materials so that the radiological and nonradiological hazards did not exceed standards established by the U.S. Environmental Protection Agency (EPA).

The *Uranium Mill Tailings Radiation Control Act* [URL Ref. No. 311] required the EPA to promulgate standards of general application for protection of public health, safety, and the environment from radiological and nonradiological hazards associated with residual radioactive material located at inactive uranium mill tailings sites and depository sites. In 1983, the EPA established standards (40 CFR Part 192 [URL Ref. No. 69 {10}], "*Health and Environmental Protection Standards for Uranium Mill Tailings*"), but the groundwater portion of the standards was remanded to the EPA [(American Mining Congress v. Thomas, 772 F.2d 617, (10th Cir. 1985); cert. denied 476 U.S. 1158 (1986)]. In 1987, the EPA published the revised proposed standards (52 FR 36000) [URL Ref. No. 67–69] (U.S. DOE, 1994e).

The DOE was authorized to perform groundwater restoration in Senate Report 100-543, which accompanies the 1988 UMTRCA amendments, where necessary to comply with 40 CFR Part 192 [URL Ref. No. 69 {10}], "*Health and Environmental Protection Standards for Uranium Mill Tailings*," Parts A, B, and C. The UMTRA Project was also required to comply with the NEPA [URL Ref. No. 308] of 1969 as implemented by the Council on Environmental Quality [URL Ref. No. 177] regulation 40 CFR Part 1500 [URL Ref. No. 69 {10}]. In addition, the DOE codified implementing procedures for NEPA [URL Ref. No. 312] under 10 CFR Part 1021 [URL Ref. No. 67–69]. DOE Order 5440.1E [URL Ref. No. 275] (*National Environmental Policy Act Compliance Program*) and DOE Supplemental Directive AL 540.1D (*DOE Albuquerque Field Office National Environmental Policy Act Compliance Program*) established DOE guidelines for implementing the NEPA.

Under the DOE, the Uranium Mill Tailings Remedial Action (UMTRA) addressed the decontamination of 24 inactive sites and adjacent properties in 10 states (U.S. DOE, 1994e). Five of the sites are on or near Native American lands.

The UMTRA Project was divided into two projects, surface and groundwater. On November 18, 1992, the DOE issued a notice of intent (57 FR 54374, 1992) [URL Ref. No. 68–69] to prepare a programmatic environmental statement (PEIS) for the UMTRA Groundwater Project. In April 1995, the DOE issued a draft document on the PEIS (U.S. DOE, 1995m).

Summary

The proposed action and active remediation to background levels alternatives are most effective at protecting human health and the environment from the contaminated groundwater at the UMTRA [URL Ref. No. 310] project sites (U.S. DOE, 1995a–1995m). When cost is factored in, the proposed action is likely to be more cost-effective than the active remediation alternatives, because it would use less costly passive remediation strategies at sites where these strategies are shown to be protective of human health and the environment. Implementation of the active remediation to background levels alternative would be the most costly because active groundwater remediation methods would be used at most sites.

Groundwater below the uranium mill tailings sites may be contaminated with uranium [URL Ref. No. 283] and chemicals used in processing. To address groundwater compliance issues at these inactive uranium processing sites, the DOE is developing a program to ensure the protection of human health and the environment and to meet the proposed EPA groundwater standards.

Mixed Waste [URL Ref. No. 291]

Mixed waste [URL Ref. No. 291] is waste that is radioactive and hazardous as defined by the Resource Conservation Recovery Act (RCRA, Public Law 94-573, Oct. 21, 1976) [URL Ref. No. 280]. Mixed waste is generated by users of radionuclides and consists of contaminated organic solvents, oils, lead shielding, and chromate solutions.

DOE has explained radioactive mixed waste to be (U.S. DOE, 1990b) waste containing radioactive and hazardous components regulated by the AEA (Atomic Energy Act) [URL Ref. No. 330] and RCRA, respectively, with the term radioactive components referring only to the actual radionuclides dispersed or suspended in the waste substance (DOE Order 5400.3: Hazardous and Radioactive Mixed Waste Program, 2/22/89) [URL Ref. No. 275].

The mixed waste land disposal problems occur because the nonradioactive components are hazardous and promote the mobility of radionuclides. They also present regulatory authority problems, since these wastes are under the authority of the EPA, the NRC [URL Ref. No. 339 {34}], and different state agencies. It is also now the responsibility of generators to identify and properly manage mixed wastes, and at the present time, disposal options do not exist for mixed wastes and they cannot be legally stored by the generator for more than 90 days unless the facility has an RCRA Part B Permit.

Although mixed waste comprises less than 10 percent of the LLRW [URL Ref. No. 290], it has been identified by states as their major concern in managing LLRW (Office of Technology Assessment, 1989) [URL Ref. No. 178]. This concern of the states continues to this date where the issues associated with the management of mixed wastes are still subject to dual, and at times conflicting, regulations governing the radioactive and hazardous components of the waste (Devarakonda and Seiler, 1995). However, there is still a need for the development of a national mixed waste strategy based on consistent guidelines that relate to all situations.

Low-Level Radioactive Waste [URL Ref. No. 290]

Low-level radioactive wastes (LLRW) (Tykva and Sabol, 1995) [URL Ref. No. 290] are defined in the Low-Level Radioactive Waste Policy Act of 1990 (LLRWA, Public Law 96-573, Dec. 22, 1980) and in its 1985 amendments (Low-Level Radioactive Waste Policy Amendments Act; LLRW-PAA; Public Law 99-240, Jan. 15, 1986) [URL Ref. No. 68–69]. LLRW includes all the radioactive waste that is not classified as spent fuel from defense [URL Ref. No. 265]—high-level radioactive activities from producing weapons, commercial nuclear power plants, or uranium mill tailings (Office of Technology Assessment, 1988, 1989). About 97 percent of all LLRW produces relatively low levels of radiation and heat, requires no radiation shielding to protect workers or the surrounding community, and the radiation decays within less than 100 years to levels that the NRC finds do not pose an unacceptable risk to public health. The remaining 3 percent of LLRW requires shielding and can remain harmful for 300 to 500 years. A small percentage of LLRW is harmful, is the responsibility of the federal government to dispose, and needs to be isolated for a few hundred to a few thousand years (Office of Technology Assessment, 1989).

Low-Level Radioactive Waste Performance Assessments (Source Term Modeling)

Low-level radioactive wastes (LLRW) [URL Ref. No. 290] generated by government and commercial operations need to be isolated from the environment for at least 300 to 500 years. Most existing sites for the storage or disposal of LLRW employ the shallow-land burial approach. However, the U.S.

Department of Energy currently emphasizes the use of engineered systems (e.g., packaging, concrete and metal barriers, and water collection systems). Future commercial LLRW disposal sites may include such systems to mitigate *radionuclide transport* through the biosphere.

Performance assessments must be conducted for LLRW disposal facilities. These studies *include comprehensive evaluations of radionuclide migration from the waste package, through the vadose zone, and within the water table.* Atmospheric transport mechanisms also need to be studied. Estimates of the release of radionuclides from the waste packages (i.e., source terms) are used for subsequent hydrogeologic calculations required by a performance assessment. Computer models are typically used to describe the complex interactions of water with LLRW and to determine the transport of radionuclides. Several commonly used computer programs for evaluating source terms include GWSCREEN, BLT (Breach-Leach-Transport), DUST (Disposal Unit Source Term), BARRIER, and SOURCE 1 and SOURCE 2 (Icenhour et al., 1995).

The disposal of low-level radioactive waste entails financial and safety risks not common to most market commodities (Bullard et al., 1998). This manifests debilitating uncertainty [URL Ref. No. 261] regarding future waste volume and disposal technology performance in the market for waste disposal services. Dealing with the publicly perceived risks of LLRW disposal increases the total cost of the technology by an order of magnitude, relative to traditional shallow land burial. A marketable disposal permit mechanism is proposed by Bullard et al. (1998) and is analyzed for the purpose of reducing market uncertainty and facilitating a market solution to the waste disposal problem.

A Case History Example of some U.S. DOE Facilities within the Albuquerque Operations Office

DOE—Albuquerque Operations
[URL Ref. No. 312]

The most active national programs pertaining to radioactive waste management in the U.S. are those administered by the U.S. Department of Energy (Thomson, 1991, 1992) [URL Ref. No. 61]. The DOE created an Office of Environmental Restoration and Waste Management that prepared a five-year plan to achieve compliance with U.S. environmental laws [URL Ref. No. 68–69] with jurisdiction over radioactive and hazardous waste (Thomson, 1991; U.S. DOE, 1990a, 1990c, 1993b; Los Alamos National Laboratory, 1990 [URL Ref. No. 97, 288]; Rail, 1992).

The DOE Five-Year Plan includes the following goals:

(a) Clean up the DOE's sites by the year 2019

(b) Comply with public health and environmental protection laws and regulations

(c) Contain contamination at inactive sites

(d) Ensure that the DOE's compliance actions reduce risk to human health and the environment

The five-year plan (U.S. Department of Energy, 1990a) updated the FY 1991–1995 Five-Year Plan and incorporated a condensed version of the Draft Applied Research, Development, Testing, and Evaluation (RDT&E) Plan and added Transportation. It began with FY 1990 budget execution and continues through FY 1991 budget requests, FY 1992 budget formulation, and outyear cost estimates through FY 1996. The Plan also reflected a new Headquarters organization, the Office of Environmental Restoration and Waste Management (EM). This organization, established in November 1989, fulfilled a major DOE Departmental commitment to create a high-level focal point for the consolidated environmental management [URL Ref. No. 61] of nuclear-related facilities and sites formerly under the separate cognizance of the Assistant Secretaries for Defense Programs and Nuclear Energy and the Director of the Office of Energy Research. Superfund sites at which DOE is considered to be a potentially responsible party continue to be included in the Plan as they are identified [URL Ref. No.61].

Sections 2–4 of the Plan (U.S. Department of Energy, 1990a) provide information on planned activities in the three compliance-related areas of Corrective Activities, Environmental Restoration, and Waste Operations (i.e., including projects to modernize certain facilities), with specific information presented by the Operations Office. The Scope of Environmental Restoration/Remediation [URL Ref. No. 61] has activities for assessment and remediation at all inactive/surplus facilities and sites contaminated with radioactive, hazardous, or mixed wastes. The program is comprised of the following elements:

- Environmental Restoration and Remedial Actions Program (ERRA) [URL Ref. No. 61]
- Decontamination and Decommissioning (D&D)
- Uranium Mill Tailings Remedial Action Program (UMTRA) [URL Ref. No. 310]

The following activities are also within the ERRA Program:

- *Remedial actions:* activities required at all inactive/surplus facilities and sites contaminated with radioactive [URL Ref. No. 304], hazardous [URL Ref. No. 266], or mixed wastes [URL Ref. No. 291]; activities to protect or restore natural resources damaged by contamination from past activities that resulted in hazardous substance releases
- *Underground storage tanks* [URL Ref. No. 251]: tanks in operation before November 1988, after they have been identified as inactive/surplus
- *Investigations:* Activities to identify, confirm, and quantify contamination, feasibility studies, remedial action plans and designs, and remedial actions

- *Multiparty agreements:* costs associated with cooperative multiparty cleanup plans and activities
- *Monitoring systems:* installation of post-closure long-term monitoring systems
- *CERCLA* [URL Ref. No. 264]: Comprehensive Environmental Response, Compensation, and Liability Act (CERCLA) assessments necessary before assessing real property assets
- *RCRA* [URL Ref. No. 280]: Resource Conservation and Recovery Act (RCRA) permit provisions associated with solid waste management units [URL Ref. No. 299] that would meet the definition of past disposal sites under CERCLA/Superfund Amendments and the Reauthorization Act (SARA)
- *NEPA* [URL Ref. No. 308]: documentation preparation of all NEPA documentation related to environmental restoration activities—site-wide NEPA documentation is not covered
- *Land units:* closure of land units in operation prior to November 1988 including underground storage tanks
- *Studies:* specific studies and support for risk assessments [URL Ref. No. 235] for hazardous waste remedial actions

Additionally, the following activities are also within the scope of the Decontamination and Decommissioning (D&D) Program:

- *Assessment:* activities to identify, confirm, and quantify contamination; feasibility studies; remedial action plans and designs; and remedial actions
- *Remedial action:* activities required at all inactive/surplus facilities and sites contaminated with radioactive, hazardous, or mixed wastes; activities to protect or restore natural resources damaged by contamination from past activities that resulted in hazardous substance releases
- *Surveillance and maintenance:* after the facility has been accepted into a D&D funded program
- *NEPA Documentation:* preparation of all NEPA documentation related to decontamination and decommissioning activities; site-wide NEPA documentation is not covered
- *Cleanup:* decontamination and decommissioning after the facility has been accepted into a decontamination and decommissioning (D&D) funded program

Decontamination and decommissioning activities are to be broken into assessment, remediation, and surveillance and maintenance.

The Uranium Mill Tailings Remedial Action Program (UMTRA) [URL Ref. No. 310] is authorized by Public Law 95-604, the Uranium Mill Tailings Radiation Control Act of 1978 [URL Ref. No. 311], which calls for such actions as necessary to minimize radiation health hazards and other environmental hazards from inactive uranium mill sites.

The U.S. Department of Energy Albuquerque Operations Office [URL Ref. No. 312] Environmental Restoration and Waste Management Five-Year Plan (FY 1993–1997; U.S. Department of Energy, 1990a), for example, includes information of potential areas of integrated demonstrations {Inhalation Toxicology Research Institute [URL Ref. No. 336 {48}], NM; Kansas City Plant, KS; Los Alamos National Laboratory, NM; EG&G Mound Plant, OH; Pantex Plant, TX; Pinelas Plant, FL; Sandia National Laboratories, NM; Canonsburg, PA; Durango, CO (UMTRA); Rocky Flats, CO; Edgemont, SD (UMTRA); Falls City, TX (UMTRA); Grand Junction, CO (UMTRA); Green River, UT (UMTRA); Lakeview, OR (UMTRA); Lowman, ID (UMTRA); Maybell, CO (UMTRA); Monument Valley, AZ (UMTRA); Mexican Hat, UT (UMTRA); Naturita, CO (UMTRA); Slick Rock, CO (UMTRA); Spook, WY (UMTRA); Tuba City, AZ (UMTRA); South Valley Site, Ambrosio Lake, NM (UMTRA); and Belfield, ND (UMTRA) [URL Ref. No. 45]}. These integrated areas (i.e., Office of Technology Development Potential Integrated Demonstrations) include the following (groundwater and soils cleanup):

- cleanup of volatile organic components [URL Ref. No. 247, 267] in saturated soils and groundwater (Savannah River, gaseous, diffusion plants)
- Pu contaminated soils (Nevada and Rocky Flats) [URL Ref. No. 179]
- U contaminated soils (Oak Ridge, Fernald) [URL Ref. No. 294]
- unsaturated soils cleanup (arid sites; Idaho, Rocky Flats, Los Alamos, Pantex, Lawrence Livermore National Laboratory, Sandia National Laboratory)
- non-Pu/U metals in soil (Oregon, Idaho, Hanford)
- toxic chemicals (Savannah River, Oregon, Fernald)
- non-VOC in saturated soils (Oregon, Savannah River, Fernald, gaseous diffusion plants, Kansas City)
- non-VOC in unsaturated soils (Idaho, Hanford, Sandia-Livermore, Los Alamos, Pantex, Tonopah Test Range)

Los Alamos National Laboratory, A Specific Case History Review [URL Ref. No. 97]

The Los Alamos National Laboratory follows the Department of Energy's annual Environmental Restoration and Waste Management Five-Year Plan (Los Alamos National Laboratory, 1990). The Site-Specific Plan (SSP) is subject to a dynamic environment affected by agreements, permits, regulations, and site activities. The plan is written to encompass all activities necessary to comply with laws and regulations applicable to protect public health and the environment. Implementation, theoretically, is a must. The DOE conducted a

comprehensive environmental survey at Los Alamos National Laboratory in 1987 (U.S. Department of Energy, 1988b) and reported no major environmental problems at the Laboratory that represented an immediate threat to human life. The identified concerns varied in terms of their magnitude and risk. Since the survey, the DOE and the Laboratory have investigated the findings, and most of them have been mitigated and possibly closed out. Any findings that are still outstanding have been placed in a Corrective Activities Plan for systematic and scheduled corrective action, and they must all be closed out eventually.

The following information (i.e., an example of what was involved at a DOE National Laboratory in the time frame of 1989, although conditions can change from year to year to the present) addresses the Laboratory corrective activities that resulted from the Environmental Survey (U.S. Department of Energy, 1988b; Los Alamos National Laboratory, 1990) conducted by the Office of DOE Environment, Safety and Health Environmental Audit [URL Ref. No. 339 {37}]. Corrective activities at the Laboratory will be discussed in more depth under a separate heading ("*Los Alamos National Laboratory—Groundwater Interactions/Concerns*") that follows this section.

- The Laboratory's NPDES [URL Ref. No. 313] permit regulates 112 treated wastewater discharges. Three violations of the permit occurred in FY89. During the year, six upgrades of wastewater treatment systems were completed, four of which were installed pursuant to a Federal Facility Compliance Agreement negotiated between the EPA and the DOE. A major modification of the wastewater treatment system was completed at Technical Area (TA) - 53. During FY90, NPDES treatment system upgrading will continue.
- A Title I engineering design was completed for the Sanitary Wastewater Consolidation System (SWCS) Project. This project will replace seven existing sanitary wastewater treatment plants and approximately 30 septic tanks [URL Ref. No. 217–218], implementing state-of-the-art sanitary wastewater treatment and improving NPDES compliance.
- Water quality data were collected on all NPDES wastewater outfalls in preparation for submission of a NPDES reapplication during FY90. Additional water quality data, including biomonitoring analyses, will be collected.
- During FY89, two product and three radioactive waste USTs [URL Ref. No. 251] were removed. Eighty-eight tanks are currently in use (35 product and 53 radioactive waste tanks). Fifteen USTs are scheduled to be upgraded by replacement or retrofit to new tank standards.
- During FY89, engineering designs were completed for all major potential spill sites, and most sites were

redressed with secondary containment structures pursuant to the Laboratory's Spill Prevention Control and Countermeasure (SPCC) Plan.

- Septic tank systems [URL Ref. No. 217–218] were upgraded at TA-9 during FY89. Additional upgrades will be required and other technical areas will be investigated for upgrading.

- Throughout FY89, numerous PCB transformers and capacitors were replaced by non-PCB equipment. At the close of FY89, the Laboratory had 118 PCB transformers and 365 PCB large capacitors in its inventory.

- Groundwater protection was augmented at TA-56, the Fenton Hill Geothermal Site [URL Ref. No. 248], by cleaning out the large drilling mud pond (EE-1 pond) and preparing a design for installation of a seepage detection system and membrane liner.

- Throughout FY89, hazardous waste was primarily managed by using facilities located at TA-50 and TA-54. The Laboratory will pursue the construction of a Hazardous Waste Treatment Facility at TA-50. This system will consolidate and improve the handling and treatment of hazardous waste and will ensure compliance with RCRA [URL Ref. No. 280] requirements.

Under the *Corrective Activities Program,* numerous waste types must be addressed. They include sanitary and industrial wastewater consisting of sewage effluent, power plant and boiler blowdown effluent, treated cooling water and noncontact cooling water effluent, high explosive processing effluent, photographic processing effluent, printed circuit board processing effluent, and radioactive wastewater treatment effluent, toxic substances such as polychlorinated biphenyls (PCBs), radioactive air emissions, such as those at Los Alamos Meson Physics Facility (LAMPF), which are primarily made up of short-lived radionuclides having half-lives of 71 seconds to 1.8 hours, and various other hazardous wastes.

Specific hazardous wastes [URL Ref. No. 279] result from various Laboratory operations and programs. For example, underground storage tanks [URL Ref. No. 251] at the Laboratory contain petroleum products such as gasoline, kerosene, dielectric mineral oil, and waste motor oil. Other USTs contain chemical products, such as acids and bases, and miscellaneous hazardous and radioactive wastes. Wastes generated at high explosive processing and testing sites can include high explosive compounds, various chemicals, such as solvents, elements such as lead, and sometimes trace amounts of radioactive solids. Other hazardous wastes are generated because of diverse research and development activities throughout the 33 active technical areas of the Laboratory.

Regardless of the type of waste, mismanagement of waste materials could cause noncompliance with federal and state environmental regulations [URL Ref. No. 68–69, 333]. If noncompliance occurs, programmatic interruptions can result, with temporary curtailment of Laboratory operations potentially occurring. A more severe environmental problem could cause discontinuance of specific Laboratory operations. Since the Laboratory covers 43 square miles, with 33 active technical areas, the Corrective Activities Program affects virtually all of the technical areas. Throughout the Laboratory are nine active sanitary wastewater treatment facilities, 76 active septic tanks, 102 active industrial wastewater treatment facilities, 88 active underground storage tanks, more than 200 active satellite or less-than-90-day hazardous waste storage facilities, 118 PCB transformers, 60 PCB-contaminated transformers, 365 large PCB capacitors, six pieces of miscellaneous PCB equipment, 26 large-volume secondary containment facilities for spill control, and one major radioactive air emission source at TA-53. Although Corrective Activities can be put into place, because of the nature of past activities at Los Alamos related to nuclear weapons production or other wartime activities (e.g., chemical, biological, etc.), a never-ending systematic monitoring and surveillance program is essential (Los Alamos National Laboratory Environmental Surveillance Report Series, 1987–1997).

Environmental Safety and Health Vulnerabilities of Plutonium [URL Ref. No. 284] *at the Los Alamos National Laboratory* (LANL) [URL Ref. No. 97]

A national effort to assess the environmental safety and health (ES&H) issues of plutonium at defense nuclear facilities included an assessment of such vulnerabilities at the Los Alamos National Laboratory (Pillay, 1995). Of the 14 major locations identified within the DOE with ES&H vulnerabilities, Los Alamos was ranked thirteenth, well below the most serious problem sites within the defense complex. However, the problems at LANL are serious enough to require immediate attention, and resources are being sought to address the most serious ES&H vulnerabilities of plutonium at LANL. About 10 percent of the problem is located at LANL, and most of this inventory is in the form of residues generated from nuclear weapons production during the last decade.

At LANL, there are nearly 10,000 containers of plutonium with a large proportion of chemically reactive residues and several unsheltered containers. Most of the reactive residues at Los Alamos originated from the plutonium metal production activities during the 1980s. They included a variety of chloride salts from molten-salt extraction, oxide reduction, and electro-refining. In addition, there are large quantities of spent crucibles, molds, and filters made of graphite; contaminated sand, slag, and magnesia crucibles from bomb reduction; and numerous other residues from machining, molding, and aqueous chemical processing. An internal assessment identified necessary technologies to correct the vulnerabilities of these residues. The corrosion and rupture of containers of plutonium experienced during the past several years is likely to accelerate, and potentials for release to the environment are identified as the primary risk [URL Ref. No. 235, 262] to workers. Groundwater interactions here could be very significant.

Soil Adsorption of Radioactive Wastes at Los Alamos National Laboratory

The disposal of radioactive wastes by discharge to the ground and eventually groundwater has been practiced at Oak Ridge, Hanford, and Savannah River (Christensen et al., 1958), and during the early years of operation at Los Alamos, NM, all wastes were discharged to seepage pits or to canyons. In 1952 this practice was stopped, and chemical precipitating treatment plants were installed. The areas receiving these known plutonium-bearing [URL Ref. No. 284] wastes have been repeatedly monitored since that time, and no appreciable movement of plutonium through the soils has been noted. It has been observed, however, that the concentration of plutonium in the soil of a canyon receiving low-level wastes had progressively moved downstream. This concentration was not high, and, although it was measurable, it was still within acceptable tolerance levels, and the movement has not been extensive and is confined within the limits of the Los Alamos Project. Because there was some movement, however, it was deemed advisable to investigate the travel of plutonium through the local soils under varying conditions, since there was a possibility that wastes containing strontium-90 and cesium-137 might be produced by the Laboratory in the near future; consequently, it was decided to investigate these isotopes within the confines of the Los Alamos National Laboratory.

The results obtained in the study (Christensen et al., 1958) were not in complete agreement with those of others that showed that cesium and hardness [URL Ref. No. 194] broke through resin columns at about the same point. It had been demonstrated that the tuff local to Los Alamos had a rather high capacity for retention of various nuclides, and this was particularly notable since this particular material had an ion exchange capacity that was about as low as any to be found in nature. ^{137}Cs was also apparently very tightly bound to the tuff and resisted leaching by any of the common agents. ^{239}Pu likewise was readily retained by the tuff, and from the actual experience at Los Alamos, plutonium in wastes discharged into the ground appeared to remain at the point of discharge. However, from what is known about the chemistry of plutonium [URL Ref. No. 284], it was entirely possible that this nuclide could be released at some future time by inadvertent discharge of solutions, such as versene, in the same area.

At Los Alamos, ^{90}Sr was not retained by the tuff nearly as well as cesium and plutonium, and it was much more easily released. Disposal of this isotope to soils is to be undertaken with extreme caution and only with fore-knowledge of the nature of the soil and its capacity for the ions known to be present in the waste. ^{90}Sr can be leached by other ions, and a disposal area receiving this isotope must be closely guarded.

Mobility of Plutonium and Americium through a Shallow Aquifer in a Semiarid Region at Los Alamos

Treated waste effluents from the Central Waste Treatment Plant at the Los Alamos National Laboratory have been discharged into Mortandad Canyon since 1963 (Penrose, et al., 1990b). The shallow alluvium of Mortandad Canyon is composed of lensed sandy to silty clay materials formed by the weathering of volcanic rocks (Bandelier Tuff) and contains a small elevated aquifer of $(20–30) \times 10^3$ m^3 storage capacity. Annual storm runoff into the canyon ranges from 25 to 125 $\times$ 10^3 m^3 per year, and treated effluents are released into the seasonal stream, flow down the canyon and, under ordinary conditions, infiltrate into the tuff with $\cong$ 2 km. Surface water may flow as far as 3.4 km beyond the waste outfall during storm events. Subsurface flow represents $\cong$ 90 percent of the water movement in the canyon. Tritium [URL Ref. No. 293] oxide tracer experiments have also shown that 85 percent of the water released from the waste plant was lost through evapotranspiration. Total transit time for the tritium from the waste outfall to a monitoring well 3,390 m down the canyon was about a year, and the water that did not evaporate was assumed to be lost by infiltration into the underlying tuff, because no continuous surface flow existed along the reach of the canyon.

The waste treatment process included the addition of iron sulfate and lime. The precipitation of iron hydroxides and calcium carbonate acted to remove almost all of the actinides from the waste, although traces of ^{238}Pu, $^{239, 240}$Pu and ^{241}Am remain in the effluent and are released to the canyon.

The shallow aquifer also contains a series of monitoring wells. Sampling of these wells has revealed that plutonium [URL Ref. No. 284] and americium are found in the groundwater at distances of 3,400 m from the outfall. Some of the actinides could be transported to the lower region of the canyon by surface flow during occasional storm events. However, the tritium oxide transit time measurements suggest that the majority of water movement takes place in the subsurface. Reports of plutonium and americium movement in groundwater over distances of even a few meters are rare. It was the purpose of the study by Penrose et al. (1990b) to determine the true mobility of these actinides in the groundwater of Mortandad Canyon and to establish the features of either the wastewater or the aquifer that might contribute to enhanced mobility.

Regular monitoring of the effluents and groundwater confirmed, based on their sampling points, that all effluents are contained within the laboratory boundary, that the concentrations of plutonium and americium have not exceeded the Department of Energy Concentration Guidelines for Controlled Areas, and that no water is derived from Mortandad Canyon for drinking, industrial, or agricultural purposes. These trace level actinides, however, act as tracers to evaluate the potential for colloidal transport of subsurface groundwater contaminants. These other contaminants include actinides and other radionuclides [URL Ref. No. 232, 304], toxic metals [URL Ref. No. 317], and toxic organic materials [URL Ref. No. 210].

Soil Adsorption of Radioactive Wastes at Los Alamos and Potential Groundwater Interactions

In late 1943, a site with the primary responsibility for the purification of Pu was established at Los Alamos, NM (Nyhan et al., 1985). Because of the urgency, limited construction time, and the lack of information on the resulting radioactive wastes [URL Ref. No. 304], it was initially decided to dispose of radioactive wastes in several ways. Untreated liquid wastes were at first discharged into canyons, underground storage tanks [URL Ref. No. 251], and absorption beds filled with gravel and cobble.

The interaction of some of these radionuclides in these liquid wastes with local soils and geologic materials was initially studied in the laboratory. Cores of Bandelier tuff collected at Los Alamos were contaminated with waste solutions of Pu, essentially all of which was retained in the top few millimeters of the core even after subsequent leaching (Christenson et al., 1958) experiments. Five 3 to 6 m deep holes were drilled in and around the absorption beds, and an effort was made to gather samples at 30 cm intervals using a pick and shovel, a driven pipe, and a drilling rig with a core bar. The results of this study indicated that the vertical migration of Pu occurred within 6 m of the surface of the absorption beds and that Pu was readily retained by the components in the bed.

These field observations were in sharp contrast with the results of the early laboratory studies of Pu solutions (with and without complexing agents) in tuff cores, which demonstrated that essentially all of the Pu was retained within the top few millimeters of the tuff core (Christenson et al., 1958). Whether or not these differences in radionuclide behavior can be explained by physical and chemical differences in the liquid waste streams is unclear at this time and not within the scope of the study by Nyhan et al. (1985).

Occurrence, Fate, and Transport
[URL Ref. No. 302]

Fate/Transport

Since radionuclides can be naturally occurring or introduced to groundwater by humans or their activities, contami-

nation is defined here as concentrations of radionuclides that pose a health risk [URL Ref. No. 235, 262]. The radionuclide contamination problem can be divided into two parts, according to Toran (1993) and Knox et al. (1993), that include the potential adverse effect on health and the environment and the need to dispose of radioactive waste underground without creating problems related to the first part. However, then one therefore needs to address two corresponding questions:

(1) Have groundwater supplies been extensively contaminated by radionuclides?

(2) Can we prevent future contamination?

An understanding of these questions and complexities of radionuclide geochemistry [URL Ref. No. 314] is, therefore, required to answer these questions. This is because radionuclides are not a uniform class of elements, but they can vary widely in their behavior, as has been discussed in the first section of this *chapter*. Since waste forms typically contain mixtures of radionuclides, predicting behavior in groundwater is complex, and disposal environments will always be complex systems with extreme or varying *Eh,* temperature, ionic strength, and hydrogeological properties [URL Ref. No. 13–14, 187], which make such studies difficult. Radionuclides also have an advantage over some contaminants in that they decay naturally (i.e., albeit some decay so slowly that in human terms they must be treated as long-term threats). Additionally, there is more extensive information on health effects of radionuclides (i.e., due to medical research and studies of known exposure to humans) than for most other groundwater contaminants.

The problem of radionuclide contamination of groundwater also becomes an emotional one involving public fears of exposure to risks [URL Per. No. 235] they cannot see or understand. Consequently, ultimately, to answer these questions, there is a need to address the scientific issues involving radionuclide geochemistry and transport in groundwater.

This is accomplished by emphasizing factors that can mobilize radionuclides such as organic complexes [URL Ref. No. 210] and colloids, high-temperature interactions, and surface complexation models that describe sorption in greater detail than was previously possible. Issues involved in evaluating radioactive waste-disposal sites and the modeling to address these issues also then need to be presented. And, it must be understood that perhaps the greatest challenges facing modelers are presented by limited data available on radionuclide migration for model calibration and the total lack of long-term (thousands of years) studies. Unsaturated flow, fracture flow, and coupled flow and geochemistry are important areas of research in radionuclide modeling, and radionuclides previously believed to be immobile (e.g., [241]Am) have now been known to have been transported significant distances, often attributable to transport of colloids. Thus far, public water supplies have not been extensively contaminated by human-made radiological sources, but adequate continued satisfactory monitoring is needed.

Remediation of contaminated groundwater in the Chernobyl 30 km evacuation zone is frequently identified as a priority by technical experts and Chernobyl site officials in Ukraine (Bugai et al., 1996). And, in order to evaluate the health risk basis [URL Ref. No. 235, 262] for the groundwater remediation, Bugai et al. (1996) estimated both on-site and off-site health risks caused by radionuclide migration to the groundwater and compared these risks with those from exposure to radioactive contamination on the ground surface. This analysis by Bugai et al. (1996) implied that, relative to other exposure pathways, there was little current or future health risk basis for the proposed complex and costly groundwater remediation measures that occurred in the 30 km zone. Therefore, these activities would be abandoned in favor of more pressing health issues caused by the Chernobyl accident.

Sources of Radionuclides in Groundwater

There are three sources of radionuclides in groundwater and they include the following:

(1) Atmospheric radiation originated in cosmogenic decay and thermonuclear bomb testing
(2) Natural sources in the subsurface
(3) Other human-made sources

Cosmogenic radiation is defined as the result of bombardment of atmospheric gases (e.g., argon and nitrogen) by cosmic rays. This bombardment interaction then produces radioactive isotopes such as ^3H, ^{14}C, ^{32}Si, ^{36}Cl, ^{39}Ar, ^{85}Kr, and ^{129}I, which also have been identified in groundwater (Toran, 1993). Natural levels of many of these radionuclides previously listed were increased by testing of thermonuclear bombs in the 1960s, but these higher human-made levels have since declined. Both natural and human-made atmospheric radionuclides (e.g., ^{14}C and ^3H) have also been used in groundwater dating, with varying success.

Radioactivity in groundwater from natural sources comes from rocks containing uranium [URL Ref. No. 283] or thorium [URL Ref. No. 339 {35}], which decay to produce products sufficiently long-lived to be detected in groundwater (e.g., ^{238}U, ^{235}U, ^{226}Ra, ^{228}Ra, and ^{222}Rn). The rock type linked with the highest groundwater concentrations is granite and its derivatives, metagranites, and arkoses (i.e., which is weathered from granite). Concentrations of radionuclides in water may be enhanced by human activities, such as uranium mining and milling [URL Ref. No. 311]. Recently, high concentrations (i.e., up to levels encountered in uranium mines) of naturally occurring radioactive radium have been described in pipe scale from oil field brines.

Human-made sources of radionuclides additionally include nuclear power reactors, processing plants for nuclear fuel (to enrich ^{238}U with ^{235}U), fission products from weapons fabrication, medical therapy and research, and various industrial categories. Industrial sources include radiography (e.g., gamma rays used to inspect metal parts for flaws), luminescent signs, and smoke detectors. The nuclear power industry is a large source of waste because of the fuel elements (i.e., although, technically, fuel elements are not considered waste, since it is possible to reprocess them). Since 1972, there has been no disposal of fuel element wastes in the United States; they are stored aboveground on reactor sites.

What will be provided next is a brief overview of the geochemical behavior of radionuclides as they relate to fate and transport [URL Ref. No. 302].

GEOCHEMISTRY [URL REF. NO. 188, 314]

The geochemistry of radionuclides is critical to determining their mobility. This is because mobility is a key factor in groundwater contamination, as it determines whether contaminants are released from a source, how far they spread, and at what concentrations they exist. Also, like other groundwater constituents, mobility is affected by solubility, complexation, sorption on immobile solids and on colloids, and oxidation state, which affects the other behaviors previously listed. The only factors distinctive for radionuclides are radioactive decay and alpha recoil. Radioactive decay is well quantified and reduces the concentration of the parent radionuclide, but it gives rise to daughter products that may also be radioactive. Alpha recoil is mobilization due to a physical rebound from decay (i.e., a factor in Rn release). Additionally, radionuclide geochemistry may seem more complex because there are a large number of elements to consider, with some radionuclides having a wider range of oxidation states than most major ions, consequently, they are uncommon constituents of natural environments and are studied less. Furthermore, their behavior must be understood in diverse environments if prediction of mobility in a radioactive repository is to be assessed.

Nonetheless, there is a wide range of studies available describing the geochemistry of radionuclides, and while some trends can be observed, specific radionuclides and geochemical environments must be studied to determine whether radionuclides will be mobile in groundwater and potentially can significantly contaminate groundwater supplies (Bugai et al., 1996; Toran, 1993).

SOLUBILITY AND OXIDATION STATE

Precipitation of radionuclides as crystalline phases, amorphous phases, and coprecipitates can also be an important factor in reducing mobility in groundwater. Any of these phases could be present in the groundwater environment because of low groundwater velocities (Toran, 1993). A common crystalline phase is made up of oxide minerals, for example TcO_2 or PuO_2. Uranium [URL Ref. No. 283] forms an oxide precipitate and coprecipitates with other oxides. Additionally, because many radionuclides have a high atomic number (and thus have more electrons), a wide range of oxidation states is possible, and determining the oxidation state is the first step in determining what complexes will form and whether a solid phase will be present. For example, $Tc(IV)$ forms TcO_2 or is readily sorbed, while $Tc(VII)$ forms water-soluble oxide and hydroxide complexes (TcO_2- and $HTcO_4$). Uranium shows the tendency of higher oxidation states to be more mobile. Eh-pH diagrams are often used to find conditions leading to mobility. However, many assumptions are used to construct diagrams. These include equilibrium thermodynamics, constant specified temperature, species concentrations, and selection of species. Any variation in these factors or addition of species requires new diagrams, and infinite combinations are possible. Thus, Eh-pH diagrams represent only the first step in making predictions of radionuclide behavior (Toran, 1993).

COMPLEXATION AND COLLOID FORMATION

Because of the large number of elements, the range in possible oxidation states, and limited occurrence in nature, there are many uncertainties in radionuclide complexes (types and equilibrium constants) that hinder predictions of radionuclide mobility. Some mobile complexes have been identified by analyzing mobile species in groundwater samples, but, unfortunately, there has not been a comprehensive determination of mobile species from field data (Toran, 1993). Instead, laboratory titration experiments have been a major source of information, which is limited by the time-consuming nature of these experiments. There are simply not enough laboratory or field studies under the range of possible conditions in waste environments to sufficiently characterize radionuclide complexes.

SORPTION

The final geochemical process to consider in immobilization of radionuclides is sorption. Sorption is a term used to describe several different processes, from ion exchange to site binding of oxides. Prediction of sorption is hindered by the need to understand the physicochemical properties of prevailing complexes, the wide range in possible geochemical conditions, as well as by the need for further detailed field and laboratory studies.

The simplest model to estimate sorption uses the distribution coefficient (Kd), which is the ratio of solute sorbed to surfaces to that remaining in solution, assuming reversible equilibrium with the surrounding solution. Kds are specific to the solute, the soil, and the ambient water chemistry [URL Ref. No. 186, 189]. Kd translates to a retardation factor (R) in the solute transport equation (Toran, 1993). Large Kds are indicative of slow transport that consequently, leads to dilution of peak concentrations by transporting water. Alternative concentrations to using Kd to examine sorption behavior involve more detailed geochemical characterization of solute and sorbent in terms of processes such as complexation, oxidation/reduction, precipitation, types and capacities of sorption sites, and competing sorbates.

MODELING AND REPOSITORY SITING [URL REF. NO. 315]

Modeling is related to the problem of siting an underground repository for disposal of radioactive waste because the regulations surrounding repository siting place unusual demands on radionuclide transport models (Toran, 1993). Specifically, these models are expected to predict, with high certainty, transport behavior in groundwater 1,000 and even 10,000 years into the future. These expectations are unrealistic given contemporary scientific knowledge about any chemical or any physical system.

Models are important for evaluating radionuclide behavior because of the large number of processes, complex geochemistry, and the need for scaling up in space and time. For high-level radioactive wastes (HLRW) [URL Ref. No. 305], model studies are typically divided between near field (near waste canisters where high-temperature, high-ionic strength geochemical reactions are important) and far field (where groundwater geochemistry and transport are important). Modeling needs for assessing radionuclide mobility are similar to those for other contaminants with a realistic conceptual model being the first step. Advection must then be carefully evaluated with flow model; then transport, including geochemical reactions and dispersion, must be considered.

Additional processes to consider when modeling radionuclide transport include radioactive decay, unsaturated flow and fracture flow (two factors in many proposed low-permeability repository environments) [URL Ref. No. 315], and parent plus daughter element transport. Sensitivity analysis to evaluate uncertainty [URL Ref. No. 261] and variability is also a factor. Model limitations are similar to those for any contaminant transport model with insufficient characterization of heterogeneity in the field and insufficient information on sorption behavior.

Coupled flow and geochemical codes are often suggested for use in radionuclide-transport problems because they can account for geochemical variability within the flow field and more complex geochemical behavior than described in transport codes alone (Toran, 1993). Gaps in geochemical database knowledge include organic complexes, high-temperature reactions, high ionic strength, amorphous solids and solution series, and associations with waste canisters.

The primary limitation in radionuclide modeling is that there are almost no opportunities to calibrate models because of the unusual environments and chemical studies that occur. Sites considered for radionuclide disposal are difficult to characterize because, by requirement, they have low accessibility and thus have been studied less.

RADIONUCLIDE MIGRATION

At Los Alamos National Laboratory [URL Ref. No. 97, 101], Pu and Am from liquid LLRW [URL Ref. No. 290] have been detected in four wells along Mortandad Canyon, 3,390 m from a disposal area having rates of as fast as 0.5 km/y based on separate releases of ^{239}Pu and ^{240}Pu (Penrose et al., 1990a, 1990b). It is likely that the radionuclides travel through Bandelier tuff. Colloidal mobilizations of Pu and Am closer to the waste pits were reported previously (Nyhan et al., 1985), and specific size fractions were identified through ultrafiltration techniques. For Pu, 85 percent was retained by the 25 to 450 nm fraction, while only 28 percent of the Am was of this size, 26 percent was associated with 2 to 5 nm size colloids, and the remainder passed through the 2 nm filters. Filtration in the aquifer of the larger colloids may have led to the observed decline in Pu concentration along the flow path, whereas, the Am in smaller fractions or ionic forms does not show this trend. The specific colloids were not identified, but inorganic colloids, dissolved organic carbon, and an unknown anionic complex of Am were detected in the groundwater samples (Nyhan et al., 1985).

PROBLEMS ASSOCIATED WITH ASSESSING RADIONUCLIDE CONTAMINATION OF GROUNDWATER

There are significant problems associated with assessing radionuclide contamination of groundwater. These include (Toran, 1993) the following:

(1) Field, laboratory, and modeling studies on complexation, colloids, and sorption are needed to improve prediction of radionuclide migration and to better estimate risks [URL Ref. No. 235].
(2) Contamination of groundwater by naturally occurring radon [URL Ref. No. 107, 233] has occurred, and further assessment is needed to determine its impact and scope.
(3) Groundwater contamination is a potential pathway to surface water contamination, which may increase environmental exposure.
(4) Siting of an underground repository [URL Ref. No. 315] for civil and defense radioactive waste raises sociopolitical issues that are difficult to address scientifically. Cleanup operations at existing disposal sites quite often face unrealistic expectations.

SUMMARY

Radionuclide contamination of groundwater presents challenges to our knowledge of radionuclide geochemistry

[URL Ref. No. 314] and transport processes, however, in the meantime, other more immediate risks [URL Ref. No. 235] to groundwater supply (e.g., numerous landfills) should not be neglected in view of the sociopolitical stigma surrounding radioactive waste disposal. Scientists need to find ways to address scientific, technological problems, and social problems associated with groundwater contamination by radionuclides in appropriate time intervals.

Transport and Fate of Contaminants in the Subsurface

Congress requires (U.S. EPA, 1989a) that the U.S. Environmental Protection Agency, as well as other regulatory entities and the regulated community, meet four interrelated objectives for the protection of groundwater quality. These include the following:

(1) Assessment of the probable impact of existing pollution on groundwater at points of withdrawal or discharge {Safe Drinking Water Act (SDWA), 1974, 1986, [URL Ref. No. 258]}
(2) Establishment of criteria for location, design, and operation of waste disposal activities to prevent contamination of groundwater or movement of contaminants to points of withdrawal or discharge {Resource Conservation and Recovery Act of 1976 (RCRA) [URL Ref. No. 280] and the Hazardous and Solid Waste Amendments of 1984 (HSWA) [URL Ref. No. 300]}
(3) Regulation of the production, use, and disposal of specific chemicals possessing an unacceptably high potential for contaminating groundwater when released to the environment {Toxic Substances Control Act (TSCA) [URL Ref. No. 295] and the Federal Insecticide, Fungicide, and Rodenticide Act (FIFRA) [URL Ref. No. 296]}
(4) Development of remediation technologies that are effective in protecting and restoring groundwater quality without being unnecessarily complex or costly and without unduly restricting other land use activities {Comprehensive Environmental Response, Compensation, and Liability Act of 1980 (CERCLA or Superfund) [URL Ref. No. 264] and the Superfund Amendments and Reauthorization Act of 1986 (SARA)}

To achieve these objectives, definite knowledge of the *transport and fate of contaminants* in the subsurface environment is essential [URL Ref. No. 302]. Without this knowledge, regulatory agencies [URL Ref. No. 19, 89, 333] run the twin risks of undercontrol and overcontrol, and regulatory undercontrol would result in inadequate prevention and cleanup of groundwater contamination. Regulatory overcontrol would result in costly preventative actions and remedial responses to contamination. However, gaining and using knowledge about the contaminant transport and fate can be difficult because of the complexity of the subsurface environment. The activities of site characterization and remediation illustrate and are an example of this complexity.

Site Characterization

Transport and fate assessments require interdisciplinary analyses and interpretations because the processes involved in these activities are naturally intertwined (U.S. EPA, 1989a), and each transport process must be viewed from the broadest of interdisciplinary viewpoints, with the interactions between them identified and understood. In addition to a sound conceptual basis, integrating information on geologic, hydrologic, chemical, and biological processes into an effective contaminant transport evaluation requires data that are accurate, precise, and appropriate at the intended problem scale.

The issues of contaminant transport and fate in the subsurface are difficult to address at Superfund sites [URL Ref. No. 264] because of the complex array of chemical wastes involved. The hydrogeologic settings of these sites are usually measured in hundreds of feet and, at this scale, are extremely complicated when characterized for a remediation plan. The methods and tools used for large-scale characterizations are generally applicable to the specialized needs at hazardous waste sites, however, the transition to smaller scale is fraught with scientific and economic problems. Some problems stem from the highly variable nature of contaminant distributions at hazardous waste sites, and other problems result from the limitations of available methods, tools, and theories.

When using a conceptual model to interpret contaminant transport processes, it is crucial that special attention be given to the spatial and temporal variations of the collected data (i.e., hydraulic conductivity) (U.S. EPA, 1989a). Additionally, to circumvent the large numbers of measurements and samples needed to reduce uncertainties [URL Ref. No. 261] in dealing with subsurface parameters, more comprehensive theories are constantly under development. The use of many developed theories, however, is also frustrating because many call for data that are not yet practically obtainable, such as chemical interaction coefficients or relative permeabilities of immiscible solvents and water. Therefore, modern contaminant transport and fate studies involve a compromise between sophisticated theories, current limitations for acquiring data, and economics.

There are several contaminant transport processes that may be responsible for the persistence of residual contamination. Releases of contaminant residuals may, for example, be slow relative to water movement through the subsurface while pumping is occurring. Transport processes that generate this kind of behavior include the following:

(1) Diffusion of contaminants within spatially variable sediments

(2) Hydrodynamic isolation

(3) Sorption-desorption

(4) Liquid-liquid partitioning

There are many misconceptions regarding the processes affecting the transport and fate of contaminants in the subsurface. Some of these can be addressed by educational efforts, while others can be studied only by applied research methods (Jury and Roth, 1990). The document that describes this specific information about known transport and fate of contaminants in the subsurface is presented by the U.S. EPA (1989a) and includes the following information:

(1) Physical processes controlling the transport of contaminants in the aqueous phase

(2) Physical processes controlling the transport of nonaqueous phase liquids in the subsurface

(3) Determination of physical transport parameters

(4) Subsurface chemical processes

(5) Subsurface chemical processes (field examples)

(6) Microbial ecology [URL Ref. No. 214] and pollutant biodegradation in subsurface ecosystems

(7) Microbiological principles influencing the biorestoration of aquifers

(8) Modeling subsurface contaminant transport and fate

(9) Management considerations in transport and fate issues

Transport of Reactive Contaminants in Heterogeneous Porous Media

The transport and fate of contaminants in subsurface will continue to be one of the major research areas in the environmental, hydrological, and Earth sciences (Brusseau, 1994). However, an in-depth understanding of how contaminants move in the subsurface is required to specifically address environmental problems. And, such knowledge is needed to evaluate the probability of a contaminant being associated with a chemical spill reaching an aquifer and contaminating groundwater. Just as importantly, knowledge of contaminant transport and fate is necessary to design pollution prevention strategies and develop and evaluate methods for cleaning up contaminated soils and aquifers.

The study by Brusseau (1994) expanded on this by including the results of theoretical studies designed to pose and evaluate hypotheses, results of experiments designed to test hypotheses and investigate processes, and development and application of mathematical models [URL Ref. No. 339 {31}] useful for integrating theoretical/experimental results for evaluating complex systems. Brusseau (1994) began with a review of the basic concepts related to contaminant transport and followed with a discussion of the results obtained from some of the few well-controlled field experiments designed to investigate transport of reactive contaminants in the subsurface. Some of the major factors controlling contaminant transport were then discussed, followed by a review of conceptual and mathematical approaches used to represent those factors in mathematical models. Brusseau (1994) then concluded with a brief overview of future needs and opportunities in contaminant transport, indicating that it is clear that a large number of physical, chemical, and biological factors and processes influence contaminant transport.

Additionally, it is well established that the original paradigm for contaminant transport in porous media, based on assumptions of subsurface homogeneity and instantaneous mass transfer, are invalid at the field scale. Despite this fact, models based on the paradigm continue to be developed and used. The continued use of this paradigm can be related to at least three factors:

(1) *Lack of knowledge:* While the invalidity of the original paradigm may be well known to those actively doing research in this field, it may not be universally known to all those using contaminant transport models or the information obtained from them. Thus, more information is needed to transfer knowledge from those developing it to those that need it.

(2) *Lack of computational resources:* In many cases, although the existence of nonideal transport may be recognized, models based on the original paradigm are developed and used because the computational resources required to accurately simulate nonideal transport are not available. The great increase in the computational power of desktop computers, the advent of inexpensive workstations, and the increasing accessibility to supercomputers should eventually eliminate this constant.

(3) *Lack of information:* The greatest impediment to the use of advanced, nonideal transport models is the lack of information available for parameterizing the models. The expense and time associated with the traditional means of field characterization (e.g., sampling wells and collecting cores) preclude the availability of the required information for most sites. Thus, the widespread use of advanced contaminant transport models is intimately coupled to advances in methods for site characterization.

The application of single-factor nonideality models to systems affected by more than one factor yields lumped parameters. Values of these lumped parameters can usually be obtained only by calibration and will be valid only for the specific set of conditions for which they were obtained. In addition, these lumped parameters cannot supply process-discrete information and, thus, are useless for elucidating the relative contributions of various nonideality factors to total nonideality. Such information can only be obtained with the use of a model that accounts explicitly for the existence of multiple nonideality factors. To date, very few such models have been presented. Given the large probability that contaminant transport at the field scale is influenced by multiple nonideality factors, the need for additional work in the development and evaluation of multifactor nonideality models is apparent.

It is probable that for most site applications, one will never be able to generate sufficient information to use fully deterministic models in the distributed parameter mode. The fact that properties of the subsurface cannot be measured at all points in the domain of interest means the parameter values used for input are uncertain. The use of stochastic-deterministic models, for which input parameters can be treated as having a mean value with associated variance and error, is a better means by which to account for this uncertainty. The fact that the input for a model is uncertain leads to uncertainty in the output. It is important that this uncertainty, therefore, also be considered in the development and application of contaminant transport models.

A major focus of the paper by Brusseau (1994) included factors that control contaminant transport in subsurface systems. A large body of work has been reported on laboratory-scale investigations of these factors. The vast majority of the work done by Brusseau (1994) involved studying a particular factor in isolation. Such work is still needed, however, to further refine the understanding of the mechanism involved along with an additional need for research on coupled processes, an area that is just beginning to receive attention. Research is needed to investigate the dynamics of systems influenced by multiple, simultaneously occurring processes wherein synergistic or antagonistic interactions may influence transport. This is especially critical for transferring the understanding from laboratory- to field-scale systems. Additional field-scale experiments are needed for studying transport of reactive contaminants in the subsurface. Especially needed are detailed investigations of the transport of contaminants that have been in contact with porous media for long times. It is critical that future experiments be designed in accordance with the current understanding of contaminant transport.

Radioactive Colloid Transport

Most studies of radionuclide migration in possible repository environments have focused on the transport of dissolved forms of the radionuclides by flowing groundwater (Chung and Lee, 1991; Hwang et al., 1991). Colloid formation, however, is not uncommon in geologic systems, and radionuclides in a colloidal or particulate form could conceivably migrate further and faster than they would in a dissolved form because of weaker sorption on stationary solids. It is therefore worthwhile to consider their role in the geologic environments surrounding future waste repositories [URL Ref. No. 315]. Chung and Lee (1991) examine this role by discussing theories that quantify colloid migration in porous media and describe certain processes and mechanisms that affect the transport colloidal particles in porous media.

Bimodal Filtration Coefficient for Radiocolloid Migration

During waste dissolution in a nuclear waste repository [URL Ref. No. 315], colloids can also be created, according to Hwang et al. (1991). The migration of radiocolloids through porous material in an advection-dominated system is described in an experiment on colloidal migration to determine an appropriate filtration coefficient. A bimodal probability distribution for the filtration coefficient is needed for colloidal transport in porous media.

Laboratory Studies for Prediction of Radionuclide Migration in Groundwater

The sorption of ^{60}Co, ^{90}Sr, and ^{137}Cs on five core samples from El-Dabaa (the site of the future Egyptian power reactor) were determined, and the migration velocities of the tested radionuclides in groundwater of the area were calculated by Aziz et al. (1994). The dependence of sorption on the solution pH [URL Ref. No. 203], the concentration of the tested cations, the presence of foreign salts, the time period of contact between the solid and aqueous phases and stability of binding of the tested cations on the rock samples were also investigated, and the results were discussed. Also, according to Aziz et al. (1994), the stability of the binding of the radionuclide to rock is one of the most important parameters for migration of radionuclides, and the rate of migration depends on the ion and the process by which it is bound to the rock particles.

Effect of Chelating Agents on the Migration of Radionuclides

It has been stated that chelate formation of radionuclides with chelating agents such as decontamination reagents (e.g., ethylenediaminetetraacetic acid) and natural organic compounds (e.g., fulvic and humic acids) observed in groundwater significantly influences the migration behavior of radionuclides (Baik et al., 1991). They form extremely strong chelates with radionuclides and mobilize

these radionuclides from radioactive waste repositories (especially from low-level waste) [URL Ref. No. 290]. In the study by Baik et al. (1991), a new retardation factor incorporating a chelation effect was introduced, and a general convection-dispersion transport equation that included a degradation of solute caused by various physicochemical reactions in porous medium was used and solved by an analytical method.

The effect of the degradation constant on transport of radionuclides showed the degradation processes affecting the migration of radionuclides in the presence of chelating agents (i.e., may be precipitation, hydrolysis, oxidation-reduction, microbial transformation, and chelating agent degradation). This mechanism conserved the performance of the natural geologic medium as a barrier for the migration of radionuclides.

Improved Technique for Estimating Parameters of Diffusion Experiments

Cement has been widely used to solidify low-level radioactive waste [URL Ref. No. 290] (Teng and Lee, 1991), and it is also used as an engineered barrier to retard the release of radionuclei into the biosphere. Since diffusion is the main transport mechanism by which radionuclei are released through the engineered barriers of a low-level waste repository, it is necessary to know the diffusion coefficients of radionuclides in cement to evaluate its retardation effectiveness.

The diffusion coefficients of radioactive nuclides in cement can usually be determined by experiments using diffusion cells. In the study by Teng and Lee (1991), cement specimens were made into circular plates and were individually installed in the middle of the cell. The specimen divided the cell into the inlet high-concentration region and the outlet low-concentration region. The diffusion of radionuclei through a cement specimen was monitored by measuring cumulative activities in the outlet region. A good estimate of the diffusion coefficient required the diffusion process to approach a steady state in which the flux of diffused radionuclei remained constant. Then, the linear portion of the cumulative activity data was fitted by the asymptotic solution of the diffusion equation. The diffusion coefficient was then obtained from the slope of the linear portion with an intercept in the time axis.

To solve these problems, a technique using the Nonlinear Chi-Square and Newton's (NCSN) method was used in the study by Teng and Lee (1991). The NCSN method was considered reliable and capable of analyzing measured data to judge the off-steady-state degree of the diffusion process. The NCSN method also offers error estimates of the fitting parameters, whereas, the contribution to the uncertainty of the overall risk assessment thereby can be evaluated.

Groundwater Remediation/Restoration of Radioactively and Chemically Contaminated Sites [URL Ref. No. 36, 48, 72, 92, 103 {8}, 130, 132, 155, 303, 336 {34, 68}, 337 {20, 27}]

General

Environmental Remediation/Restoration in the United States

Environmental remediation/restoration of radioactively and chemically contaminated sites [URL Ref. No. 303] represents complex challenges that are currently a problem at nuclear weapons sites in the United States, however, as the civilian nuclear industry everywhere continues to deal with decommissioning and decontamination, the lessons learned will be influential in correcting deficiencies (Muntzing and Person, 1994). Standards governing remedial action are complex and are constantly evolving, and unless contaminated material is to be stabilized in place, it must be removed and sent to another facility for storage and ultimate disposal. The task is technically demanding, and those who undertake the challenge must be technically sophisticated, creative, and innovative. Additionally, those who seek to remediate past contamination may find themselves exposed to expanding and unfair allegations of liability for that very contamination, because there is often a basic crisis of public confidence regarding remediation efforts. This was determined by conducting a public attitude survey in neighborhoods adjacent to a radioactively contaminated site where remediation was under the auspices of the U.S. DOE FUSRAP (Formerly Utilized Sites Remedial Action Program) (Feldman and Hanahan, 1996). The survey's purpose was to ascertain levels of actual and desired public involvement in the remediation process; to identify health, environmental, economic, and future land-use concerns associated with the site; and to solicit remediation strategy preferences. Surface water and groundwater contamination, the desire for public involvement, and potential health risks [URL Ref. No. 235] were summarized in this study to be the most highly ranked site concerns. Preferred remediation strategies in the survey showed favor toward treatment of contaminated soil and excavation with off-site disposal. Respondents were concerned with protecting future generations, better assessing of risks to health and the environment, and avoiding generation of additional materials that would be contaminated and add to the volume. Also, according to Crowley (1997), the survey focused on three issues: disposal of spent fuel, treatment and disposal of high-level waste [URL Ref. No. 305], and cleanup of soil and groundwater contamination at U.S. defense sites.

Remediation [URL Ref. No. 36, 48, 72, 92, 103 {8}, 130, 132, 145, 155, 285, 303, 336 {34, 68}, 337 {20, 27}]

A major issue in cleaning up groundwater contamination is determining when remediation is complete and has been successful. Generally, in remedial actions, the level of contamination measured at monitoring wells may be dramatically reduced after a moderate period of time, but low levels of contamination always seem to persist. In parallel, the contaminant load removed by extraction wells, for example, declines over time and gradually approaches a residual level in the latter stages. Based on this information, then, a decision must be made to continue or to end remediation. By continuing remediation, efforts will be made to clean up small amounts of residual contamination. However, if remediation is ended prematurely, an increase in the level of groundwater contamination may follow through time.

Flow through the zones of highest hydraulic conductivity results in rapid cleansing of these zones by extraction wellfields, but cleanup of contaminants in low permeability zones can occur only after the slow process of diffusion takes place. The situation is similar, though reversed, for *in situ* remediations [URL Ref. No. 246] that require the injection and delivery of nutrients or reactants to the zone of intended action. Because of the surface area of low-permeability sediments, greater amounts of contaminants accumulate on them. Hence, the majority of contaminant reserves may be available only under diffusion-controlled conditions in many heterogeneous settings.

For remediation efforts involving compounds that readily sorb to aquifer materials, the number of pore volumes to be removed depends not only on the sorptive tendencies of the contaminants, but also on whether flow rates during remediation are too rapid to allow contaminant levels to approach equilibrium. If insufficient contact time is allowed, the affected water is advected away from sorbed contaminants prior to reaching equilibrium and is replaced by upgradient freshwater. This method of removal generates large volumes of mildly contaminated water where small volumes of highly contaminated water would otherwise result.

When nonaqueous phase liquid (NAPL) [URL Ref. No. 99 {11}, 103 {4}, 316] residuals, such as gasoline, are trapped in pores by surface tension, diffusive liquid-liquid partitioning controls dissolution of the toxic compounds within the NAPLs into the groundwater. And, as with sorbing compounds, flow rates during remediation may be too rapid to allow saturation levels of the partitioned contaminants to be reached, and large volumes of mildly contaminated water will be generated.

The practical use of remediation wellfields and other groundwater cleanup technologies are highly dependent on site-specific knowledge and the influence of transport processes on contaminant levels. Although, there is still much to be learned about highly specific and cost-effective remediations. However, far more could be accomplished if the processes that govern the behavior and treatability of contaminants would be actively investigated at each site. In general, conventional field characterization efforts have not led to satisfactory remediations, although recent transport-process-oriented approaches of characterization are resulting in more permanent and cost-effective remediations.

Sampling of Uranium-Contaminated Groundwater at the Fernald Environmental [URL Ref. No. 294] *Management Project/Environmental Remediation Site*

The Fernald Environmental Management Project (FEMP) is a United States Department of Energy (DOE) facility, located in southwestern Ohio, that formerly produced uranium [URL Ref. No. 283] metal products. Environmental restoration activities began at the 1,050-acre FEMP complex, and adjacent areas were impacted by contamination releases (Shanklin et al., 1995). Groundwater transport was the primary exit pathway of concern for contamination from FEMP radioactive and nonradioactive pollutants. For example, a uranium plume was identified as early as 1981 migrating off-site, and this discovery prompted a removal action under the Comprehensive Environmental Response, Compensation, and Liability Act (CERCLA) [URL Ref. No. 264]. Groundwater remediation at the FEMP will continue to require long-term commitment of resources, and groundwater monitoring programs during past and future periods will be vital to assess the performance of remedial programs.

Past practices of regulatory-driven groundwater sampling necessitate the collection of large volumes of purge water in order to obtain representative samples. During a typical three-month monitoring period at the FEMP, 90 to 110 monitoring wells were sampled (following regulatory-approved sampling practices), and more than 26,500 liters of purge water were generated. Additionally, several hundred liters of decontamination water were generated from the cleaning of purging and sampling equipment during the same period. Contaminated sample waters were then handled, stored, and treated prior to appropriate disposal. Because contaminant concentrations were unknown until laboratory analyses were conducted, all purged groundwater was handled and treated at considerable expense for treatment and labor.

Micro-purge low-flow sampling has been demonstrated to improve groundwater sampling efficiency, to minimize the generation of wastewater, and to better ensure the collection of representative groundwater samples from narrow diameter wells with short-screened intervals. The review of regulatory guidance for purging methodologies prompted the experiments conducted by Shanklin et al. (1995).

Removal of Plutonium from Low-Level Process Wastewaters by Absorption

Wastewater containing small concentrations of plutonium [URL Ref. No. 284] was generated during processing operations at the Plutonium Finishing Plant (PFP) on the Hanford Site in Washington State (Barney et al., 1991). This waste-water consisted mainly of cooling water and was generated at a rate of about 6 to 11 million liters per month. It was thought that plutonium will be adsorbed on ion exchange resins or other types of adsorbents by passing the wastewater through columns containing beds of the adsorbent.

The objectives of the study by Barney et al. (1991) were to evaluate potential adsorbents for use in the treatment facility. The adsorbent, therefore, had to have a high affinity for plutonium species dissolved or suspended in the wastewater over the range of pH [URL Ref. No. 203] values expected.

Plutonium removal from low-level wastewater effluents at the Hanford Site [URL Ref. No. 292] was faster and more complete using a bond char adsorbent rather than using the other commercially available adsorbents tested. Equilibrium distribution coefficients (K_d values) were high (8,000 to 31,000 ml/g) for plutonium adsorption on bone char over the range of pH values expected in the wastewater (5 to 9).

Groundwater Contamination, Optimal Capture, and Containment

Gorelick et al. (1993) summarized a set of techniques that can be used to design efficient and cost-effective capture and containment systems for groundwater remediation. The contaminants found in groundwater at many sites are representative of virtually all major industrial by-products, and although the characteristics of groundwater contamination and the hydrogeologic conditions may be unique to each site, the design techniques presented were quite general and should be applicable to a large number of these sites.

Systems based on a pump-and-treat [URL Ref. No. 337 {27}] approach are an essential alternative of choice for treatment of many groundwater contamination problems. However, for many sites, the pump-and-treat strategy must be exercised for decades as a means to capture the contaminated groundwater and does not always result in complete aquifer remediation. Because pump-and-treat systems often evolve into *long-term* operations, the potentially high costs dictate the need to attempt to maximize their efficiency.

The recommended approach to maximize efficiency (i.e., mathematically formalizing checks and balances that might be used to ensure that the design is optimal) is one based on a combination of simulation and optimization. Simulation is carried out with the usual types of groundwater models for flow and transport. Optimization is based on standard linear programming techniques. Gorelick et al. (1993) presented the mechanics of the suggested approach, sets the framework for the problem, and placed the simulation-optimization technique into the context of other available solution technologies. However, there were limitations on the application of the simulation-optimization methodology, and these are recognized throughout the manuscript by Gorelick et al. (1993).

Particle Methods to Reliable Identification of Groundwater Pollution Sources

An alternative strategy for identifying sources of contamination in groundwater systems is presented by Bagtzoglou et al. (1992). Under the assumption that remediation costs are affected by the level of contamination, the scheme provides probabilistic estimates of source locations and spill-time histories. Moreover, the method successfully assessed the relative importance of each potential source. The proposed methodology by Bagtzoglou et al. (1992) provides crucial information for the design of monitoring and data-collection networks in support of groundwater pollution source identification efforts. The network must be designed with minimum inter-sampling distances of about one to two correlation lengths, λ.

Remediation of Toxic Particles from Groundwater

The presence of radioactive colloids (e.g., radiocolloids) in groundwater has been documented by Nuttall and Kale (1994), and there is significant evidence to indicate that these colloids may accelerate the transport of radioactive species in groundwater. But, because field experiments are often fraught with uncertainties [URL Ref. No. 261], colloid migration in groundwater continues to be an area of active research, and thus, the role and existence of radiocolloids is still being investigated. The publication by Nuttall and Kale (1994) describes an ongoing study that characterizes groundwater colloids to understand the geochemical factors affecting colloid transport in groundwater and to develop an *in situ* colloid remediation process. The colloids and suspended particulate matter used in the study by Nuttall and Kale (1994) were collected from a perched aquifer site (i.e., located at Los Alamos National Laboratory's Mortandad Canyon in Northern New Mexico) where the radiation levels at several hundred times the natural background have shown the presence of radiocolloids containing plutonium and americium. At this site, radionuclides were spread over several kilometers, and the inorganic colloids collected from water samples were characterized with respect to concentration, mineralogy, size distribution, electrophoretic mobility (e.g., zeta potential), and radioactivity levels.

Facilitated transport of radioactive waste in groundwater presents an environmental threat and a challenging restoration problem. Mortandad Canyon at Los Alamos was the chosen investigation site because plutonium and americium migrated over 2 km in a perch aquifer at this location. Groundwater colloids from the Mortandad Canyon site were characterized and studied in a series of packed column experiments under conditions that simulated the natural perched aquifer. Also, spherical silica colloids were used as a simu-lant to represent the natural colloids and provide a better controlled system in which to study colloid transport.

Mortandad colloids were composed of feldspar, quartz, clay, and contained trace concentrations of plutonium and americium. The source or colloids was from the surrounding rock, which has the same mineral composition. They have a zeta potential of -18 mV indicating that these colloids were electrostatically stabilized. Their average size was approximately one micron in diameter, and the batch flocculation and column experiments confirmed the relative stability of the Mortandad colloids. Because of electrostatic stabilization, both Mortandad and silica colloids transported relatively unretarded through the untreated 20 cm quartz-packed columns. Treatment of the quartz packing with the polyelectrolyte CATFLOC effectively stopped colloid migration within these laboratory columns, and post-analysis of the packing material using an ESEM showed a monolayer absorption coating of colloids on the quartz packing. The mass concentration of the loaded column was x gr of colloids per gram of quartz packing. The study by Nuttall and Kale (1994) additionally showed the existence of Mortandad colloids, their chemistry, and their charge stabilization properties, as well as their ability to be transported through a porous media.

Sharp-Interface Model for Assessing NAPL Contamination [URL Ref. No. 316] and Remediation of Groundwater Systems

A numerical model was presented for areal analyses of three-dimensional (3-D) flow behavior of nonaqueous-phase liquids (NAPLs) in groundwater systems in a study conducted by Huyakorn et al. (1994). The model presented was designed for specific application to chemical and petroleum spills and leaks and remedial design and evaluation of NAPL-contaminated sites. The mathematical formulation was based on vertical integration of the 3-D two-phase flow equations and incorporation of the modified concept of gravity-segregated vertical equilibrium (GSVE) that yields sharp interfaces separating zones of mobile NAPL and groundwater. History-dependent pseudoconstitutive relations were developed for LNAPLs and DNAPLs (light and dense NAPLs) scenarios, taking into account the effects of residual saturations. Owing to the sharp-interface assumption, the soil capillary pressure and relative permeability curves were not needed in the evaluation of pseudofunctions. Efficient and mass-conservative nonlinear numerical techniques were adopted for solving the governing equations and treating practical boundary conditions that included injection and recovery wells and trenches. Simulation and application examples additionally were provided to demonstrate verification and utility of the model. Numerical results obtained using the sharp-interface modeling approach were compared with analytical solutions and rigorous multi-phase numerical solutions that accounted for vertical flow components/capillary effects. The verification results showed the validity of the FSVE modeling assumptions and accuracy of the proposed formulation and computational schemes predicting the NAPL recovery. The numerical study by Huyakorn et al. (1994) indicated that the present model is highly efficient and is, thus, suitable for preliminary analyses of site-specific problems that have limited data and personal computer resources.

Continuous Monitoring for Tritium [URL Ref. No. 293] *in Aqueous Effluents*

Effluents from selected facilities that routinely handle or produce large quantities of tritium are continuously monitored for tritium concentrations at the Savannah River Site (SRS) [URL Ref. No. 61] as reported by Hofsetter (1993). The tritium [URL Ref. No. 293] effluent waste monitors (TEWMs) developed at the SRS are placed at strategic locations in these facilities to sample and analyze the aqueous effluents in real time. The TEWMs include a water purification system, a flow cell containing a solid scintillator, coincidence electronics, alarms, and interfaces to plant systems. The main purpose of the TEWM is to alert cognizant personnel to an upset condition that might result in an unplanned release of tritium to the environment.

The TEWM installed on the outfall from the K reactor remains in operation to support decontamination and decommissioning of the reactor facility. The removal of the heavy water moderator from the reactor and supporting facilities has the potential to release tritium to building sumps and drains that are discharged to the effluent canal. During these activities, SRS management committed to continuous tritium monitoring of aqueous discharges to mitigate the consequences of an unplanned release of tritium to the Savannah River. Under the current discharge flow-rate conditions from the K reactor, monitored by the TEWM, only $\approx$ 0.1 Bq (3Ci) of tritium would be released before an alarm would be received in the reactor control room.

The installation of continuous tritium monitors on effluent streams that can contain significant quantities of tritium is part of the active program at the SRS to reduce the quantity of tritium released to the environment. As tritium is the most significant radioisotope emitted from SRS operations in terms of off-site dose consequence, reduction of tritium releases is paramount in improving public perception of the SRS and is proactive in anticipating possible changes in regulatory limits.

Can Groundwater Restoration Be Achieved?

Alternative source and plume control remediation technologies have been developed and evaluated to enhance or replace the pumping and treatment technology (Olsen and Kavanaugh, 1993; Austin, 1995; Fine II, 1991). And, source control technologies, such as soil vapor extraction including injection of chemicals into saturated zones, focus on the removal or destruction of subsurface contamination sources, and plume control technologies, such as reinjection of treated groundwater, pulsing, air sparging, *in situ* biodegradation, and permeability enhancement, focus on control and removal of the dissolved plume (Olsen and Kavanaugh, 1993). In summary, the following should be considered:

(1) Soil vapor extraction: Soil vapor extraction should be considered to reduce cleanup times by removing contamination sources from the unsaturated zone (source control). As volatile organic compounds evaporate or volatize from the groundwater into the unsaturated zone to reestablish equilibrium, contamination concentrations decrease, increasing the rate of cleanup.

(2) Injection of chemicals into saturated zones: Chemical enhancement can also increase contaminant removal efficiency. Cosolvents can be used to increase solubility and decrease adsorption of organic compounds. Surfactants can be used to decrease interfacial tension, which increases the mobility of NAPLs.

(3) Reinjection of treated groundwater: Upgradient reinjection can decrease cleanup times by increasing the hydraulic gradients, saturated thicknesses, and amounts of water flushed through the aquifer. Reinjection into the aquifer with treated water can decrease cleanup times by 30 percent, however, the present-worth costs of direct discharge to surface water are about equal because of the costs of infiltration galleries.

(4) Pulsing: If diffusion controls the release of contaminants, intermittent operation (pulsing) of the pumping and treatment system should recover additional mass. When the system is turned off, contaminants diffuse into the mobile zone. Modeling studies of site cleanups using pulsed pumping and treatment systems indicate that cleanup takes longer, but the overall costs may be lower because less water is treated.

(5) Air sparging: Injecting air into the saturated soil zone using horizontal or vertical wells enhances the release of contaminants by stripping volatile compounds from the groundwater via biodegradation. Modeling studies have indicated that air sparging may decrease remediation times from one-half to two-thirds. Air sparging can be cost-effective in many cases.

(6) In situ bioremediation: The addition of nutrients, oxygen, methane, or other chemicals may enhance *in situ* bioremediation and decrease the remediation time. However, recent studies indicate that biodegradation occurs only in the dissolved phase. Therefore, if contamination release is controlled by diffusion from the solid phase, biodegradation will not increase cleanup rates

because the cleanup time is still controlled by the diffusion from the immobile zone.

(7) *Permeability enhancement:* If diffusion controls the release of contaminants, methods used to increase the permeability or access the immobile zone should also decrease cleanup times. Hydraulic jetting of clay materials increases contaminant recovery and fracturing bedrock by pneumatic methods greatly increases mass recovery.

(8) *Slurry wall containment:* This technology reduces the quantity of water requiring treatment, and dewatering a saturated zone inside a slurry wall can increase the amount of unsaturated soil that can be treated with the soil vapor extraction method. Chemical and physical slurry wall enhancement results in reactive barrier walls.

The difficulty of cleaning up groundwater, as indicated, depends on the source and type of contamination and on the nature of the subsurface (Macdonald and Kavanaugh, 1994). And, in theory, the goal of cleaning up contaminated groundwater is not contrary to any fundamental principle of science, however, the inherent complexities of the Earth's surface and the types of contaminants often found in groundwater make what should be possible in theory impossible in practice at many sites. Also, the chief technical reasons for the difficulty of cleanup according to Macdonald and Kavanaugh (1994) include the following:

(1) *Physical heterogeneity:* The Earth's subsurface is highly heterogeneous. Groundwater is stored in aquifers consisting of layers of sand, gravel, and rock having vastly different properties, and because of this variability, determining the pathways by which contaminants will spread is very difficult, complicating the design of cleanup systems.

(2) *Presence of nonaqueous-phase liquids (NAPLs):* Many common groundwater contaminants are NAPLs that, like oil, do not dissolve readily in water. Light NAPLs (DNAPLs), such as the common solvent trichloroethylene, are denser than water, and as NAPLs move underground, they leave small immobile globules trapped in the porous materials of the subsurface; these globules, in turn, cannot be flushed out of the subsurface with conventional groundwater cleanup systems.

(3) *Diffusion of contaminants into inaccessible regions:* Contaminants may also diffuse into very small pore spaces in the geologic materials making up the aquifer. These small pores are difficult to flush with conventional groundwater cleanup systems, and at the same time, contaminants in the pores serve as long-term sources of pollution as they slowly diffuse from the pores when the contaminant concentration in the groundwater decreases.

(4) *Adherence of contaminants to subsurface materials:* Many common contaminants adhere to solid materials in the subsurface by physical attraction or chemical reactions (e.g., sorption), and if a contaminant sorbs strongly to the aquifer solids, a large volume of water is required to flush it out.

(5) *Difficulties in characterizing the subsurface:* The subsurface cannot always be viewed in its entirety but is usually observed only through a series of drilled holes. Without knowing the subsurface characteristics, it is difficult to design an effective cleanup system. A degree of uncertainty will always be present.

Additional Examples of Groundwater Remediation/Restoration Techniques/Methods for Radioactive and Hazardous Waste Sites

Sun Fuels Groundwater Remediation

A field test of a solar photocatalytic process for water detoxification, conducted at Tyndall Air Force Base, Florida, successfully destroyed benzene, toluene, ethylbenzene, and xylenes (BTEX) in fuel-contaminated groundwater (Crittenden et al., 1995). Destruction of BTEX compounds to less than 0.001 mg/l for total BTEX concentrations of 2.25 mg/l and solar irradiances of 0.32 mW/cm^2 (rainy conditions) and greater was achieved using a contact time of 2.5 minutes in a 1.3-cm-dia (0.5-in.-dia) reactor.

The primary advantage of using photocatalysis for water and wastewater treatment is that photocatalysis mineralizes many organic compounds into nontoxic forms, simple mineral acids, carbon dioxide, and water. Many conventional air-stripping and adsorption technologies transfer contaminants from one medium to another. Other advantages included degradation of nuisance color and odor compounds, destruction of disinfection by-product precursors, and on-site treatment without the risk of transporting hazardous wastes. The preliminary estimated treatment cost of $1.38/1,000 l ($5.22/1,000 gal) for the site was competitive with conventional technologies. The preliminary cost analysis included a large contingency factor and could be reduced based on the selection of pretreatment technology and long-term pilot testing. For example, less expensive alternatives to pretreatment using ion exchange may be feasible.

The following were the considerations and assumptions used for the estimation of treatment costs (Crittenden et al., 1995):

(1) Capital costs were converted to an annual cost using a fixed charge rate of 13.4 percent based on an assumed plant life of 20 years with an interest rate of 12 percent.

(2) Photoreactors were assumed to be operational for nine hours each day.

(3) The photocatalyst was 1.0 percent Pt-TiO$_2$ supported on silica gel (the life of the photocatalyst was assumed to be five years).

(4) Plastic tubes 1.3-cm-dia (0.5-in-dia.) were used for the construction of the photoreactors (the life of the photoreactors was expected to be 10 years).

(5) The life of ion exchange resin was expected to be five years.

(6) The total land required for the system was set at 1.6 ha (4 ac) at $20,000/ha ($100,000/ac).

(7) The cost of administration, including site preparation, design, and permitting; assembling and installation; contractor fees; and contingencies were 18 percent, 30 percent, 12 percent, and 15 percent of the total system capital cost, respectively.

Removing Groundwater Contaminants Through Irrigation

A University of Nebraska-Lincoln (UNL) research team was successful in implementing a technique to clean up contaminated groundwater that could save communities millions of dollars and irrigate crops at the same time (Hurst, 1995). Sprinklers, hallmarks of irrigated agriculture [URL Ref. No. 226], were used to safely clean up groundwater, according to researchers. An interdisciplinary UNL research project showed that treatment removed the contaminants from groundwater during irrigation, with the project focused on cleaning up groundwater at two subsites of the Hastings Groundwater Contamination Site, the Far-Mar-Co and North Landfill locations (i.e., Hastings is on the U.S. Environmental Protection Agency's Superfund National Priorities list) [URL Ref. No. 19]. Savings to communities, companies, and individuals have the potential to be very large, and the average cleanup cost for a Superfund site is $27 to $30 million. Use of the sprinkler technique may reduce costs to $500,000 or less with this process capturing, containing, and removing the contamination.

The 62-acre experimental site was a furrow irrigated cornfield on the eastern edge of Hastings, and contaminated groundwater occurred underneath the site at 120 feet below the surface. It contained trace levels of solvents, or degreasing agents, trichloroethylene (TCE), trichloroethane (TCA), and tetrachloroethylene (PCE), as well as trace levels of two fumigants, carbon tetrachloride (CT) and ethylene dibromide (EDB).

The contaminants allegedly originated from two sites. A grain elevator was allegedly the source of fumigants, and the solvents supposedly came from the abandoned north landfill and industrial sites located southwest of the elevator, according to Hurst (1995). The contaminants were volatile organic compounds [URL Ref. No. 247, 267], which means they are easily vaporized into the atmosphere. Contaminated water at the sites was pumped through a well to a sprinkler and sprayed through nozzles against a pad, and impact turned the water into a thin film from which small droplets

emerged, releasing contaminates into the atmosphere, where some degradation occurs.

Risk [URL Ref. No. 235] to the public and/or environment during the water-to-air exchange, according to Hurst (1995), was minimal. Compounds were emitted into the air at levels far below those requiring an emission permit from the Nebraska Department of Environmental Quality. Also, since predicted volatile emissions would amount to 500 tons per year, a state permit was required for emissions (i.e., a permit is required if 2.5 tons per year or more are emitted). Health-risk-base models indicated that there was no increase of cancer risk for field workers, at these concentrations, at the Hastings site.

Another benefit of the technique was that it was based on irrigation, a farming practice that is vital not only to small grains production in central Nebraska but also to the agricultural economy of Western states. This is a positive note, because groundwater is the source of irrigation water for 56 percent of the 46.2 million irrigated acres in the contiguous United States and in the Western states (75 percent of cropland is irrigated with groundwater).

Remediation using the sprinkler irrigation treatment [URL Ref. No. 226] is a beneficial use for the treated water and eliminates the costly disposal of the discharged water.

Use of Plants in the Remediation of Soil and Groundwater Contaminated with Organic Materials

The use of plants in remediation of soil and unconfined groundwater contaminated with organic materials [URL Ref. No. 210] is appealing for a variety of reasons that follow (Shimp et al., 1993):

(1) Plants provide a remediation strategy that utilizes solar energy

(2) Vegetation is aesthetically pleasing

(3) Plant samples can be harvested and tested as indicators of the level of remediation

(4) Plants help contain the region of contamination by removing water from soil

(5) Rhizosphere microbial communities are able to biodegrade a wide variety of organic contaminants

(6) Many plants have mechanisms for transporting oxygen to the rhizosphere

However, before effective plant remediation strategies can be developed, an understanding is needed of the physical, biological, and chemical relationships that determine the fate of each organic contaminant in the rhizosphere. Shimp et al. (1993) presents an overview of some factors required to understand and model the complex processes that determine the fate of the organic contaminants in plant remediation strategies. Planning and management criteria for the development of practical plant remediation strategies are also presented in the overview by Shimp et al. (1993).

Review of In Situ Air Sparging for the Remediation of VOC-Contaminated Saturated Soils and Groundwater

Many hazardous waste sites that exist in the United States today are the result of accidental surface spills, leaking underground storage tanks, uncontrolled waste disposal, and leaking landfills, to mention some of the sources. The remediation of these hazardous waste sites is a top priority of local, state, and federal government agencies as well as of those who were directly responsible for the contamination [i.e., the potentially responsible parties (PRPs)]. The contamination of unsaturated soils and the subsequent migration of these contaminants into the groundwater can cause adverse effects to human health and the environment. And, as a result, researchers continue to be challenged to develop new and innovative techniques to clean up hazardous waste sites faster, more effectively, less expensively, and with a greater degree of safety.

Subsurface contamination from volatile organic compounds (VOCs) [URL Ref. No. 247, 267] is a widespread problem across the United States, and the contamination of groundwater with VOCs can create large contaminant plumes that have the potential to migrate rapidly, both vertically and horizontally. A number of *in situ* techniques have been developed and implemented in order to remediate VOC-contaminated sites; however, these techniques have shown limited success when used to remediate saturated soils and groundwater. The most commonly employed remediation in situ techniques and their specific limitations are outlined as follows (Reddy et al., 1995):

- *Pump-and-treat systems* have been extensively used for groundwater remediation, but these systems require pumping of relatively large volumes of water with relatively low contaminant concentrations, although this method has significant limitations for remediating VOCs sorbed onto saturated soils, due to soil heterogeneity, contaminant distribution, and the kinetic limitation of the mass removal process. Additionally, this technique is expensive, and thus, priorities usually shift to other remediation methods.

- *In situ bioremediation* is commonly used to remediate saturated zones. This method can be economical and desirable; however, its past performance and effectiveness have been significantly limited by several biological parameters and the requirement of intimate mixing between the contaminated groundwater and the microorganisms that are injected into the subsurface. In addition, the long-term effectiveness of this remediation method has not yet been established.

- *In situ chemical treatment* requires the injection of chemicals in order to transform the contaminants in place into nontoxic substances. However, the difficulty with this method arises in selecting the appropriate chemicals and delivering them to the low-permeable zones. In addition, chemical reactions may adversely affect subsurface areas.

- *In situ soil vapor extraction* is a popular technique that has been shown to be a successful and cost-effective remediation technology for removing VOCs from unsaturated soils or the vadose zone, however, it is not always applicable for remediating saturated soils and groundwater.

In situ air sparging is a developing remediation technique that has significant potential for use in VOC-contaminated [URL Ref. No. 247] saturated soils and groundwater (Reddy et al., 1995). This technique consists of injecting air below the contaminated area to partition the dissolved, sorbed, and free phase VOCs into the gas phase and to enhance the aerobic biodegradation of the VOCs. Because of the buoyancy effect, the VOCs in the gas phase are transported by air to the vadose zone where they are removed and subsequently treated by a soil vapor extraction system. The design, operation, and monitoring of air sparging systems is based mainly on an empirical approach that has been subjected to limited field experiments. Extreme care must be exercised in designing and implementing the air sparging system so that the contaminants are removed efficiently and without adverse effects on the subsurface environment, particularly when related to the spread of groundwater contaminants to clean areas.

Reddy et al. (1995) outlined the fundamentals of air sparging and presented an overview of previous air sparging field and laboratory investigations. Reddy et al. (1995) detailed a critical assessment of modeling studies that predicted contaminant transport during the air sparging process and were involved in an ongoing comprehensive research program that developed the most efficient and economical air sparging systems that performed laboratory aquifer simulation tests to characterize the basic mechanisms of air sparging, a contaminant transport model to optimize the different design variables in a typical air sparging system, and a field demonstration of optimal conditions.

Groundwater Remediation Using the Simulated Annealing Algorithm

The contamination of groundwater supplies, as has been described, poses widespread and important environmental problems. Different solutions strategies have been proposed to solve such problems, and a great deal of research is still in progress in this area. Some of this research includes the placement of pumping wells and selection of the pumping rates that constitute the most important factors in solving pump-and-treat strategies (i.e., a simulation model with known hydraulic data is utilized to predict the contamination aquifer behavior, and an optimization methodology is devised to place the pumps and assign pumping rates to clean up a plume with the minimum capital cost or minimum cleanup time).

As indicated, the problems of the placement of pumps and the selection of pumping rates are important issues in designing contaminated groundwater remediation systems using a pump-and-treat strategy (Kuo Chin-Hwa et al., 1992). Kuo Chin-Hwa et al. (1992) proposed three nonlinear optimization formulations to address pump-and-treat problems. The first problem formulation considered hydraulic constraints and reduced the plume concentration to a specified regulation standard value within a given planning time, while it minimized capital cost. The second formulation was similar to the first formulation; however, in this formulation, the number of pumps was prespecified by using the results from the initial formulation. The inclusion of well installation costs in the first problem formulation resulted in non-smooth local solutions that related to the use of conventional nonlinear optimization techniques (i.e., the simulated annealing algorithm was used to overcome difficulties, and specific simulation studies indicated that the method advanced herein was promising because it involved acceptable computation times). It is anticipated that in the near future, more information will be available concerning *Simulated Annealing Algorithms.*

Remediation of Groundwater Polluted with Chlorinated Ethylenes by Ozone-Electron Beam Irradiation Treatment

A serious concern facing a large number of water suppliers is the appearance of chlorinated ethylenes, such as trichloroethylene (TCE) and perchloroethylene (PCE) in groundwater (Gehringer, 1992). The removal of them by carbon adsorption simply transfers the problem from the water to the adsorbent.

Water treatment oxidation processes are attracting a growing interest because of their capability to mineralize organic substances. PCE, for instance, has been shown to be completely mineralized by oxidation, and mineralization of the dissolved chlorinated ethylenes would solve the whole problem (i.e., water cleanup and detoxification) of the pollutants in a single step.

TCE and PCE contained as micropollutants in water can be effectively oxidized by the attack of OH radicals only. This conventional method of OH radical production in water is based on the decomposition of dissolved ozone with *UV* light or hydrogen peroxide as the promoter. However, the efficacy of the process is limited by the low water solubility of ozone, and for groundwater remediation on a technical scale, oxidation of TCE and/or PCE by OH radicals has not yet been carried out.

In the article by Gehringer (1992), information was presented that OH radicals formed in water radiolysis may be effectively used for the oxidative decomposition of tricholorethylene and perchloroethylene contained as micropollutants in groundwater. The addition of ozone to the water before irradiation caused the reducing species of the water radiolysis to be converted into OH radicals.

Remediation of Contaminated Soil and Groundwater Using Air-Stripping and Soil Venting Technologies

The public's heightened awareness of environmental issues over the past several years has served as a catalyst for remediating contaminated sites (Fine II, 1991). Additionally, the expanding array of environmental legislation has continued to increase the level of effort necessary to satisfy cleanup criteria. Consequently, to address the regulatory impetus imposed during the remediation of soils and groundwater, a variety of technologies has been employed. Some of these are relatively unique to the environmental field, while others are a spin-off of more traditional unit operations within the chemical process industry. Two of the more common remedial technologies that have been frequently utilized during the cleanup of soils and groundwater contamination are commonly referred to as *soil venting* and *air stripping.* Fine II (1991) discusses these in detail.

Recovery of Toxic Heavy Metals from Contaminated Groundwaters

The work by Rayson et al. (1994) specifically addresses Section 2.5.5.2 in the Department of Energy's Applied Research, Development, Demonstration, Testing, and Evaluation (RDDT&E) Plan for Environmental Restoration and Waste Management that describes that to protect public health and the environment, the DOE must provide reduction or elimination of radioactive, heavy metal, and/or inorganic contamination in groundwater through extraction and *in situ* technologies.

Initial work, therefore, has begun regarding the use of nuclear magnetic resonance spectrometry for probing the mechanisms of binding of other metals to *Datura innoxia* (angel's trumpet) cell material (Rayson et al., 1994). The operating parameters for measuring the local chemical environments of cadmium (Cd) nuclei have been determined, and initial measurements of the metal bound to the cell material in a suspension have been made.

The objective of the research as reported by Rayson et al. (1994) was to characterize biologically produced materials for the recovery of toxic heavy metals [URL Ref. No. 317] from contaminated groundwater. Specifically, components of the cells of the organism *Datura innoxia* were investigated. Several approaches to understanding the interactions involved in the binding process were employed and ranged from the phenomenological to the fundamental Measurement of the binding capacity of cadmium ions onto both the free *Datura innoxia* cells and immobilized cells were also undertaken. Immobilization of the cell material was then observed to result in a significant increase in the amount of cadmium ion bound to the material. A methodology that provided the binding coefficient and the binding capacity of the metal ion had also been developed and was based on adsorption isotherms. A methodology for the rapid characterization of the binding of a metal pollutant with minimal generation of waste was also investigated by Rayson et al. (1994).

Cost Components of Remedial Investigation/Feasibility Studies

The first step in conducting any type of remedial action at a hazardous waste site is the remedial investigation/feasibility study (RI/FS) process (Cressman, 1991). The RI is conducted concurrently with the FS and emphasizes data collection and site characterization. The data collected during the RI supports the analysis and decision-making activities of the FS as well as remedial alternative evaluations through treatability investigations. The overall RI process includes the following:

(1) Scoping process
(2) Sampling/analysis plan development
(3) Site characterization
(4) Treatability investigation
(5) Data analysis and report
(6) Data management
(7) Health/safety planning
(8) Community relations

The U.S. Department of Energy (DOE) [URL Ref. No. 45, 61], in its major environmental restoration projects of active and inactive sites throughout the United States, developed a list of environmental problems at its sites and probable cleanup technologies and techniques that could be used.

Some projects were identified that were common to many or all cleanup projects (Cressman, 1991). The article by Cressman (1991) focused on cost components of the RI and the FS processes.

Superfund and Groundwater Remediation, Another Perspective

Don't pollute groundwater resources because contamination plumes have no quick fix. This was underscored over 10 years ago when earth scientists at the U.S. Geological Survey described that deterioration in groundwater quality constituted a permanent loss of water resources because treatment of the water or rehabilitation of the aquifers was presently generally impractical, and solutions rested largely in changing land and water management practices to take into account the susceptibility of groundwater resources to degradation (Rowe, 1991). To support this, Rowe (1991) discussed a list of recommendations for modifying the Superfund approach to groundwater remediation. The six recommendations that were made, included the following (Rowe, 1991):

(1) Start groundwater actions early at sites
(2) Make the remedial goal plume containment and contaminant mass reduction
(3) Focus on source control actions at sites
(4) Classify plumes on the basis of remediation priority by taking aquifer water quality and use into account
(5) Admit that aquifer restoration is unachievable in some instances
(6) Abandon chemical-specific cleanup goals for remedial action

Groundwater contamination at many sites has no quick fix and usually constitutes a permanent loss in water resources. What remains ahead is a difficult task of dealing with residual groundwater contamination, and in spite of many active restoration efforts, residual contamination persists at many sites, requiring a management strategy that balances active restoration, waivers, alternate concentration limits, and source control efforts. Where contamination is not a problem, the proverb is a familiar one—*prevention is still the best fix.*

FIVE

Technical Evaluations of Groundwater and Groundwater Protection Plans Related to Contamination [URL Ref. No. 53, 55, 73, 97, 101, 339 {12–13}]

Technical Evaluations of Groundwater

The Mid Rio Grande Area of New Mexico

Albuquerque-Belen Basin Area

The Albuquerque-Belen groundwater basin [URL Ref. No. 55, 339 {12–13}], located in central New Mexico, is dependent on ground and surface waters for irrigation and municipal use [U.S. Geological Survey (U.S.G.S.), 1988] [URL Ref. No. 53]. The general area also encompasses two of the U.S. Department of Energy National Laboratories. These include the Sandia National Laboratory [URL Ref. No. 73] and the Los Alamos National Laboratory [URL Ref. No. 97]. The City of Albuquerque [URL Ref. No. 327] and Bernalillo County [URL Ref. No. 89] are part of the Albuquerque-Belen basin. The geochemistry [URL Ref. No. 314] of groundwater in the basin has been studied by the U.S.G.S. as part of the Southwest Alluvial Basins Regional Aquifer-Systems Analysis (U.S. Geological Survey, 1988). The purpose of the U.S.G.S. study was to define the areal distribution of different water qualities, to define the groundwater flow system, and to determine groundwater quality in the Albuquerque-Belen basin.

The Albuquerque-Belen basin [URL Ref. No. 55, 339 {12–13}] contains as much as 18,000 feet of basin-fill sediments of the Santa Fe Group, which forms the principal aquifer in the basin. The majority of groundwater inflow to the principal aquifer occurs as infiltration of surface water through river channels, infiltration of surface inflow from adjacent areas, infiltration of excess irrigation water, groundwater inflow from adjacent bedrock units, and groundwater inflow from the upgradient Santo Domingo basin. In general, groundwater flows from the margins of the basin toward the basin center and then southward to the adjacent Socorro basin. The majority of groundwater outflow is evapotranspiration, groundwater pumpage, and groundwater outflow to the Socorro basin.

The chemistry of inflow water to the aquifer has the largest effect on the distribution of different water quality parameters in the Albuquerque-Belen basin. In the southeastern area of the basin, inflow is derived from Paleozoic and Mesozoic rocks that contain gypsum, that have specific conductance ranges from 1,000 to 1,200 microsiemens per centimeter at 25° Celsius, and that have calcium [URL Ref. No. 192] and sulfate [URL Ref. No. 197] as the dominant ions. On the eastern side of the basin, inflow is derived from Precambrian and Paleozoic rocks, and groundwater in this area of the basin has a specific conductance that is usually less than 400 microsiemens per centimeter, and calcium [URL Ref. No. 192] and bicarbonate are the dominant ions. Along the southwestern margin of the basin, groundwater enters the basin from adjacent Paleozoic rocks and from the infiltration of surface water from adjacent areas. The inflow from adjacent bedrock units has a specific conductance generally greater than 20,000 microsiemens per centimeter. This groundwater also contains large concentrations of sodium [URL Ref. No. 191] and chloride [URL Ref. No. 198]. The mixing of this groundwater and the infiltration of surface water from adjacent areas, which generally has a small specific conductance, results in groundwater with a large range of specific conductance values. Sodium [URL Ref. No. 191] and sulfate [URL Ref. No. 197] are the dominant ions in groundwater inflow from Cretaceous rocks in this area, which is along the western margin of the basin. In the northern area of the Albuquerque-Belen basin, groundwater inflow from the Jemez geothermal reservoir [URL Ref. No. 248] mixes with local recharge water and exhibits large concentrations of silica and chloride.

In a large area west of Albuquerque, NM, sodium is the dominant cation in groundwater, and in this area of the basin, the exchange of calcium and magnesium [URL Ref. No. 193] for sodium is a dominant process affecting groundwater quality. This is also the same area of the basin that is underlain by relatively fine-grained sediments as has been indicated by well drillers' geophysical logs.

Groundwater in the Rio Grande valley is affected by the infiltration of excess irrigation water [URL Ref. No. 241], and excess water generally has a larger specific conductance than other groundwater in the Rio Grande valley. Consequently, the mixing of these waters then results in shallow groundwater having a generally larger specific conductance than deeper groundwater from the same area.

Simulation of Groundwater Flow in the Albuquerque Basin, Central New Mexico

This report (U.S. Geological Survey, 1995) describes a three-dimensional finite-difference groundwater flow model of the Santa Fe Group aquifer system in the Albuquerque basin, which comprises the Santa Fe Group (late Oligocene to middle Pleistocene age) and overlying valley and basin-fill deposits (Pleistocene to Holocene age). The model was designed to be flexible and adaptive to new geologic and hydrologic information [URL Ref. No. 187] by using a geographic information system (GIS) [URL Ref. No. 163] as a database interface. The aquifer system was defined and quantified in the model consistent with the current (July 1994) understanding of the structural and geohydrologic framework of the basin.

The model simulates groundwater flow over an area of about 2,400 square miles to a depth of 1,730–2,020 feet below the water table with 244 rows, 178 columns, and 11 layers. Of the 477,752 cells in the model, 310,376 are active, and the top four model layers approximate the 80-foot thickness of alluvium in the incised and refilled valley of the Rio Grande basin to provide detail of the effect of groundwater withdrawals on the surface-water system. Away from the valley, these four layers represented the interval within the Santa Fe Group aquifer system between the computed predevelopment water table and a level 80 feet below the grade of the Rio Grande River. The simulations included in the model involved initial conditions (steady-state), the 1901–1994 historical period, and four possible groundwater withdrawal scenarios depicted from 1994 to 2020.

The model also indicates that for the year ending in March 1994, net surface-water loss in the basin resulting from the City of Albuquerque's [URL Ref. No. 327] groundwater withdrawal totaled about 53,000 acre-feet. The balance of the about 123,000 acre-feet of withdrawal then came from aquifer storage depletion (about 67,800 acre-feet) and captured or salvaged evapotranspiration (about 2,500 acre-feet).

Additionally, in the four scenarios projected from 1994 to 2020, the City of Albuquerque annual withdrawals ranged from about 98,700 to about 177,000 acre-feet by the year 2020, and the range of resulting surface-water loss was from about 62,000 to about 77,000 acre-feet. The range of aquifer storage depletion was estimated to be from about 33,400 to about 95,900 acre-feet, if the captured evapotranspiration

and drain-return flow remained nearly constant for all scenarios. Maximum projected declines in hydraulic head in the primary water-production zone of the aquifer (mode layer 9) for the four scenarios ranged from 55 to 164 feet east of the Rio Grande and from 91 to 258 feet west of the river. Average declines in a 383.7 square-mile area around Albuquerque ranged from 28 to 65 feet in the reduction zone for the same period.

Other Studies

Many hydrologic studies, qualitative and quantitative, have been conducted in the Albuquerque basin [URL Ref. No. 55, 339 {12–13}] and date back to the late nineteenth century. Recent investigations (i.e., within the last five years) of the Albuquerque-Belen basin, in particular in the Albuquerque area, indicate that the zone of highly productive aquifer material was less extensive and thinner than previously thought. Consequently, on the basis of these and other investigations, officials with the City of Albuquerque decided that a better understanding of the hydrologic system of the Albuquerque Basin must be developed so that present and future water demands can be met for all basin residents. Therefore, in July 1992, the U.S. Geological Survey [URL Ref. No. 53] in cooperation with the City of Albuquerque Public Works Department [URL Ref. No. 160] began a long-term investigation designed to reevaluate the geohydrology of the Albuquerque Basin in central New Mexico, with emphasis on the greater Albuquerque area.

The study goals and objectives described in the report by the U.S. Geological Survey (1995) were to be the third of a three-phase study to quantify groundwater resources in the Albuquerque Basin. The first phase, conducted by the New Mexico Bureau of Mines and Mineral Resources [URL Ref. No. 339 {40}] in cooperation with the City, described the hydrogeologic framework of the Albuquerque Basin on the basis of recent data (Hawley and Haase, 1992). The second phase of the study resulted in a description of the geohydrologic framework and hydrologic conditions in the Albuquerque Basin (Thorn et al., 1993).

The U.S. Geological Survey (1995) report described a three-dimensional finite-difference groundwater-flow model of the Albuquerque basin, with emphasis on the Albuquerque area. The model incorporated recent information on geologic and hydrologic data about the Albuquerque basin. The model simulated initial conditions and historical responses to groundwater withdrawals for 1901–1994 and projected responses to selected possible future conditions to the year 2020. The hydrogeologic framework for the model was based on material presented by Hawley and Haase (1992). Additionally, geohydrologic characteristics of the basin were based on those presented by Thorn et al. (1993).

The modeling effort previously presented differs from previous modeling efforts in the Albuquerque Basin:

(1) The database and data extraction system can be dynamically updated and used for enhancements to the model as updated information on the geohydrologic system becomes available.

(2) The model simulates detailed surface-water/groundwater interaction.

(3) The disagreements between measured and simulated conditions are used to identify areas where more information is needed to improve the understanding of the geohydrologic system.

Previous Investigations

Previous investigations involving groundwater concerns in the Albuquerque Basin are described in Thorn et al. (1993). Although groundwater flow modeling investigations within the basin are few in number, some of the first groundwater modeling efforts performed in the Albuquerque area were completed by Bjorklund and Maxwell (1961) and Reeder et al. (1967). Kernodle and Scott (1986) developed a three-dimensional simulation of steady-state conditions in the Santa Fe Group aquifer system underlying the Albuquerque Basin. Transient groundwater flow, also in the Santa Fe Group aquifer system in the Albuquerque Basin, was additionally discussed in Kernodle et al. (1987), and bibliographies that provide other useful references concerning the hydrogeology of the Albuquerque Basin were reported by Kelley (1977), Borton (1978, 1980, 1983), Wright (1978), and Stone and Mizell (1979).

The Sandia National Laboratories Site-Wide Hydrogeologic Characterization Project
[URL Ref. No. 73]

Land Use

Sandia National Laboratories (SNL) [URL Ref. No. 73], a U.S. DOE Laboratory Facility [URL Ref. No. 45], consists of five technical areas and several additional test areas with each area having its own distinctive operations. SNL facilities located on Kirtland Air Force Base (KAFB) [URL Ref. No. 327] operate under a complex series of land-use agreements among the DOE Albuquerque Operations (DOE/AL) [URL Ref. No. 312], KAFB, and the U.S. Forest Service (USFS) [URL Ref. No. 339 {20}]. An additional 14,920 acres (60.3 km^2) are provided by land-use permits from KAFB, the USFS, the State of New Mexico [URL Ref. No. 333], and the Isleta Pueblo Indian Reservation (Sandia National Laboratory, 1992).

Demographics

The largest and closest population center to KAFB [URL Ref. No. 327] is the City of Albuquerque, which bounds KAFB on the north and northwest. The 1990 census lists the population of the City of Albuquerque [URL Ref. No. 327] as 384,734 and the population of Bernalillo County [URL Ref. No. 89] as 480,577, which includes permanent residents of KAFB living on base in KAFB housing areas. Isleta Pueblo, south of KAFB, is the next nearest population center, with a 1993 population of 4,538.

History and Mission

SNL was established in 1945 and was operated by the University of California [URL Ref. No. 339 {23}] until 1949, when President Truman asked American Telephone and Telegraph to assume the operation as an opportunity to render an exceptional service in the national interest. Designated by Congress as a National Laboratory in 1979, SNL is one of the DOE's most diverse laboratories and one of the nation's comprehensive research and development facilities. SNL's main responsibility is national security programs in defense and energy, with primary emphasis on nuclear weapons research and development. SNL also does work for the Department of Defense [URL Ref. No. 265] and other federal agencies on a noninterference basis.

Current activities at SNL include process development, environmental testing, radiation research, combustion research, computing, and microelectronics research and development. Over SNL's four decades of existence, its mission has changed. From an original focus of nuclear weapons research and development, its activities have expanded to include research on other advanced military technologies, energy programs, arms verification, control technology, and applied research in numerous scientific fields, including an extensive program in materials research. Energy efforts include combustion research, integrated geosciences research, and solar and wind power programs. SNL's environmental projects include programs in waste reduction and research for environmentally conscious manufacturing and environmental restoration.

Calendar Year 1992 Characterization Activities Relevant to SNL/KAFB

During the 1992 reporting period, a number of characterization activities were undertaken within the SNL/KAFB region that contributed to current understanding (Sandia National Laboratory, 1992). One of these activities was designed solely to increase the understanding of the regional hydrogeologic system. The others were related to contamination characterization studies for particular Environmental Restoration (ER) sites within the confines of the area.

South Fence Road Hydrogeologic Wells

Given the paucity of data near the south boundary of KAFB straddling the basin-bounding normal fault system, a gravity geophysical survey was performed, and a series of hydrogeologic characterization wells were installed along South Fence Road (SFR) adjacent to the Isleta Pueblo Indian Reservation. The gravity survey was intended to provide an estimate of depth to Paleozoic bedrock and, thus, help locate the buried fault system. On the basis of the interpreted bedrock depths, four boreholes were advanced at four different locations. Two of the boreholes were completed with well screens at two different depth intervals to ascertain vertical hydraulic gradients in the uppermost aquifer(s). One of the boreholes was completed in a single permeable zone, and the fourth remains uncompleted. Neel and McCord (1993) presented a detailed report of the field activities and data collected as part of this drilling program.

Contaminant Characterization Activities at the SNL Chemical Waste Landfill

Work performed at the Chemical Waste Landfill (CWL) for calendar year 1992 consisted of characterization of a volatile organic compound (VOC) [URL Ref. No. 247, 267] contaminant plume in the vadose zone. These activities began in November of 1992 and continued into calendar year 1993. Soil and soil gas samples were taken from depths of up to 275 ft (83.8 m) in the vicinity of the landfill. Although data collection efforts focused on contaminant characterization, some samples were analyzed for *in situ* moisture content.

Contaminant Characterization Activities at the SNL Mixed Waste Landfill

Work performed at the Mixed Waste Landfill (MWL) [URL Ref. No. 291] for calendar year 1992 included an infrared thermographic surface geophysical survey to aid in locating buried wastes and moisture content anomalies. This information was used to aid in siting an angled monitoring well, which was begun in calendar year 1992. This well was drilled approximately 6.75 degrees from vertical and was designed to be completed beneath a trench suspected of receiving tritiated water discharges. Continuous core was collected using the sonic drilling technique, with lithologic and borehole geophysical logging to be performed by U.S. Geological Survey personnel. The well was to be completed in two zones to assess vertical gradients and hydraulic conductivity anisotropy in the aquifer beneath the MWL.

Contaminant Characterization Wells at the Technical Area V Liquid Waste Disposal System

Work performed at the TA-V Liquid Waste Disposal System (LWDS) for calendar year 1992 included the installation

of eight contaminant characterization boreholes and one monitoring well. The sonic technique was used to drill all of the holes. Five of the boreholes were advanced to a total depth of 100 ft (30.5 m) around the perimeter of the LWDS surface impoundment, and three were angle-drilled to a depth of 50 ft (15.2 m). Continuous cores were taken from all of the boreholes, and lithologic logging was performed by U.S. Geological Survey personnel. The monitoring well was advanced to a depth of 525 ft (160 m), intersecting the water table at a depth of about 510 ft (155.5 m). Cores were collected during drilling of the monitoring well, with lithologic and borehole geophysical logging performed by U.S. Geological Survey personnel.

Contaminant Characterization Well at Technical Area II

Work performed at TA-II consisted of the installation of a single borehole advanced to a depth of 330 ft (106.1 m); the borehole was located just outside the TA-II fence. Samples were collected at regular intervals for contaminant characterization and for determination of *in situ* moisture content, cation exchange capacity, and percent organic carbon. Lithologic and borehole geophysical logging were performed by U.S. Geological Survey personnel. Water was encountered at 309 ft (94.2 m), and a monitoring well was completed at this depth. The completion interval was presumed to be in the perched zone.

Geologic Mapping of the Sanitary Sewer Line Trench

From 1990 through 1991, a sewer line [URL Ref. No. 222] extension was constructed from just south of Tijeras Arroyo south eastward to the Inhalation Toxicology Research Institute (ITRI) [URL Ref. No. 336 {48}] facility. During excavation, the ER Program mapped the geologic materials exposed in trench walls. A report was issued in calendar year 1992 (International Technology Corporation, 1992) that described the geologic conditions encountered. Along its entire 30,000 + ft (9,146 m) length (including laterals), the 3 to 18 ft deep trench exposed alluvial fan deposits. In the last quarter of 1992, a geostatistical analysis of trench map data was initiated by McCord et al. (1993) to quantify the spatial correlation characteristics of alluvial fan materials.

Geologic Mapping of the Travertine Hills

The Travertine Hills, located on the eastern portion of KAFB, represent the western outcropping of Precambrian and Paleozoic rocks in this part of the Albuquerque Basin. The

Sandia National Laboratory, New Mexico (SNL/NM), Site-Wide Hydrogeologic Characterization (SWHC) project has been implemented as part of the SNL/NM Environmental Restoration (ER) Program to develop the regional hydrogeologic framework and baseline for the approximately 100 square miles of Kirtland Air Force Base (KAFB) [URL Ref. No. 327] and adjacent withdrawn public lands upon which SNL has performed research and development activities (McCord et al., 1993). Additionally, the project investigated and characterized generic hydrogeologic issues associated with the 172 ER sites owned by SNL/NM across its facilities on KAFB. Examples of generic issues exist in all components of the hydrogeologic system, including surface water (e.g., erosion), the vadose zone (e.g., groundwater recharge), and the saturated zone (e.g., regional groundwater flow to receptors).

As called for in the Hazardous and Solid Waste Amendments (HSWA) [URL Ref. No. 300] to the Resource Conservation and Recovery Act (RCRA) [URL Ref. No. 280], Part B permit agreement between the U.S. Environmental Protection Agency (EPA) [URL Ref. No. 19] as the permitor and the U.S. Department of Energy (DOE) [URL Ref. No. 45] and SNL [URL Ref. No. 73] as the permittees, an annual report was prepared by the SWHC project team that served two primary purposes:

(1) To identify and describe the conceptual framework for the hydrogeologic system underlying SNL/NM

(2) To describe characterization activities undertaken in the preceding year that add to the understanding (reduce the uncertainties) regarding the conceptual and quantitative hydrologic framework

This SWHC project annual report focused primarily on the first purpose, providing a summary description of the current state of knowledge of the SNL/KAFB hydrogeologic setting. This summary description included information obtained from all appropriate reports that SNL is currently aware of, as well as from ER Program studies and projects, and summarized the current understanding, including the associated uncertainties [URL Ref. No. 261] of the occurrence, movement, and interaction of water in the geosphere. Additionally, this first annual report is an appropriate vehicle for communicating the rationale SNL will employ to identify and prioritize field characterization activities.

All SNL/NM facilities are situated on KAFB and are adjacent to Cibola National Forest lands that were withdrawn as part of an agreement between the U.S. DOE [URL Ref. No. 45] and the U.S. Forest Service [URL Ref. No. 339 {20}]. The SNL/KAFB area is located on a high, semiarid mesa and adjacent foothills, about 5 mi (8.0km) east of the Rio Grande. The mesa is cut by the east-west-trending Tijeras Arroyo, which drains into the Rio Grande. The eastern side of KAFB, north of Tijeras Arroyo, is bound by the southern end of the Sandia Mountains, and, south of Tijeras Arroyo, by the northern end of the Manzano Mountains.

Most of the area is relatively flat, sloping gently westward toward the Rio Grande. However, the eastern portions of KAFB and SNL extend into the canyons of the northern Manzano (or Manzanita) Mountains.

Geologically, SNL/KAFB is located in the east-central Albuquerque Basin, a major structural feature of the Basin and Range Province. The site sits on a partially dissected bajada built by three alluvial fan systems, Tijeras Arroyo Fan, Arroyo de Coyote Fan, and the Travertine Hills Arroyo Fan at the eastern margin of the Sandia-Manzanita mountains. The site straddles the major basin-bounding normal fault system (the Sandia and Hubbell Spring Faults), which trends north-south, as well as the Tijeras strike-slip fault, which cuts oblique across the site along a northeast-southwest alignment. Because few deep boreholes have been drilled within the SNL region and because the distribution of subsurface geologic data is sparse, the subsurface hydrogeology of the SNL/KAFB site is, at present, poorly defined. The hydrogeologic framework across SNL/KAFB is deformed by structural boundaries (e.g., the positions of major faults) and the spatial distribution of hydrostratigraphic units, lithotypes, and interlithotype heterogeneity.

The hydrogeologic setting at SNL also includes the meteorological environment, surface water runoff, percolation through the vadose zone, and saturated groundwater flow (Sandia National Laboratory, 1992). Refining this understanding helps establish a quantitative basis for understanding the potential pathways for transport of contaminants from SNL/KAFB sites to receptors that could lead to adverse impacts to human health and safety. Of additional concern, was the possible erosion and subsequent surface transport and redistribution of contaminants at SNL (Sandia National Laboratory, 1992).

At SNL, the vadose zone provides the link between surface water hydrology (i.e., which deals with surficial processes such as precipitation, snow melt, runoff, infiltration, overland flow, and evapotranspiration) and groundwater hydrology concerned with the flow and transport processes in aquifer systems. The vadose zone is an important part of the hydrologic system in the SNL/KAFB area, and in this semiarid climate, the vadose zone thickness is generally quite large [from 50 to > 500 ft (15 to > 150 m)], consequently, most contaminants released near the ground surface must travel a long distance before reaching the water table. The majority of environmental restoration (ER) sites are located at or near the land surface, and contaminant concentrations that reach the water table are always a concern with respect to the RCRA [URL Ref. No. 280] maximum contaminant limit (MCL). Dispersive effects in the vadose zone could dilute contaminant concentrations to the point that when and if contaminants reach the water table, concentrations might be less than the MCL. The regional areal recharge rate, which controls the upper boundary condition of the saturated zone, is also affected by vadose zone characteristics.

With respect to regional saturated zone hydrology, the basin-fill Santa Fe Group deposits are contributors to the primary aquifer in the Albuquerque basin [URL Ref. No. 55, 339 {12–13}]. This is because the basin-fill aquifer consists of interbedded gravel, sand, silt, and clay and is part of a complex stream-aquifer system that has been extensively developed in parts of the basin for irrigation and domestic and municipal water supplies. The aquifer properties have a considerable range of values because of the large variations in the lithology of basin-fill deposits. Groundwater is generally assumed to be unconfined in the upper part of the aquifer, although, however, in the deeper parts of the aquifer, the water can be semiconfined or confined, and accordingly, the depth to groundwater is quite variable in the basin.

SNL/KAFB is also situated in an area that includes two very different geologic environments separated by an assemblage of fault systems. This melange of geologic elements contributes to a complex saturated zone hydrogeologic framework that is divided into three distinct hydrogeologic regions (HR) based on local geology. This hydrogeologic framework establishes the basis for the conceptual model that identifies four subareas within these three HRs defined by a mix of hydrologic characteristics that strongly impact the local saturated zone hydrology. Two of these subareas (subareas 1 and 2) are distinguished by the local transients caused by water-supply pumping wells. The other two subareas (subareas 3 and 4) include differences in flow systems (porous media flow and fracture flow), the type(s) of aquifers (unconfined/perched and confined), the state of the flow system (steady state and transient), flow system boundaries, and flow system heterogeneities and anisotropies. Elements involved in future saturated zone characterization include developing subarea flow models (planning and reporting), applying a quantitative approach to identify critical data gaps (numerical model application), and acquiring parameters to fill these gaps (drilling and testing). These elements previously listed are intended to establish a strong, site-wide hydrogeologic understanding that will be used to support ER Program characterization and remediation projects.

With respect to saturated zone water chemistry, the springs were located in Tijeras Arroyo, the mouth of Coyote Canyon, and Hubbell Spring on the Isleta Pueblo Indian Reservation. In general, the wells and the springs in Tijeras Arroyo, reflect water compositions derived from runoff from the Sandia and Manzanita mountains, with calcium-bicarbonate-rich waters and lesser amounts of chloride, sulfate, and sodium. The water samples from Coyote Springs differ from samples obtained from other wells and springs in that they have significantly higher total dissolved solids and boron concentrations. Groundwater quality data indicate, among other things, that the groundwater east and west of the Sandia/Hubbell fault line is characterized by statistically different chemistries.

A site-wide, subsurface conceptual hydrogeological model (CM) was also developed by constructing the hydrogeologic framework and evaluating the spatial distribution of geologic features that control hydrologic and contaminant transport parameters. This CM accounted for all relevant hydrogeologic processes, from surficial processes to vadose zone processes to saturated zone processes. This CM will be implemented, as required, by mathematical models [URL Ref. No. 339 {31}] that will permit SNL to make quantitative predictions regarding the behavior of the total hydrogeologic system and will be used to help guide site-specific field characterization activities.

However, there are still many aspects of the hydrogeologic system that are not fully understood; thus, the current conceptualization of flow and transport processes and the conceptual model(s) are limited. Future characterization work is planned and prioritized according to the methodology described, which is driven by critical uncertainties [URL Ref. No. 261] and identified performance measures. The critical uncertainties are largely dictated by data needs for the ER Program operable units. This strategy will reflect close coordination with all ER Program task leaders. Once processes in need of characterization are identified, SNL will implement an iterative stochastic simulation/field characterization procedure to define particular field characterization activities and refine SNL understanding of the site-wide hydrologic system.

Hydrogeologic Conceptual Model for the SNL/KAFB

Developing a Conceptual Model (CM) that identifies important transport processes [URL Ref. No. 302] and the interrelationships between those processes is a vital first step in building a predictive quantitative model of flow and contaminants in water systems. Knowing the likely fate and transport behavior of contaminants of concern through the regional hydrogeologic system is prerequisite to assessing the total risk [URL Ref. No. 235] posed by each of the ER sites. Although one would never be able to precisely know a contaminant's ultimate fate and transport behavior, mathematical models [URL Ref. No. 339 {31}] might be used to develop an appropriately quantitative understanding. Thus, an overall CM for the SNL/KAFB area must be developed to permit a defensible risk assessment for each of the ER sites. In addition, a well-defined CM is required in order to identify and select field characterization activities of the regional hydrological system.

Integrated Conceptual Model (CM)

The integrated CM includes the following:

(1) Precipitation will contribute to surface water flow and recharge into and possibly through the vadose zone into

the underlying aquifer. Overland flow and diffuse recharge are potential contaminant mechanisms.

(2) The majority of the SNL/KAFB area is overlain by an alluvial cover that will play host to vadose zone processes [estimated to range from 50 to 150 ft (15 to 46 m)].

(3) In addition to this thin alluvial cover, the vadose zone in subareas 1 and 2 extends deep into the underlying Santa Fe Group. In these subareas, the vadose zone is generally greater than 300 ft (91.5 m) in thickness.

(4) Subarea 4 extends from the Manzanita Mountain-front area west to the Tijeras/Hubbell Spring fault complex and is characterized by complex bedrock geology. This subarea might include local unconfined alluvial aquifers and confined porous media and fractured rock aquifers. Fracturing might contribute high-permeability pathways for groundwater flow and contaminant transport.

(5) Subarea 3 includes the Sandia, Tijeras, and Hubbell Spring fault complex. There is significant uncertainty on the impact of faulting on groundwater flow. Four scenarios on the impact of the fault complex include the following:

   (a) Faulting might create a hydraulic discontinuity, leading to a very high lateral hydraulic gradient between one or more subarea 4 aquifers and the uppermost Santa Fe Group unconfined aquifer in subarea 2.

   (b) Faulting might provide a conduit for upward flow from deep, confined aquifers into the shallow Santa Fe Group aquifer.

   (c) Faulting might result in deep recharge from the subarea 4 aquifers into a deeper Santa Fe Group confined aquifer.

   (d) Faulting might have a minor impact on the hydraulic connection between subareas 2 and 4.

(6) The Santa Fe Group underlying subareas 1 and 2 is characterized by a complex hydrostratigraphic architecture. This complexity will result in highly variable aquifer characteristics. These characteristics might include the opportunity for perched aquifers, locally confined conditions, high-permeability flow paths, and significant vertical gradients between distinct Santa Fe Group aquifers.

(7) Subarea 1 is within the radius of influence of the KAFB water supply well field. The local gradients associated with groundwater pumping will have a strong impact on the flow field in this subarea. This well field is a major groundwater receptor.

Uncertainties in the Conceptual Model

In each component of the hydrogeologic system (e.g., surface water, vadose, and saturated zones) are a number of un-

certainties that detract from the ability to predict fate and transport of contaminants. The only way to reduce these uncertainties is through well-focused field characterization activities. Characterization of individual ER sites provides the focus of upcoming activities and develops information that reduces uncertainties with respect to generic, site-wide hydrologic processes.

Surface Water and Vadose Zone Uncertainties

The surface water regime at SNL can affect contamination transport by erosion and sediment, and subsequent infiltration into the vadose zone. The net effect then can be a surficial redistribution of waste sources that can greatly complicate the ability to predict transport, although one knows that as many as six of the ER sites lie within the flood-prone areas, as identified by the U.S. Army Corps of Engineers (1979). It is also possible (if not likely) that some of the ER sites with surficial contamination that lie outside the flood-prone areas might be subject to source redistribution by surface processes.

Additionally, because of the spatial and temporal variability in the climatic conditions (i.e., the surface boundary for the vadose zone) and the surface hydrologic redistribution of precipitate, total water flux in the near-surface soil is highly variable. And, presumably, as the water moves deeper into the vadose zone, the natural lateral variability in moisture fluxes should be attenuated. Locations of near-surface spills (e.g., leaking tanks) and concentrated effluent loading (e.g., septic tank leachfields) [URL Ref. No. 217–218] can also cause the development of moisture plumes, although both the mode of fluid flow and its magnitude can be highly uncertain. For instance, one cannot rule out the existence of preferential flow paths through the vadose zone (i.e., which would significantly reduce contaminant travel time and attenuation), particularly beneath effluent sources. And, because large uncertainties exist regarding the natural recharge rates as well as the dispersion/dilution characteristics of thick vadose zones, the possibility of contaminant interaction and movement in separate phases (i.e., gas phase or separate liquid phase) must also be recognized and addressed.

Saturated Zone Processes

The saturated zone underlying the SNL/KAFB area is also highly complex. This complexity, in turn, contributes uncertainty to characterizing groundwater flow and transport, although subdividing the SNL/KAFB area helps to reduce area-wide complexity. However, even within the subdivided individual subareas, the local hydrogeologic framework still retains significant complexity and uncertainties [URL Ref. No. 261] that can include aquifer-type identification, quantification of effective hydraulic parameters (e.g., permeability, porosity, and compressibility), lateral and vertical hydraulic gradients, identification of high-permeability flow paths, aquifer transport characteristics (e.g., dispersivity dispersion, distribution coefficient, and contaminant degradation), and boundary conditions associated with the fault complex.

Hydraulic Gradients

The mix of different aquifer types and the heterogeneous lithologies lead to uncertainty in horizontal and vertical hydraulic gradients. Horizontal gradient uncertainty is related to correlating water levels from different aquifers to define a characteristic potentiometric surface (i.e., like mixing apples and oranges). The potential for locally significant vertical gradients in the saturated zone underlying the S14 has been demonstrated, however, the controls on vertical gradients (lithologic and/or structural) have not been identified, and there is significant uncertainty in the three-dimensional distribution of vertical gradients.

Transport Parameters

The variability in the hydrogeologic framework, and the range of different types of contaminants create many uncertainties in transport parameters. The magnitude and mix of these parameters could vary significantly at individual locations as well as between locations.

Boundary Conditions at the Fault Complex

The impact of the fault complex on site-wide horizontal and vertical groundwater flow is currently uncertain, that is, does the fault complex constitute a no-flow boundary, a constant head boundary, or no boundary at all?

Summary of SNL/KAFB Hydrologic Modeling

A CM has been developed for the SNL/KAFB hydrogeologic system. This CM accounts for all relevant hydrogeologic processes from the surficial processes to the vadose zone processes to the saturated zone processes. This CM will be implemented as required by mathematical models that will permit SNL to make quantitative predictions regarding behavior of the total hydrogeologic system. The CM will also be used to help guide field characterization activities to support the overall ER Program.

As previously stated, the SNL hydrologic project focuses on investigating the subsurface water system underlying SNL/NM facilities. And, it is important to note that over most of this area, the DOE is a tenant of lands controlled by KAFB/USFS, that many U.S. Air Force facilities are inter-

spersed with the DOE/SNL facilities, and that the relevant hydrogeologic system affecting and affected by SNL extends beyond the political boundaries shown. Therefore, the region to be investigated, in coordination with the appropriate controlling organization, must include all lands within the boundaries of KAFB, adjacent withdrawn buffer lands, and areas such as the adjacent Indian Reservations.

Los Alamos National Laboratory [URL Ref. No. 97] Groundwater Protection Plan

Groundwater Protection Plan

The Groundwater Protection Management Plan (GW-PMPP) that follows provides a detailed framework for consolidating and coordinating groundwater protection activities at Los Alamos National Laboratory (Los Alamos National Laboratory, 1990, 1995; Rogers and Gallaher, 1995) [URL Ref. No. 97]. The purpose of the groundwater protection plan is to monitor and protect the main aquifer underlying the Pajarito Plateau from contamination or other adverse impacts resulting from Laboratory operations and to preserve the quality of water for Los Alamos and surrounding communities in northern New Mexico for future generations.

The GWPMPP addresses the following concerning the groundwater situation at the Los Alamos National Laboratory:

(1) Hydrogeological characterization
(2) Potential contamination
(3) Groundwater monitoring network
(4) Water supply
(5) Information management
(6) Quality assurance
(7) Regulatory compliance

These previously listed issues have been discussed in audits, reports, assessments, and various deficiencies regarding current Los Alamos National Laboratory operations. Additionally, to remedy problems that have been detected, hydrologists, geologists, and consultants, as well as representatives of the New Mexico Environment Department [URL Ref. No. 333] and the Environmental Protection Agency [URL Ref. No. 19], have examined and reviewed issues/concerns related to activities at the Laboratory and have recommended solutions to prevent groundwater contamination, as they relate to past, present, and future on-site operations.

The primary solutions, therefore, to protect groundwater from contamination included the expansion of the current groundwater monitoring network. By increasing the number of monitoring wells and boreholes and by constructing the wells at select locations across the Los Alamos Pajarito Plateau, hydrologists, through time, will be able to collect sufficient data to adequately characterize the (aquifer) groundwater of the area.

This will allow the Laboratory to centralize, collect, and report actual groundwater information on an organized and systematic basis, since groundwater collection and analyzing efforts are presently spread among several organizations at the Laboratory. Consequently, to remedy problems, the GWPMPP details the development of a computer database network to ensure that timely groundwater-related information will be accessible to internal and external organizations and other interested parties. Consistency of sampling procedures, well construction and abandonment techniques, and other procedures will also be implemented so that the sharing of groundwater-related information continues to be facilitated among all concerned.

The GWPMPP also presents a business plan detailing the organizational hierarchy, roles and responsibilities, accountability and authority, funding allotments, and other financial considerations involved in implementing the groundwater plan. The core of the business plan includes a prioritized list of groundwater activities that ranks the activities according to their cost and overall importance. The final list of priorities, in the plan, then, represents the best attempt to enhance Laboratory environmental operations within budgetary constraints.

In *summary,* the GWPMPP at Los Alamos, strives to present an organized written approach to managing and protecting groundwater in the Los Alamos area through a dynamic process of coordinating its activities by implementing the following objectives meant to ensure the long-term protection of the local and regional groundwater supply:

(1) Consolidating the activities of different LANL environmental groups to ensure a unified approach to groundwater protection and to prevent duplication of effort

(2) Establishing an information system in which all groundwater-related data will be stored and that will be accessible to different LANL groups and outside customers

(3) Addressing the requirements of the HSWA Permit [URL Ref. No. 300], Module 8, Task III

(4) Providing enhanced groundwater documentation to support a Laboratory-wide Environmental Impact Statement as requested by the DOE under the National Environmental Policy Act (NEPA) [URL Ref. No. 308]

(5) Maintaining ongoing groundwater protection activities and addressing new issues of concern as they occur

Groundwater Protection Laws, Regulations, Statutes, and a Case Study Groundwater Protection Plan for Bernalillo County, New Mexico [URL Ref. No. 89]

Federal, State, and Local Laws and Regulations

The following paragraphs briefly discuss Federal and State regulations and local ordinances applicable to groundwater protection in the Albuquerque, New Mexico area (Albuquerque Public Works Department et al., 1995).

Federal Regulations [URL Ref. No. 68–69]

(1) The Resource Conservation and Recovery Act (RCRA) [URL Ref. No. 280]

(2) The Hazardous Materials Transportation Act [URL Ref. No. 328]

(3) The Toxic Substances Control Act (TSCA) [URL Ref. 295]

(4) The Comprehensive Environmental Response, Compensation, and Liability Act (CERCLA) [URL Ref. No. 264] and the Superfund Amendments and Reauthorization Act (SARA)

(5) The Water Pollution Control Act [URL Ref. 329]

(6) The Safe Drinking Water Act [URL Ref. No. 258]

(7) The National Environmental Policy Act (NEPA) [URL Ref. No. 308]

(8) Applicable portions of the Occupational Safety and Health Act (OSHA) [URL Ref. No. 332]

(9) The Federal Insecticide, Fungicide, and Rodenticide Act (FIFRA) [URL Ref. No. 296]

(10) The Atomic Energy Act [URL Ref. No. 330]

(11) The Clean Air Act [URL Ref. No. 331]

Resource Conservation and Recovery Act (RCRA)

The RCRA [URL Ref. No. 280] is the most comprehensive federal regulation specifically dedicated to managing the generation, storage, treatment, disposal, and transportation of hazardous wastes. RCRA also contains provisions for solid and medical wastes. The hazardous waste regulations are extensive. They are a combination of design specifications and performance standards, including permitting and reporting requirements. Congress gave the Environmental Protection Agency [URL Ref. No. 19] strong enforcement powers under this act and authorized funding for implementation of State programs in the EPA's budget. The New Mexico Environment Department (NMED) [URL Ref. No. 333] has "Primacy" or authorization to implement RCRA in New Mexico, with the exception of mixed (hazardous and radioactive) waste regulations, however, the EPA maintains some oversight and influence on the State's implementation.

The major deficiencies in RCRA are the exemptions for small quantity and conditionally exempt small quantity generators. Enforcement is difficult because small quantity generators are so numerous relative to large quantity generators. Also, RCRA does not regulate all "hazardous" wastes, only those listed and characterized by definition.

The solid waste (Subtitle D) provisions of RCRA cover solid waste management including sanitary landfills [URL Ref. No. 220], which typically accept only "nonhazardous" solid waste, with the exception of some household hazardous wastes [URL Ref. No. 334]. Subtitle J covers the management and tracking of some medical wastes.

Subtitle I of the RCRA regulates underground storage tanks (USTs) [URL Ref. No. 251] and the storage of certain hazardous substances, including gasoline and oil. Subtitle I does not regulate transportation [URL Ref. No. 250], disposal, or treatment of some hazardous substances (i.e., those defined by RCRA Subtitle C) stored in USTs.

Hazardous Materials Transportation Act [URL Ref. No. 328]

The Hazardous Materials Transportation Act, as the name implies, regulates the transport of hazardous materials throughout the United States. The regulations specify appropriate packaging and labeling requirements but do not cover storage, siting, disposal, or treatment of hazardous materials.

Toxic Substances Control Act (TSCA)
[URL Ref. No. 295]

The TSCA regulates the manufacture and use of toxic chemicals. TSCA also regulates storage, transport, and disposal of polychlorinated biphenyls (PCBs), fully halogenated chlorofluoroalkanes, and asbestos. Under this act, substances are registered with the EPA [URL Ref. No. 19] prior to manufacture and use. The registration process includes an evaluation of product toxicity. The act does not specify storage and site requirements for substances other than PCBs.

Comprehensive Environmental Response, Compensation, and Liability Act (CERCLA) and Superfund Amendments and Reauthorization Act (SARA)

The CERCLA (also known as Superfund) [URL Ref. No. 264], as amended in 1986 by the Superfund Amendments and Reauthorization Act (SARA), regulates the investigation, cleanup, and release of hazardous substances to the environment. It requires the following:

(1) Notification of releases of hazardous substances
(2) Reporting of manufacture and use of hazardous substances and toxic chemicals (under SARA Title III)

The regulation does not specifically consider management of hazardous materials, except that listed materials stored in excess of a predetermined amount must be reported to the Local Emergency Planning Committee and the State Emergency Response Commission. Using these reports, local and state agencies can assess the amounts and types of materials stored in an area, however, acceptable storage practices are not mandated under this regulation. CERCLA addresses primarily the remediation of contamination. Some observers have concluded that its focus on fixing the blame for the contamination actually impedes remediation efforts.

Water Pollution Control Act

The Water Pollution Control Act (the Clean Water Act) [URL Ref. No. 329] regulates point discharges of wastewaters to navigable waters, discharges of dredge and fill into navigable waters, disposals of sewage sludge [URL Ref. No. 236], and discharges of wastewater to publicly owned treatment works (POTW) through industrial pretreatment standards. The only provisions of the act that deal specifically with hazardous materials or wastes are the discharge

limitations of the industrial pretreatment standards. These standards limit the concentrations of toxic pollutants permitted in a discharge to a POTW.

The 1987 Water Quality Act amended the Clean Water Act to include a program to manage pollution in municipal storm sewer systems [URL Ref. No. 249]. The EPA has subsequently published proposed rules that will require municipalities to obtain permits for discharges to municipal storm sewers from industries and hazardous waste TSD facilities. The proposed rules will require the municipal permit applicant to certify and monitor such discharges and to describe a program to assist and facilitate the proper management of used oil and toxic materials. These proposed regulations, therefore, impact discharges of hazardous wastes or materials to municipal storm sewer systems and require the City of Albuquerque to obtain a system-wide discharge permit for storm sewers in their jurisdiction.

Safe Drinking Water Act

The Safe Drinking Water Act [URL Ref. No. 258] regulates the quality of the water in a municipal water supply for public consumption. It also regulates the use of injection wells for the disposal of hazardous wastewater under the Underground Injection Control (UIC) program. The UIC regulations contain design specifications and performance standards for the operation of permitted injection wells. The Sole-Source Aquifer provisions of the act allow EPA to intervene when federally-funded projects pose a threat to a designated sole-source aquifer. The act also calls for states to develop wellhead protection strategies, but it provides no funding.

National Environmental Policy Act (NEPA)

The NEPA [URL Ref. No. 308] requires the preparation of an environmental impact statement to analyze the effects of major federal projects.

Occupational Safety and Health Act (OSHA)

The Hazard Communication Act, under the OSHA [URL Ref. No. 332], requires training and employee notification for chemical manufacturing, importing, and laboratories. The act does not consider storage or siting of hazardous materials or wastes, or their treatment, disposal, or transportation. Other OSHA provisions are not applicable to hazardous materials and waste storage and siting or groundwater protection in general.

State Regulations [URL Ref. No. 333, 339 {42}]

Within the State of New Mexico [URL Ref. No. 339 {42}], handling, storage, and transportation of hazardous materials and wastes are regulated by statutes and regulations. These include the following:

(1) The Water Quality Act
(2) The Hazardous Waste Act
(3) UST Regulations and the Ground Water Protection Act
(4) The New Mexico Mining Act
(5) The Emergency Management Act
(6) The Environmental Improvement Act
(7) The Radiation Protection Act
(8) The Public Nuisance Provision
(9) The Oil and Gas Act
(10) The Pipeline Safety Act
(11) The Motor Carriers Act
(12) The Flammable Liquids Statute
(13) The Pesticide Control Act
(14) The Municipal Health Act
(15) The Solid Waste Act
(16) The Subdivision Act
(17) The Surface Mining Act
(18) The Environmental Services Gross Receipts Tax Acts

Water Quality Act

The State Water Quality Act delegates primary responsibility for regulating water pollution to the Water Quality Control Commission (WQCC). The act defines the makeup of the WQCC. The responsibilities given to the WQCC include the following:

(1) Adopt water quality standards as a guide to water pollution control
(2) Adopt, promulgate, and publish regulations to prevent or abate water pollution
(3) Assign responsibility for administering its regulations to constituent agencies

The WQCC has delegated authority partly to NMED [URL Ref. No. 333] and partly to the Oil Conservation Division (OCD). The WQCC has promulgated surface water quality standards, with which dischargers must comply. WQCC has also promulgated effluent limitations that apply only to dischargers who are not subject to a federal NPDES [URL Ref. No. 313] permit or who have been notified that they are violating a NPDES permit. (The EPA may authorize a state to administer the NPDES program.)

The WQCC has promulgated standards for groundwater quality. Facilities may not discharge contaminants that may move into the groundwater without submitting a discharge plan to the New Mexico Environment (NMED) Secretary (or Oil Conservation Department Director) and obtaining their approval. The standards do not apply to saline groundwater (total dissolved solids [URL Ref. No. 190] greater

than 10,000 milligrams per liter). The regulations allow groundwater to be contaminated up to the standards that, in some cases, allow contamination above the concentrations permitted in public water supplies. The WQCC has also promulgated special regulations for effluent disposal (injection wells).

Although the WQCC regulations provide a comprehensive and flexible framework, historical effectiveness of the enforcement may have been hindered by resource limitations. The Water Quality Act does not provide regulatory coverage to hazardous materials storage or siting. In addition, discharge plan regulations exempt most agricultural applications, constituents that are subject to effective and enforceable effluent limitations in a NPDES permit, discharges resulting from flood control systems, or discharges covered by other state regulations (Solid Waste, OCD, and Surface Coal Mining Commission).

Hazardous Waste Act

The State regulates hazardous wastes [URL Ref. No. 266] primarily through its authorization to enforce RCRA [URL Ref. No. 280] and through the State Hazardous Waste Act (RCRA equivalent) and associated regulations. These regulations cover a wide range of requirements for hazardous waste generators, including treatment, storage, and disposal facilities, and transporters. The regulations provide some exemptions for conditionally exempt and small quantity generators based on types and volumes of wastes generated. Household hazardous wastes [URL Ref. No. 334] are not regulated. These deficiencies are important because the number of small quantity generators and households producing hazardous wastes is greater than the numbers of large quantity generators and treatment, storage, or disposal (TSD) facilities that are covered by most of the RCRA and Hazardous Waste Act regulations.

Underground Storage Tank Regulations/ Ground-Water Protection Act

The Ground Water Protection Act and Underground Storage Tank Regulations provide a regulatory framework consistent with the federal underground storage tank regulations [URL Ref. No. 251] included in RCRA. Regulations require owners to register USTs, meet performance standards for new USTs (double-wall tanks for hazardous substances other than petroleum products, leak detection, corrosion, and spill/overflow protection), upgrade existing USTs, and report, investigate, and correct release.

The Ground Water Protection Act established a corrective action fund for the investigation, mitigation, containment, and remediation of contamination resulting from releases from underground storage tanks. As such, the act generally deals with pollution after the fact rather than with preventing it. Under 1992 amendments to the act, owners and operators of underground storage tanks can be reimbursed by the state corrective action fund if the owner or operator performs the cleanup action at a contaminated site. However, the owner/operator must first pay for up to $10,000 of the required minimum site assessment.

Federal Insecticide, Fungicide, and Rodenticide Act (FIFRA)

The FIFRA [URL Ref. No. 296] regulates the manufacture and use of pesticides [URL Ref. No. 239]. Like the TSCA and the OSHA, this act does not consider storage or siting of hazardous materials or wastes or their treatment, disposal, or transportation.

Atomic Energy Act

The Atomic Energy Act [URL Ref. No. 330], which includes the Radiation Protection Standards, regulates radioactive wastes [URL Ref. No. 304]. The Nuclear Waste Policy Act of 1982 established the Nuclear Power Operations Regulations, which regulate radioactive wastes, including releases. The Energy Reorganization Act of 1974 established the Standards for Protection Against Radiation, which also covers releases. The Standards for Protection Against Radiation include performance standards for releases, shipments, disposal, and treatment of radioactive wastes. The Radiation Protection Standards cover management and disposal of radioactive wastes, including spent nuclear fuel, high-level radioactive wastes [URL Ref. No. 305], and transuranic radioactive wastes [URL Ref. No. 289]. These standards also specify requirements for disposal site location and site monitoring.

Clean Air Act

The Clean Air Act [URL Ref. No. 331] regulates hazardous and nonhazardous air emissions. It does not directly affect sources of groundwater pollution.

National and Industry Standards

Various industry standards reviewed include the following:

- Uniform Fire Code
- National Fire Code
- Steel Tank Institute
- Underwriters Laboratories
- American National Standards Institute
- American Petroleum Institute
- American Society for Testing and Materials
- National Association of Corrosion Engineers

The Uniform and National Fire Codes, which the city and county have formally adopted, recommends storage and siting practices for hazardous materials and wastes. The other national standards reviewed did not specifically address hazardous material and waste storage, siting, and disposal.

New Mexico Mining Act

Effective in June 1993, the New Mexico Mining Act established procedures that are intended to promote the responsible use and reclamation of lands affected by mineral exploration and extraction. The act covers all minerals except for construction aggregate, wastes regulated by the Nuclear Regulatory Commission [URL Ref. No. 339 {34}], wastes regulated by the Federal RCRA regulations, coal, oil, natural gas, and geothermal wastes. A Mining Commission, made up of representatives of various state agencies, will adopt regulations governing the implementation of the act.

The act prescribes that the regulations protect water resources from degradation caused by mining activities. Specifically, the regulations require permit applicants to do the following:

(1) Describe all watersheds [URL Ref. No. 257] that may be affected by the mining activity
(2) Determine the hydrologic consequences [URL Ref. No. 187] of the new mining operation and reclamation with respect to surface and groundwater quality and quantity
(3) Show cross sections or plans depicting the locations of aquifers and springs and the estimated position of the water table and flow characteristics

Emergency Management Act

The Emergency Management Act provides a mechanism for coordinating response to hazardous materials incidents and specifies reporting requirements for facilities storing hazardous materials. The act does not specify hazardous materials storage or siting practices or specific groundwater protection measures.

Environmental Improvement Act

The Environmental Improvement Act [URL Ref. No. 333, 339 {42}] is general legislation that established both the Environment Improvement Board (EIB) and the (former) Environmental Improvement Division (EID) of the Health and Environment Department (HED). (The Environment Department Act reorganized the HED, creating the current Environment Department.) The Environmental Improvement Act authorizes the Board to promulgate regulations and standards for water supplies and liquid waste.

The EIB promulgated the Water Supply Regulations that specify general operating requirements, maximum contaminant levels (consistent with the federal SDWA), and siting and construction requirements. However, the Water Supply Regulations focus on protecting the quality of the drinking water delivered through public water systems, not on protecting groundwater.

The State Liquid Waste Disposal Regulations cover septic tanks [URL Ref. No. 217–218] receiving 2,000 gallons or less of liquid waste daily. The regulations cover permitting, installation, and modifications. They include state-wide minimum lot sizes based on design flows.

Radiation Protection Act

The Radiation Protection Act established a mechanism for dealing with the health and environmental aspects of radioactive material [URL Ref. No. 232]. The EIB, with the advice and consent of the Governor-appointed Radiation Technical Advisory Council, has the authority to promulgate rules and regulations. The state returned its uranium [URL Ref. No. 283] mill tailings responsibilities to the Nuclear Regulatory Commission (NRC) [URL Ref. No. 339 {34}] in 1986. The NRC regulates uranium mills and coordinates water quality issues with the state. The Radia-tion Protection Act regulates health and environmental aspects of radioactive materials and equipment. The act is not applicable to other potential sources of groundwater pollution.

Public Nuisance Provision

The New Mexico Criminal Code incorporates a common law provision prohibiting activities causing a nuisance to other citizens. The act of polluting water is defined as constituting a public nuisance and, therefore, it is a misdemeanor governed by the general rules of civil procedure.

Oil and Gas Act

The state's Oil and Gas Act regulates surface and underground injection of oil and gas development wastes including hazardous wastes related to the oil and gas industry [URL Ref. No. 254]. The Oil and Gas Act provides guidance on the design and construction of waste/storage disposal pits and injection wells, along with operating standards. It requires closure plans, inspection and maintenance plans, and disposal plans. The Oil Conservation Division has authority to implement the regulations, in addition to the underground injection control regulations under the Safe Drinking Water Act [URL Ref. No. 258], for which the state has primacy.

Pipeline Safety Act

The Pipeline Safety Act regulates transmission of liquid petroleum products through pipelines in the state.

Motor Carrier Act

The intent of the Motor Carrier Act is to regulate transportation of hazardous materials and wastes vehicles within the state. It requires licensing of vehicle operators.

Flammable Liquids Statute

The Flammable Liquids Statute of the State Insurance Code regulates the use, storage, handling, and vehicular transport of flammable and combustible liquids, which may include some hazardous materials and wastes. This provision, however, is not applicable to nonflammable or noncombustible hazardous materials relative to storage or siting.

Pesticide Control Act

The Pesticide Control Act regulates pesticide products [URL Ref. No. 239] or products classified for restricted use according to federal FIFRA [URL Ref. No. 296] legislation and according to the state's list of restricted products. The use, storage, and disposal of registered pesticides and pesticide containers are regulated under this act. The act also requires commercial applicators of restricted pesticides to be licensed. Funding for implementation is provided through the collection of license fees. The act prohibits local regulation of pesticides. While applicable to pesticides, the act is not directly applicable to other hazardous material and waste storage or siting.

Municipal Health Act

The Municipal Health Act is general authorizing legislation that allows local governments to adopt rules or ordinances to protect health.

Solid Waste Act

The 1990 Solid Waste Act sets into motion a much more aggressive solid waste management program in the State of New Mexico [URL Ref. No. 339 {42}]. Implementing regulations have been promulgated, and a comprehensive state-wide management plan is required by the act. The act mandates a 25 percent diversion of the solid waste stream away from landfill disposal and a 50 percent diversion by the year 2000. The regulations appear to be as stringent as

RCRA Subtitle D. The regulations provide for some ground-water protection through the use of engineering controls (such as liners) and groundwater monitoring wells. Enforcement of the new regulations is in the early stages, but the regulations would appear to allow considerable discretion by individual regulators.

Subdivision Act

The New Mexico Subdivision Act stipulates that, in regulating subdivisions, each county's Board of County Commissioners adopt regulations setting forth the county's requirements for water supply and liquid waste disposal (among other things). For subdivisions having 25 or more parcels, the act requires that the subdivider furnish liquid waste disposal facilities to fulfill the liquid waste provisions proposed by the subdivider in his disclosure statement. The act states that in reviewing a disclosure statement, the county shall request opinions from the Environmental Improvement Division (now the Environment Department) as to whether or not the subdivider can fulfill his proposals regarding liquid waste disposal.

Surface Mining Act

Surface coal-mining wastes are regulated under the state's Surface Mining Act. This act covers coal processing wastes, including acid-forming or toxic materials, acid and toxic drainage, and tailings. Surface coal mining is not conducted in Bernalillo County.

Environmental Services Gross Receipts Tax Acts

The Municipal Environmental Services Gross Receipts Tax Act allows municipalities to enact (without local referendum) up to a one-sixteenth of one percent gross receipts tax for the acquisition, construction, operation, and maintenance of solid waste facilities, waste facilities, wastewater facilities, sewer systems, and related facilities. The County Environmental Services Gross Receipts Tax Act allows any county to enact a similar gross receipts tax, up to one-eighth of one percent.

Local Ordinances

Applicable ordinances under the jurisdiction of the following city and county departments were reviewed:

- Fire Departments
- Environmental Health Department
- Public Works Departments
- Planning Department

The reviewed ordinances include the Fire Code, the City Refuse Collection Ordinance, the County Liquid Waste Disposal Ordinance, the City Sewer and Wastewater Ordinance, the City and County Zoning Ordinances, and the City and County Subdivision Ordinances.

Fire Code

The most applicable ordinances relative to the storage and siting of hazardous materials in Bernalillo County are the fire codes. The 1991 Uniform Fire Code contains requirements for the prevention, control, and mitigation of dangerous conditions related to hazardous materials and is also designed to provide information needed by emergency response personnel. The code includes detailed requirements for containers, tanks, and cylinders in addition to practices for dispensing, using, and handling hazardous materials.

The city adopted the code with many of the amendments suggested in the HMWS Policy in 1993, suggesting that this will be a good tool to regulate hazardous material and waste storage and siting.

The county also adopted the code in 1993 but did not incorporate amendments. Adoption of the amendments suggested in the HMWS Policy would enhance its groundwater protection effectiveness.

Refuse Collection Ordinance

The City Solid Waste Department maintains jurisdiction over the Refuse Collection Ordinance. The Refuse Collection Ordinance defines refuse and hazardous wastes, and it expressly prohibits the collection and disposal of hazardous wastes. However, refuse that is acceptable for collection and disposal by the city includes petroleum products, such as gasoline, kerosene, oil, and grease. Industrial wastes are also included in the definition of refuse, indicating that materials that may be hazardous (not necessarily by definition) may be collected and disposed of with regular household solid waste.

Liquid Waste Disposal Ordinance

The Bernalillo County Liquid Waste Disposal Ordinance enforced by the County Environmental Health Department regulates the use of liquid waste disposal systems discharging less than 2,000 gallons of domestic liquid waste per day. The ordinance specifies the permit process and fees; the requirements for installation, modification, and use of liquid waste systems; variances; and other administrative requirements. The ordinance focuses on assuring proper operation of septic-tank systems from a waste-disposal standpoint to prevent the surfacing of minimally treated sewage. Other than generic county-wide lot-size restrictions, it contains no specific restrictions on area-wide discharge of contaminants. Calculations suggest that the lot-size restrictions are not adequate to prevent contamination where the number of septic-tank systems along a groundwater flow path is large. Enforcement of the ordinance has improved recently, but widespread and localized contamination from septic-tank systems occurs throughout the county.

Sewer and Wastewater Ordinance

The Sewer and Wastewater Ordinance regulates industrial discharges (disposal) to city sewers. It specifies limitations on pollutant concentrations for discharges and influent pollution concentrations requiring action. The ordinance prohibits the discharge of pollutants, including toxic pollutants, to the Public Owned Treatment Works (POTW), which will interfere with its operation or contaminate the sewage sludge. The ordinance also prohibits the discharge of persistent pesticides or herbicides, PCBs, and other toxic refractory organic chemicals.

Zoning Ordinances

The city and county have broad land-use authority, which is implemented through the City and County Zoning Ordinances. These ordinances are among those most applicable to groundwater protection. Both sets of ordinances specify permissive and conditional uses of land for development within their jurisdictions. These uses typically include development and land use that may be associated with some storage of hazardous materials (for example, gasoline sales, swimming pools, utility structures, and manufacturing facilities) or wastes.

The zoning ordinances do not currently specify acceptable management of hazardous materials in terms of storage, use, or transportation practices. The zoning ordinances recognize permissive and conditional land uses. Both ordinances provide for application to their respective Planning Commission for zone designation changes and subsequent revision of land use. Application for zone designation changes does not imply that the change will be acceptable to the Planning Commission. Zone designations may change over time, and uses once considered unacceptable may be sought through application or may be considered permissive or conditional uses based on new zone designations.

Zoning changes in unincorporated areas of Bernalillo County may be more likely than in the city, because much of the city has already been developed. The county ordinance notes that all territory that may hereafter become a part of the unincorporated area of Bernalillo County shall automatically be classified in the A-1 Zone until appropriately reclassified. In the event the Zone Maps do not show the zoning of any area within Bernalillo County, the area is automatically classified in the A-1 Zone. The A-1 Zone allows one house or residence per acre. The county is currently using its zoning authority to regulate some of the private landfills in the county, no conforming uses or special use permits allow the county to place certain additional requirements, such as groundwater monitoring, on the user.

Subdivision Ordinances

The city and county have authority for the regulation of subdivisions within their jurisdictions. The City Subdivision Ordinance prescribes minimum standards for water and sanitary sewer systems. They must conform to adopted facility plans and current city policy on water and sanitary sewer line extensions. The County Subdivision Ordinance deals more directly with matters relating to groundwater contamination by setting requirements for large subdivisions to obtain approved water supply and liquid waste management plans. The minimum lot size requirements for a subdivision wishing to use individual liquid waste disposal systems are those defined by the County's Liquid Waste Ordinance (described previously) and, as such, present the same problems with respect to pollution prevention.

Bernalillo County Groundwater Protection Policy and Action Plan [URL Ref. No. 89, 160, 327]

This document sets forth the city's and county's Groundwater Protection Policy that includes a Hazardous Materials [URL Ref. No. 266] and Waste Storage and Siting Policy and an Action Plan to implement the policy (Albuquerque Public Works Department et al., 1995). The Groundwater Protection Policy and Action Plan (GPPAP) consist of the following:

(1) General policy statements, which define the desired results

(2) Protection measures, which are specific activities to protect groundwater from contamination

(3) The action plan to implement the policy and protection measures

The action plan defines the work that needs to be completed, provides an implementation schedule, identifies intergovernmental coordination needs, and identifies funding options.

It is intended that the policy statements and protection measures be detailed enough to convey the intent, rationale, and general implications of their adoption. Following adoption of the policy, detailed regulations and ordinances will then need to be developed to implement it. The action plan lays out the framework for the development of the details.

Policies cover three resource-planning areas: Bernalillo County, the Albuquerque groundwater basin [URL Ref. No. 55], and the Upper Rio Grande drainage basin [URL Ref. No. 53].

The Groundwater Protection Policy applies primarily to the City of Albuquerque and unincorporated portions of Bernalillo County. Although neighboring jurisdictions are not subject to the policy's regulatory intent, the policy calls for coordination activities with them.

This background discussion does the following:

- Describes the need for a Groundwater Protection Policy and Action Plan
- Summarizes the City and County Resolutions that called for the GPPAP
- Reviews the planning process and technical basis of the policy
- Sets forth the mission and goals of the Groundwater Protection Policy
- Relates the Groundwater Protection Policy and Action Plan to the Albuquerque/Bernalillo County Comprehensive Plan and activities

Groundwater, the County's Sole Source of Public Drinking Water

Groundwater is the sole source of public drinking water in Bernalillo County. Because the quality of the groundwater in much of the County is quite good, it does not require expensive treatment before use. At the same time, a seeming paradox exists: in many parts of the county, groundwater may not be a reliable drinking-water supply. In some places (parts of the East Mountain area for example), wells produce only small amounts of water and sometimes dry up. In other areas such as the western part of the county, the water quality is poor and unfit to drink. And in areas where we presently have an adequate supply of good quality groundwater, officials are becoming increasingly aware of its vulnerability to contamination. The shallow aquifer is highly vulnerable, and any cases of existing groundwater contamination have already affected the vulnerable aquifer system. Ten-year capture zones of public water supply wells either overlie or are quite near the polluted groundwater. Additionally, because recent work by the New Mexico Bureau of Mines and Mineral Resources [URL, Ref. No. 339 {40}] confirms the high vulnerability of the deep aquifer due to the high degree of hydraulic connection between the shallow and deep systems, these wells are also threatened by existing and potential contamination. It is clear that the adequate supply and the good quality may not last forever. So, it is essential that the quality of the county's groundwater be protected.

Vast Areas Already Contaminated

In Albuquerque and Bernalillo County, about 200 documented groundwater contamination events have contaminated vast amounts of groundwater, its quality degraded to an extent that affects its usefulness as drinking water. Of these documented cases, the U.S. Environmental Protection Agency will investigate more than 20, and some of these may reach the Superfund National Priorities List. New cases of groundwater contamination are being reported all the time. In fact, investigators have identified about 100 additional cases since 1989.

The New Mexico Environment Department (NMED) [URL Ref. No. 333] estimates that, so far, this pollution has affected about 25 public supply wells and as many as 600 private wells in Bernalillo County. More than 30 square miles of land area may overlie contaminated groundwater supplies, including the South Valley Superfund site where two contaminated city wells are now out of service. Septic tank systems, underground storage tanks, landfills, industrial facilities, and releases of hazardous materials from other sources caused this pollution. Septic systems are a major contributor to groundwater contamination, in fact, the State Environment Department says they are the major nonpoint source of groundwater pollution in New Mexico.

Magnitude of the Problem Unknown, but Likely to Worsen

Moreover, authorities know that they have probably not identified all of the existing pollution. The New Mexico Environment Department is currently investigating potential groundwater contamination at about 25 sites in the county. And, those are just the sites at which officials know or suspect past practices have caused contamination.

It is important to note that the 200 cases of groundwater contamination have come to the attention of officials not through proactive field investigations, but rather through complaints, chance encounters, or as a result of other activities such as underground storage tank removals, drilling projects, etc. Organized proactive field investigations will undoubtedly find numerous additional instances of groundwater contamination, however, local government has never had the resources to conduct this type of much-needed work.

In addition to the existing contamination-pollution that authorities know about and pollution yet to be found, a very high potential exists for even more groundwater to be polluted by future activities. Some types of contamination sources that caused the contamination we know about occur throughout the county. Some of the potential contamination sources (e.g., hazardous materials) are not even regulated with regard to groundwater protection. And, some of the regulations that exist are not being adequately enforced, in part because the governments' resources have been inadequate given the magnitude of the problem.

Existing Regulatory Framework May Not Have Prevented the Historic Groundwater Contamination Cases

In the past decades, there has been a great increase in the number of federal and state environmental laws and programs. Previous studies and government agency databases suggest that there are about 200 cases of documented groundwater contamination in Bernalillo County. Many of these cases undoubtedly occurred before the recent proliferation of environmental legislation and regulations. Although a definitive answer is not possible, the following paragraphs consider to what extent additional contamination events will be prevented under the existing regulatory framework. Previous work identified the three highest-priority sources of contamination: leaking underground storage tanks, hazardous materials and waste storage facilities, and on-site liquid waste disposal systems.

LEAKING UNDERGROUND STORAGE TANKS

Contamination from leaking underground storage tanks [URL Ref. No. 251] accounts for about 100 of the cases. For many of these cases, it is not known when the release occurred. Clearly then, the existing regulations represent enormous improvement. Many of the cases were, in fact, discovered as underground storage tank (UST) owners began complying with the leak detection and upgrade and retrofit requirements of the new regulations. However, additional leaks may have occurred even at new installations. Despite the improved regulations, remediation efforts are slowed by the high workload assigned to the NMED staff. Consequently, the GPPAP calls for enhancing enforcement and accelerating retrofit requirements within wellhead protection areas.

HAZARDOUS MATERIALS AND WASTE STORAGE FACILITIES [URL REF. NO. 266]

Of the 46 contamination cases attributed to industrial facilities' aboveground storage tanks, it is not clear how many occurred before enactment of applicable Resource Conservation and Recovery Act (RCRA), Superfund Amendments and Reauthorization Act (SARA), and New Mexico Water Quality Control Commission (WQCC) regulations. Certainly, waste management practices at existing facilities were much improved as a result of these regulations. Nonetheless, without frequent inspections, violations of these rules are likely, due sometimes to ignorance

of appropriate best management practices or sometimes to the high cost of proper disposal of hazardous wastes. In fact, recent years have seen several investigations into allegations of violations. So again, the present concern appears to be one of enforcement: conditionally exempt small quantity generators are essentially self-regulated, and officials do not know their numbers or locations. As many as 500 nonexempt facilities may operate county-wide, with few available NMED inspectors. In addition, stockpiles of hazardous materials are, except for the recently adopted 1991 Uniform Fire Code (UFC), virtually unregulated. Given this context, the GPPAP recommends enforcement of the 1991 UFC with the necessary modifications, enhanced enforcement of existing state and federal regulations, and technical assistance to help businesses that want to do the right thing.

ON-SITE LIQUID WASTE DISPOSAL [URL REF. NO. 217–219]

According to the 1990 U.S. Census, about 17,800 conventional septic-tank systems are in Bernalillo County. These discharge minimally treated sewage to the soil. The county database contains permit information for only about one-third of these systems. The county has recently enhanced the enforcement of the existing Liquid Waste Disposal Ordinance, but the structure of the ordinance may not be adequate to prevent the continued degradation of groundwater quality from on-site disposal. The current ordinance focuses on preventing public health emergencies (for example, the surfacing of raw sewage due to leach-field failures or the transport of viral or bacterial pathogens). That this ordinance may not be adequate to deal with area-wide loading of contaminants to the drinking water supply is demonstrated by the numerous documented cases of toxic contaminants such as nitrate in the groundwater, and the widespread taste and odor problems in private water supplies resulting from overly dense concentrations of septic-tank systems. The particularly vulnerable fractured sedimentary rocks of the East Mountain area and the problems already identified there, clearly show that the county-wide minimum lot-size restrictions are not adequate everywhere.

The Policy Coordinating Committee developed the Groundwater Protection Policy and Action Plan after extensive evaluation of the existing regulatory framework. The evaluation considered the adequacy of the laws and regulations and their enforcement. Based on that evaluation and extensive public comment, the GPPAP was developed with the objective of avoiding duplication of effort and unnecessary new layers of government. Wherever possible, it draws on enhancing the enforcement and effectiveness of existing programs and regulations, rather than on creating new ones. Where necessary to fill recognized gaps in the regulatory fabric, a few new programs are clearly necessary.

Each Pollution Event Can Be Devastating

Each pollution episode undermines the quality of life and the economic vitality of the community. Potential cleanup costs from just one contamination event can reach to millions of dollars. However, immediate action can substantially reduce these costs. Recognizing the critical nature of the problem, the Albuquerque City Council and Bernalillo County Board of County Commissioners called for action.

Comprehensive Groundwater Protection Policy and Action Plan

In the summer of 1988, the Albuquerque City Council and the Bernalillo County Board of County Commissioners passed a resolution (R-143 and R-49-88) calling for comprehensive Groundwater Protection Policy and Action Plan (GPPAP) and a Hazardous Materials and Waste Storage (HMWS) and Siting policy.

The resolutions noted that the City Council had authorized the city's Public Works Department to develop a Comprehensive Water-Resources Management strategy. They called for development of a GPPAP as an integral part of comprehensive water-resources management, paramount to the protection of public health, safety, and welfare.

The resolutions also called for a Policy Coordinating Committee (PCC) representing the County Environmental Health Department and the City Environmental Health, Planning, and Public Works Departments to develop the policy and action plan and for a citizen member technical advisory committee made up of: individuals who are professionals or community leaders in the field of environmental planning and/or water quality, with the remainder being selected to represent a broad range of community interests such as development, local government, academia, neighborhood, and business.

This technical advisory committee came to be known as the Groundwater Protection Advisory Committee (GPAC). As directed by the resolutions, the GPAC formed in September 1988. Since that time, a total of 45 citizens have served on the GPAC including Councilors, Commissioners, professors, attorneys, neighborhood group representatives, development interests, businessmen, environmental advocates, and representatives of pueblos, agencies, and public-interest groups.

Between September 1988 and August 1989, GPAC met 16 times to advise PCC on what was needed to be done to develop sound policies. This advice focused on producing a work plan that defined the planning process and detailed all of the tasks included in the process. During 1990 and 1991, they met over 30 times to help PCC review this policy and the technical aspects of the investigations called for in the planning process. During 1992, GPAC met seven more times to review the second draft of this policy and to review public comments.

PCC and its consultant, CH₂M HILL, with the advice and counsel of GPAC, implemented the work plan for the GPPAP, including the Hazardous Materials and Waste Storage and Siting policy and action plan. This led to the following reports, which should be used by the city and county as reference sources when existing ordinances are reviewed and updated or new ordinances are being developed to protect groundwater resources of the county.

The "Groundwater Report" sets forth the geologic and hydrologic conditions that must be taken into account to develop groundwater protection policies and locates on maps the cases of groundwater contamination in Bernalillo County.

A "Summary of Hazardous Materials and Waste Storage and Siting Practices" characterizes the hazardous materials and wastes in Albuquerque and Bernalillo County that need to be dealt with including locations of hazardous materials use; types and volumes stored; handling, storage, and transportation practices; and inventory and monitoring practices.

The "Vulnerability of Bernalillo County Groundwater Resources" report assesses the probable vulnerability of the groundwater resource based on depth to groundwater, recharge, aquifer media, soil media, topography, impact of the vadose zone, and hydraulic conductive.

Reports concerning "Review of Other State and Local Programs Related to Hazardous Materials and Waste Siting and Storage" show how other communities dealt with hazardous materials

The "Threat Characterization Report" identifies, maps, and describes nearly 10,000 potential threats of sources of groundwater contamination in the city and county.

The "Protective Measures to Address Hazardous Materials and Waste Storage and Siting," report suggests measures that the city and county might adopt to deal with hazardous materials and waste siting and storage.

The "Identification and Evaluation of Groundwater Protection Measures and Implementation Mechanisms" report identifies and evaluates 94 groundwater protection measures and a broad range of implementation mechanisms.

The report in Bernalillo County, New Mexico, entitled "Septic-Tank Systems and Their Effect on Groundwater Quality" brings together much of the data demonstrating the adverse impacts to groundwater quality caused by septic tanks.

The "Effect of Lot Sizes on Potential Groundwater Contamination from Conventional Septic-Tank Systems: Numerical Modeling" report describes the calculations made to evaluate potential septic-tank contamination.

The "Delineation of Crucial Areas for Groundwater Protection in Bernalillo County" report shows how the crucial areas were determined.

The "Executive Summary of the Documentary Basis for the Groundwater Protection Policy and Action Plan" report summarizes the received reports.

In addition to the reports listed above, the Policy Coordinating Committee and CH$_2$M HILL have prepared the following technical fact sheets:

(1) Alternatives to Conventional On-Site Sewage Disposal Systems February 1992

(2) A Summary of the Draft Groundwater Protection Policy and Action Plan

(3) Summary of the Hazardous Materials and Waste Storage Policy

(4) The Need for a Policy to Protect Our Groundwater

(5) Delineation of Crucial Areas for Groundwater Protection

(6) Controlling the Density of Conventional Septic-Tank Systems

Each report benefited from the review, advice, and counsel of GPAC before PCC adopted it, and these reports provide the basis for policies.

As specified in the resolutions, the planning process included a component for public constituency development. This component included work to educate and involve the public in formulating the policy. To inform the public, PCC translated the key technical documents into jargon-free citizen summaries that they distributed to a contact list of about 1,500 people. They prepared fact sheets, newspaper and water-bill inserts, a portable public display, and an informational video called "Groundwater —Our Future." They organized a speakers' bureau that gave talks and presentations to interested community groups and worked with Albuquerque Public Schools to involve teachers and students. In addition to these information items, PCC sought public comment at a series of seven focus groups and nine briefing meetings and public workshops held throughout the county. PCC continued to solicit public comment on an initial draft GPPAP at an open house/public workshop in November 1991. Summary reports document public comments at these meetings. These constituency-development activities provided insight into the public's concerns about groundwater protection issues and helped formulate the mission and goals. The public's concerns helped the PCC revise the initial draft, which also benefited from extensive public review.

The Policy Coordinating Committee also presented policy overviews and solicited additional comment at numerous public, civic, and professional meetings, including the following:

(1) Environmental Planning Commission

(2) County Planning Commission

(3) Environmental Tax Advisory Board

(4) Middle Rio Grande Aquifer Water Quality Forum

(5) Middle Rio Grande Council of Governments Board of Directors

(6) New Mexico Hazardous Waste Management Society

(7) Albuquerque Board of Realtors, Governmental Affairs Committee

(8) Development Process Manual Steering Committee

(9) Albuquerque Chamber of Commerce Government Academy

(10) National Association of Industrial and Office Park Developers

(11) Albuquerque Chamber of Commerce, Quality of Life Committee

(12) Cuidad Soil and Water Conservation District

(13) Local Emergency Planning Committee

(14) Albuquerque Geological Society

(15) American Water Resources Association

(16) Albuquerque/Bernalillo County Goals Commission

(17) Department of Energy/Sandia National Laboratories

(18) New Mexico Environmental Health Association

(19) Committees of the League of Women Voters

(20) East Mountain Area Builders and Realtors

(21) Tijeras Neighborhood Watch

(22) Water Well Drilling Contractors

(23) Septic-Tank Installers

(24) Underground Storage Tank Owners

(25) South Valley Chamber of Commerce

(26) Albuquerque Mortgage Bankers Association

(27) East Mountain Building and Development Association

(28) New Mexico Environmental Improvement Board

(29) New Mexico Water Quality Control Commission

(30) Albuquerque Geological Society

(31) St. Paul's Methodist Church

(32) Albuquerque Economic Forum

(33) New Mexico Conference on the Environment

(34) Middle Rio Grande Council of Governments

The policy overview presented at these briefings led to follow-up discussions and meetings with representatives of several groups. These included the New Mexico Environment Department, the East Mountain Building and Development Association, Sandia National Laboratories, and the Department of Energy.

In addition, the policy and planning processes have received recognition from two of New Mexico's professional societies:

(1) New Mexico Chapter, American Planning Association that gave the policy an award: An Outstanding Contribution to Planning in New Mexico

(2) Consulting Engineers Council/New Mexico that recognized the technical quality of the planning process re-

sults by awarding the 1993 Grand Conceptor Award of Engineering Excellence to CH$_2$M HILL and Albuquerque/Bernalillo County

The formal approval process for the GPPAP was initiated in February 11, 1993, with a joint hearing of the County Planning Commission (CPC) and the Environmental Planning Commission (EPC). In a second joint hearing on April 22, the Environmental Planning Commission approved the GPPAP, recommending that an advisory Groundwater Protection Board be established. The County Planning Commission continued to review the plan until approval was recommended in their public hearing on August 4, 1993. The CPC recommended numerous editorial changes, which have been incorporated into this document. The CPC also recommended that the Board be advisory rather than regulatory. Both Commissions recognized the possibility of the City Council and County Commission granting additional authority to the Board in the future, and the EPC called for a reevaluation of this issue three years after approval of the GPPAP.

Final approval of the GPPAP was the responsibility of the Albuquerque City Council and the Bernalillo County Board of County Commissioners, and this GPPAP will be submitted to these elected bodies.

A number of documents were created just prior to or during the approval process to address specific issues or questions. These reports, which are described below, serve as reference sources for understanding the GPPAP.

- Staff Report to the Environmental Planning Commission and the County Planning Commission—this summarizes and responds to public and agency-review comments of the Second Draft GPPAP and describes modifications and revisions that were incorporated in the Draft Final Policy
- Preliminary Fiscal Impact Analysis—this provides the Policy Coordinating Committee's initial estimate of the potential costs to local government and the private sector to implement the Draft Final GPPAP over a three-year period
- Groundwater Contamination in Bernalillo County—this provides the current understanding of the nature and extent of documented groundwater contamination cases in Bernallio County
- Evaluation of Alternative Groundwater Protection Board Structures—this considers dozens of state and local groundwater protection programs already in place across the country and, based on these and consideration of the local situation, formulates and evaluates three conceptual alternatives
- Supplemental Staff Report to the Environmental Planning Commission and the County Planning Commission—this was prepared in response to

questions raised at a joint Environmental Planning Commission/County Planning Commission hearing on February 11, 1993

- Legal Analysis of the October 1992 Draft Final Groundwater Protection Policy and Action Plan—this summarizes the legal analysis of a number of aspects of the proposed plan, including the extent of the proposed Board's legal authority, completed by the Institute of Public Law at the University of New Mexico
- Fiscal Impact Analysis—this provides PCC's final recommendations for initial implementation cost; with the advice and counsel of GPAC, and based on public comments obtained from focus groups and workshops, PCC developed a statement of the mission, goals, and objectives of the Groundwater Protection Policy and Action Plan

The mission of the GPPAP is to ensure the quality of our groundwater resources so that the public health, quality of life, and economic vitality of this and future generations are not diminished. The goals of the GPPAP are to do the following:

- protect the groundwater resource
- find and clean up the contaminated groundwater
- promote the coordinated protection and prudent use of the groundwater resource throughout the region

Additionally, the GPPAP promoted the concept that after the groundwater protection policy had been presented, public review would occur where applicable and be incorporated into the Albuquerque/Bernalillo County Comprehensive Plan and/or other county plans as an integral component for planning and zoning activities for the area.

The policies, and the action plan to implement them, which were set forth here, expanded on and extended the goals and policies expressed in the Albuquerque/Bernalillo County Comprehensive Plan. The sections that follow use a format similar to the Comprehensive Plan to relate the policy, protection measures, and action plan to the goals of the GPPAP and in addition to the Comprehensive Plan, the PCC, in developing this GPPAP, considered and benefited from several other key planning efforts. These planning efforts include the Water-Resources Management Plan, the Water Conservation Task Force, the Wastewater-Regionalization Study, the Southwest Valley Service Options Evaluation, the City of Albuquerque's Five Year Goals and One Year Objectives, and activities of the Middle Rio Grande Council of Governments.

Specific Policies

This section identifies specific policies associated with each of the three goals formulated to satisfy the GPPAP mission. For each policy statement, this section lists the protection measures required to implement the policies. Some protection measures could apply to more than one policy, but are only described once. The three goals are as follows:

- to protect the groundwater resource
- to find and clean up contaminated groundwater
- to promote the coordinated protection and prudent use of the groundwater resources throughout the region

In addition to policies and protection measures directed at achieving each of these goals, this section also identifies some general policies that will help achieve all three goals.

Policy Statement: The City and County shall prohibit or control the releases of substances having the potential to degrade the groundwater quality.

To implement this policy, the city and county must take these measures:

- establish an Albuquerque/Bernalillo County Groundwater Protection Advisory Board to recommend threat control regulations
- prohibit or restrict certain activity in crucial areas to minimize the potential for contamination of groundwater
- prohibit the release of hazardous materials and hazardous waste to the groundwater by requiring best management practices and engineering controls at hazardous materials and waste storage (HMWS) facilities that use or store hazardous materials or hazardous wastes
- limit the quantity of other contaminant discharges to maintain groundwater quality above drinking-water standards
- enhance enforcement of and compliance with local, state, and federal environmental regulations
- provide education and technical assistance to the public and regulated entities to make them aware of the groundwater protection policy and to help them meet groundwater protection goals
- promote the management of household hazardous waste
- establish wellhead protection areas surrounding the immediate vicinity of public water-supply wells, with which additional restrictions apply
- promote recycling, source reduction, waste minimization, and product substitution throughout the production, handling, and management of hazardous materials and wastes
- monitor groundwater quality associated with known or suspected sources of groundwater contamination
- identify parties responsible for groundwater contamination and seek the expeditious remedy of the contamination they caused

- advocate the use of federal or state funds to clean up sites that pose immediate threats to public health, safety, or welfare and the recovery of cleanup costs from responsible parties
- prioritize areas of known or potential septic-tank contamination and aggressively pursue expansion of water collection and treatment facilities

Policy Statement: The City and County shall promote the vigorous enforcement of laws and regulations related to groundwater protection throughout the Upper Rio Grande drainage basin planning area

To implement this policy, the city and county must take these measures:

As part of intergovernmental coordination and cooperation:

- develop and maintain an interdepartmental, interagency regional database that catalogs the locations, types, amounts, pollution-prevention controls, and related information for hazardous materials and wastes and other substances, allowing the involved agencies to actively share data generated by their efforts
- establish a Memorandum of Understanding (MOU) with federal facilities located in the county to assure regulatory compliance and foster intergovernmental cooperation
- augment with local regulation if enforcement of existing laws and regulations prove ineffective
- identify and recognize compliance by industry, business, government, and community organizations
- issue City Council and Bernalillo County Commission proclamations of good citizenship to private and public sector organizations and industrial and business concerns who demonstrate exemplary support of the GPPAP
- acknowledge community organizations and governments that contribute to improvement of the groundwater resource and compliance with the GPPAP

Policy Statement: The City and County shall undertake comprehensive water resource management planning.

To implement this policy, the city and county must take these measures:

- establish an Albuquerque/Bernalillo County Groundwater Protection Advisory Board
- continue and expand regional groundwater monitoring
- develop procedures to assure that adequate resources (funding and manpower) are available to support groundwater protection and water-resources management
- implement water conservation
- implement the Water-Resources Management Plan work plan that will guide the conservation, use, pro-

tection, acquisition development, and management of the region's water resources within the Albuquerque groundwater basin

- promote coordinated water-resources management
- periodically review and update the GPPAP, the Water-Resources Management Plan, and technical tools developed for them

Policy Statement: The City and County shall encourage, facilitate, and acknowledge public participation.

To implement this policy, the city and county must take these measures:

- inform the public of existing and potential ground-water problems, hazardous materials and waste releases, progress made in protecting groundwater, and lessons learned in the implementation of the GPPAP
- establish a public involvement program to encourage public participation in the continuing development, updating, and implementation of groundwater protection policies

Protection Measures

The rationale for each of the nine Policy A protection measures follows:

- RATIONALE: Successful implementation of the Groundwater Protection Policy throughout the city and county requires the formation of an advisory board to recommend uniform local groundwater protection regulations, to monitor implementation of the GPPAP, and to periodically review and recommend updates to the GPPAP when justified.
- RATIONALE: Groundwater underlying crucial areas must be protected to assure its quality for human consumption and economic uses. Potential short-term economic gains associated with hazardous materials, septic tanks, and other pollution threats cannot begin to offset the long-term environmental and economic costs to clean up polluted groundwater.
- RATIONALE: Pollution prevention costs much less than pollution remediation, which in many cases may be technically or economically infeasible. Best management practices and appropriate engineering controls reduce the potential for releases of hazardous substances to groundwater.
- RATIONALE: In addition to prohibiting the release from hazardous materials or hazardous waste storage facilities, to maintain groundwater quality above drinking-water standards, the city and county need to limit discharges of other contaminants to groundwater.
- RATIONALE: Local, state, and federal laws that deal with releases to groundwater are not being adequately enforced, in part because the governments' resources

have been inadequate given the magnitude of the problem. Groundwater protection must have a high priority and must be supported with adequate funding.

- RATIONALE: Making the public and businesses aware of the critical importance of our groundwater resource is essential. Public education and technical assistance will facilitate the protection effort.
- RATIONALE: Household hazardous wastes pose a unique threat to groundwater when improperly disposed of.
- RATIONALE: Federal law, state policy, and the city and county resolutions calling for the development of this groundwater protection policy recognize the need for additional groundwater protection measures within the immediate area surrounding public water supply wells.
- RATIONALE: Recycling, source reduction, waste minimization, and product substitution are cost-effective means to decrease amounts of hazardous materials or waste posing a pollution risk to groundwater.

Create a Citizen-Member Albuquerque/Bernalillo County Groundwater Protection Advisory Board

A local Groundwater Protection Advisory Board is necessary to provide more consistency in the city's and county's regulation of and approach to groundwater protection. Board members will be expected to have some technical proficiency in groundwater issues. The local Board will be more immediately concerned with local problems and can address them with more speed. The Board is not intended to duplicate or overlap existing state or federal authority relative to groundwater protection.

The Board will be given responsibility in an advisory capacity to do the following:

- recommend groundwater threat control regulations to the city and county
- monitor the enforcement of regulations and assess their effectiveness
- designate wellhead protection area and crucial area overlap
- oversee the groundwater monitoring program and interpret the results
- oversee technical assistance and public education programs
- assess the effectiveness of well-driller and septic-tank-installer certification and permitting programs and recommend changes where justified
- implement the recognition awards program
- assess adequacy of groundwater protection staffing, budgets, and progress
- review and make recommendations to update the GPPA every five years or more often, if warranted

- consider variances and hear appeals as defined in the Groundwater Protection Ordinances and City/County Joint Powers Agreement
- promote and facilitate region-wide groundwater protection (working with other jurisdictions)
- promote activities to update databases relevant to groundwater quality and quantity in Bernalillo County

Because the emphasis of government shall be on managing the quality of groundwater in an efficient manner, groundwater protection ordinances and regulations recommended by the Board shall be necessary and fairly applied. The following paragraphs describe appropriate application of these protection measures to specific types of groundwater contamination threats. The highest priority threats are as follows:

- underground storage tanks
- hazardous materials and waste storage facilities, including hazardous materials stockpiles and small quantity generators (SQGs)
- on-site liquid waste disposal systems

Threats with a moderate priority include the following:

- large quantity generators of hazardous waste (LQGs) and hazardous waste treatment, storage, or disposal (TSD) facilities
- landfills
- household hazardous waste [URL Ref. No. 334]
- abandoned wells and improperly constructed wells
- groundwater discharge plan permitted threats (surface impoundments, land application, injection wells, large-flow septic-tank systems, and mining

Threats with a lower priority include the following:

- NPDES-permitted discharges (National Pollutant Discharge Elimination System)
- urban runoff
- agricultural practices
- deicing salt storage and application
- sewer exfiltration
- pipelines

Restrict Activities in Crucial Areas

Require secondary containment with release detection when upgrading or installing new tanks or lines in crucial areas.

Use Additional Restrictions in Wellhead Protection Areas

Consistent with the Environmental Improvement Board's Water Supply Regulations, prohibit new underground storage tanks within 200 feet of public water-supply wells. Require replacement of existing uncoated steel tanks within one year. Require double-wall systems with interstitial leak detection for all new and upgraded tanks. Determine best available technology for leak detection at existing tanks and adopt it as part of best management practices for underground storage tanks. Require the implementation of best management practices for all existing underground storage tanks within one year.

Enhance Enforcement of Existing Regulations

Work with the New Mexico Environment Department to enhance compliance with the terms of the state underground storage tank regulations requiring corrective action for contaminated soil and water.

Provide Education and Technical Assistance

Provide technical assistance and education to owners and operators on regulatory compliance issues (such as the new requirements in crucial areas) and the consequences of leaks.

Restrict Activity in Crucial Areas and Wellhead Protection Areas

The Hazardous Materials and Waste Storage (HMWS) Policy requires local operating permits for facilities' crucial areas and wellhead protection areas.

Adopt Best Management Practices and Engineering Controls

To protect against threats of pollution from hazardous material stockpiles, small and large quantity generators of hazardous waste, and hazardous waste treatment, storage, and disposal facilities invoke the HMWS policy. The city and county have adopted the 1991 Uniform Fire Code (UFC). The city incorporated most amendments suggested in the HMWS policy. Bernalillo County needs to adopt the minor amendments identified. Under the authority of the UFC, the city and county should implement permitting, inspection, spill control, drainage control, and secondary containment. They will take care to avoid unnecessary duplication of existing adequate regulations.

Limit Contaminant Discharges

Three methods are available to limit the discharge of contaminants from conventional septic-tank systems:

(1) Wastewater collection and treatment
(2) Alternative on-site liquid waste disposal systems

(3) Limitations on the density of conventional septic-tank systems

WASTEWATER COLLECTION AND TREATMENT

The city and county will prioritize and aggressively pursue the expansion of facilities to collect and treat wastewater now discharged through conventional septic-tank systems. The prioritization should consider areas of known or potential septic-tank contamination.

ALTERNATIVE ON-SITE LIQUID WASTE DISPOSAL SYSTEMS

On lots unsuitable for conventional septic-tank systems, require the use of alternative on-site liquid waste disposal systems. These alternatives may include, but are not limited to split-flow systems, composting or incinerating toilets, nondischarging systems, package treatment systems, and constructed wetlands. Develop appropriate performance standards that the alternative systems must meet. Require a Professional Engineer, registered in New Mexico, with expertise in wastewater, sanitary, or environmental engineering, to design and supervise installation of alternative systems that require site-specific design. Variance procedures may consider cases where alternative systems cannot be used.

CONVENTIONAL SEPTIC-TANK SYSTEMS

To limit discharges from new conventional systems, establish minimum site-specific hydrogeologic criteria, and limit overall contaminant loading rates.

Site-Specific Hydrogeologic Criteria

Minimum site-specific hydrogeologic criteria (i.e., which shall be established to assure proper subsurface hydraulic disposal and adequate soil filtration to remove pathogenic bacteria and viruses) must be met. These criteria should include soil texture, soil profile, percolation rates, susceptibility to flooding, depth to bedrock, depth to cemented pan, depth to seasonal high water table, slope, and percent of large stones in soil.

Limit Contaminant Loading Rates

The density of conventional septic-tank systems must be limited to prevent further groundwater contamination. Guidelines should be developed to determine appropriate maximum densities and minimum lot sizes required for conventional septic-tank systems, provided the location meets minimum site-specific criteria to assure proper subsurface hydraulic functioning and adequate soil filtration of bacteria

and viruses. Where the lot-size or density criteria cannot be met, alternatives can be used.

However, because site-specific conditions may vary within the areas, the city and county shall consider other available information, such as groundwater quality and other hydrogeologic data (i.e., which may be provided by an applicant) in the permitting process. They will not permit new septic-tank systems in areas where available information shows that new systems will create a potential or actual health hazard or in those areas where drinking water standards are already exceeded.

Phased Implementation

Because some areas of Bernalillo County have limited site-specific information and because widespread use of alternative on-site systems will be such a dramatic change in Bernalillo County, protection measures for septic-tank systems need to be phased-in as new data are collected.

Phased implementation of the on-site liquid waste disposal measures will include the following:

(1) A two-year program collect and analyze groundwater quality data, particularly from the East Mountain and North Albuquerque Acres/Sandia Heights areas
(2) A concurrent two-year effort to test, demonstrate, and develop performance criteria and operating and maintenance requirements for alternative on-site disposal systems
(3) A one-year effort (following the studies above) to reevaluate lot-size guidelines based on the additional data and to complete a master plan for county-wide wastewater treatment solutions
(4) An ongoing effort as needed to collect and analyze additional groundwater quality information and evaluate alternative systems

Larger Developments

County-wide, for new developments of 25 dwelling units or more, developers and builders must provide either connection to a regulated sewer system; centralized collection and treatment of wastewater, including nitrogen removal; approved alternative on-site nondischarging systems; or conventional systems that meet lot-size and density requirements based on an acceptable hydrogeologic report.

To assure that on-site liquid waste disposal systems are installed and repaired in a manner that will protect groundwater, the city and county will work with the state to develop a program to certify contractors and regulatory personnel. Only certified contractors will be permitted to install or repair onsite waste disposal systems in Bernalillo County. Additionally, the sale of toxic septic-tank additives will be banned in support of the state regulation prohibiting introduction of such additives to on-site liquid waste disposal systems.

Consistent with the Environmental Improvement Board's Water Supply Regulations, new septic tanks within 100 feet of public water-supply wells will be prohibited. Prohibit new septic-tank drain fields within 200 feet of public water-supply wells.

The existing County On-Site Liquid Waste Disposal Ordinance (88-1) will be strictly enforced, specifically, hookup requirements for on-site liquid waste disposal systems within 200 feet of existing sewers, with the first priority being within wellhead protection areas. Funding to assist those who cannot afford the hookup fee will also be continued. Additionally, variances will only be allowed for households able to demonstrate the proper functioning of an approved alternative system.

The public will be educated about on-site liquid waste disposal permit requirements, alternative systems, water conservation, the effects of garbage disposals on nitrogen loading, and the use of toxic septic-tank additives and maintenance issues.

Activities will be restricted in crucial areas that include wellhead protection sites, and the location of new hazardous waste disposal facilities in these same areas will be prohibited. Also, working with the State to identify additional resources (when necessary) for enforcement of existing State Hazardous Waste Management Regulations will be paramount.

Prohibition of expansion of or creation of new municipal or privately-owned landfills in crucial areas and wellhead protection areas must be implemented. Relocation of existing landfills out of crucial areas will also be encouraged.

The County Special Use Permit of a landfill (operating in a crucial area) that is not in compliance with the County Special Use Permit and New Mexico Environmental Improvement Board's Solid Waste Management Regulations will also be revoked.

Developing and implementing landfill-monitoring programs for operating and closing landfills and prioritizing based on threat and risk to groundwater will limit contaminant discharge. The monitoring will include vadose zone monitoring and groundwater monitoring. If the monitoring detects contamination above action levels, the following corrective measures may be required:

- cover the landfill with suitable low-permeability material and minimize the subsequent application of supplemental water through irrigation to reduce infiltration of moisture and, consequently, additional leachate generation
- use appropriate landfill-gas migration controls, groundwater containment and treatment actions, additional monitoring, and erosion controls as required

Government should enhance enforcement of existing regulations to protect groundwater from risks of contamination posed by new landfills outside of crucial areas and rely principally on the New Mexico Solid Waste Management Regulations. As a matter of policy, the city and county will use the state's public participation program to actively participate in new permit applications. And, enforcement efforts will be increased to prohibit illegal open dumping and target those areas frequented by illegal dumpers. Identify the responsible parties and encourage responsible-party cleanup in lieu of prosecution.

The city and county will ask the state to develop a program to train and certify well drillers to ensure adherence to construction and abandonment requirements. In the meantime, the city and county will develop a local program that will end when an appropriate state program is in place. As part of the certification, require well drillers to complete an education program covering the new requirements (local program may be required as an interim measure).

Well owners will be educated about the importance of properly abandoned unused or dormant wells and the importance of properly constructed wells. A suitable training program for well drillers will be developed that requires them to complete the training program to be certified (i.e., threats requiring a NMED Groundwater Discharge Plan Permit include surface impoundments, land application sites, injection wells, large-flow greater than 2,000 gallons per day septic-tank discharges, and mining).

Additionally, a state-approved Groundwater Discharge Plan will be required prior to city or county issuance of local building or operating permits to assure that the planned release meets all state requirements prior to initiating construction or facility operation. Work will be conducted with the state to identify resources needed to perform detailed permit review, conduct frequent inspection, and review compliance data, thus enabling the state to take enforcement action when necessary.

As currently provided by NMED policy, the city and county will also actively participate in proposed and renewal plan reviews for discharges in the county. Prior to the state issuing the permit, the city and county will recommend modifications, additions, or denials when necessary.

As currently provided by NMED policy, the city and county will also review permit applications for discharges located within the county. Prior to the U.S. Environmental Protection Agency issuing the permit with state certification, the city and county will comment on the permit and recommend denial when necessary to prevent groundwater contamination.

Inventory and monitoring possible groundwater impacts resulting from stormwater runoff in coordination with the existing city program to implement stormwater discharge regulations is essential. French drains and detention basins will be identified to ensure that they are not adversely impacting groundwater.

Cooperation with Other Government Agencies

Work will be conducted cooperatively with other government agencies to assure compliance with Federal Clean Water Act stormwater regulations (i.e., the comprehensive Clean Water Act stormwater regulations require municipalities and industries to identify, monitor, and limit urban runoff that may enter arroyos and rivers, thus potentially affecting groundwater quality).

Working with the New Mexico State University Cooperative Extension Service and the New Mexico Agriculture Department is essential to provide assistance and education focusing on urban (residential) and open space or greenbelt (turf grass, parks, and golf courses) application to pesticides, herbicides, fertilizers, and irrigation water, including the importance of water conservation. Working with the Ciudad Soil and Water Conservation District and New Mexico Agriculture Department to provide on-site technical assistance in the application or management of soil, water, nutrients, and pesticides on individual farms and ranches will also be necessary. Septic-tank owners will not heavily irrigate above their leach fields.

Underground Storage Tanks and HMWS Facilities Within Wellhead Protection Areas

Consistent with the Environmental Improvement Board's water supply regulations, prohibit new underground storage tanks within 200 feet of public water-supply wells. Require replacement of existing uncoated steel tanks within one year. Require double-wall systems with interstitial leak detection for all new and upgraded tanks. Determine best available technology for leak detection at existing tanks and adopt as part of best management practices for underground storage tanks. Require implementation of best management practices for all existing underground storage tanks within one year. Assure that HMWS facilities have implemented best management practices or engineering controls as required by the HMWS policy.

Waste Minimization

Adopt established standards such as those described in the Federal Resource Conservation and Recovery Act (RCRA) for waste minimization and product substitution in an effort to reduce dependence on hazardous materials and to reduce solid and hazardous waste generation.

Encourage industry, businesses, and government agencies to offer employee incentives for developing cost-effective means to incorporate waste minimization and product substitution that benefits the employer and the environment.

Draw on local expertise, such as from the New Mexico Hazardous Waste Management Society, to provide the tech-nology transfer needed to effect recycling, source reduction, waste minimization, and product substitution.

RATIONALE: The earlier pollution is found and cleaned up, the less it will cost.

In coordination with regional monitoring efforts, implement a program to monitor groundwater quality associated with known and suspected sources of contamination. Detection of groundwater contamination should trigger the implementation of appropriate control and preventive measures.

RATIONALE: Polluters should mitigate contamination they cause.

The City and County will strive to identify the parties responsible for all point sources of groundwater contamination. Once parties are identified, the City and County will seek the expeditious remedy of the pollution caused by the responsible parties. Although the City and County do not have (nor do they seek) the authority to enforce federal and state laws requiring corrective action, they can expedite remedy of contamination by bringing information to the attention of those that do. The City and County will not accept cleanup responsibility for contamination caused by others. In extreme occurrences where there is a threat to the public health and welfare, which cannot be dealt with by the established federal and state programs, the City and County will take corrective action and pursue cost recovery.

RATIONALE: Because of the large expenditures required to clean up or contain even one contamination event, and the scarcity of funds to do so, the city and county will facilitate the use of federal and state funds in Bernalillo County. The polluters may be unknown, lack the funds to clean up, or be negligent in cleaning up. Even so, groundwater pollution should be cleaned or contained as soon as possible.

Groundwater remediation needs to be continually monitored to ensure that it is effective and taking advantage of innovative a state-of-the-art technologies.

Possible funding sources for clean up and containment include the following:

- Federal Comprehensive Environmental Response, Compensation, and Liability Act (CERCLA)
- State Groundwater Protection Act
- Federal Leaking Underground Storage Tank Trust Fund

Because existing federal and state funds are not adequate to remedy all serious groundwater contamination events, the city and county shall promote the establishment of a state fund to remediate contamination from sources other than leaking underground storage tanks.

In addition, the city and county will facilitate identification and remediation of contamination problems caused by these threats: landfills, groundwater discharge plan permitted releases, hazardous materials stockpiles, small and large quantity generators of hazardous waste, hazardous waste treatment, storage, or disposal facilities, underground storage tanks, and pipelines.

RATIONALE: Where high densities of existing conventional septic-tank systems threaten or have caused groundwater contamination, the city and county must pursue appropriate wastewater collection and treatment solutions to replace existing systems or to reduce the threat to an acceptable level.

As a matter of policy, the City and County will endeavor to remove the source of potential or existing groundwater contamination from septic-tank systems. Remedy of the existing contamination from these sources, however, may have to rely on the natural processes of dilution and mixing. To remove the source, the City and County will prioritize and aggressively pursue the expansion of utilities to collect and treat wastewater now discharged through conventional septic-tank systems. The prioritization should consider areas of known or potential septic-tank contamination. The expanded treatment facilities include:

* conventional urban sewers
* semi-urban sewers, which transmit effluent from septic tanks to a central treatment facility, but retain septic tanks to collect solids prior to routine pumping
* smaller, community-level sewer service, using self-contained collection and treatment systems each serving several homes

As part of sewer-service expansion, requiring elimination of conventional septic-tank systems and seeking financial aid for hookup fees where appropriate, the city will expand water service in coordination with expansion of its wastewater collection service and mandate connection to city sewer as a condition for receiving city water service. In cases where an immediate public health threat exists, a safe water supply should be provided as soon as practical. In addition, city sewers should be pursued as soon as practical to replace septic tanks causing the groundwater contamination. This may require changing the city's existing service extension and annexation policies. Evaluation of the annexation policy should identify alternatives, assess costs and benefits, and consider the possibility of waiving annexation requirements to mitigate a clear public-health danger or threat to the regional aquifer.

Where installation of a wastewater collection and treatment system will clearly mitigate a threat to the municipal water supply, customers of the municipal water system should share in these new system construction costs.

RATIONALE: Data collection and data management constitute a challenging part of pollution control, but have the most potential for defining additional future needs.

Establish agreements with regional, federal, and state agencies to formalize exchange of data and information about groundwater quality, potential groundwater threats, and other activities that can result in groundwater contamination. Establish a central information repository and mechanisms to keep the data current and accessible.

RATIONALE: The Federal government operates facilities in Bernalillo County that impact groundwater quality. Federal facilities are not subject to local regulations. The city and county must work closely with these Federal agencies to provide adequate groundwater protection.

Establish a Memorandum of Understanding with Federal agencies located in the County to foster intergovernmental cooperation to assure that these Federal facilities do not further contaminate groundwater.

RATIONALE: Groundwater needs protection throughout the Upper Rio Grande drainage basin. If existing laws and regulations cannot accomplish this need, additional local regulations must fill the gaps.

Establish formal agreements with regional counterparts to encourage full enforcement of environmental regulations, including those dealing with surface water (because surface water recharges the County's groundwater). Augment as necessary with additional local regulations.

RATIONALE: Business and industry that readily comply with and go beyond the requirements of the GPPAP deserve public recognition for their efforts.

To recognize compliance, the City Council and the County Commission will issue, proclamations of good citizenship to industrial, business, and governmental concerns that demonstrate exemplary of the GPPAP.

RATIONALE: Community organizations that promote the protection and prudent use of groundwater deserve public recognition for their efforts.

To recognize exceptional contributions, the City Council and the County Commission will acknowledge community organizations who contribute to the improvement of the groundwater resource and who support the GPPAP.

RATIONALE: Successful implementation of the Groundwater Protection Policy throughout the City and County requires the formation of an advisory board to recommend uniform local groundwater protection regulations, to monitor implementation of the GPPAP, and to periodically review and recommend updates to the GPPAP when justified.

Create an Albuquerque/Bernalillo County Groundwater Protection Advisory Board to recommend groundwater protection regulations, to monitor implementation of the policy and regulations, and to periodically review and make recommendations to update the GPPAP. The Board should also formulate a 20 year protection and management strategy, developed with public review, require five year budget plans updated annually as part of the budget process, also developed with public review and updated yearly.

RATIONALE: Regional groundwater monitoring is essentially the only measurement tool available to determine the existing status of the resource and to observe the inexorable changes that will occur.

In addition to monitoring associated with specific sources of contamination, develop, in concert with the Water-Resources Management Plan, a County-wide monitoring program. The program should build on and extend the efforts already underway. Phased implementation of the monitoring program should include development within one year of a detailed plan for staging implementation with complete implementation by the end of the decade.

Every five years thereafter, thoroughly analyze all data, assess the adequacy of the monitoring network, and modify the program and network, as necessary.

Program implementation will be adequately flexible so that the design of each monitoring stage will take full advantage of knowledge gained during the previous stages.

RATIONALE: Without adequate resources (funding and staff support groundwater protection and water-resources management, the needed policies and strategies cannot be implemented.

Assure the availability of adequate resources to achieve groundwater protection and water-resources management.

RATIONALE: Conserving water leads to less resource demand and prolongs the availability of the resource. Water conservation may also reduce contaminant migration and the need for water treatment.

Implement the following recommendations of the Water Conservation Task Force (Resolution 49-1992):

- Promote low water use landscaping.
- Encourage water conservation through building and plumbing codes.
- Evaluate modification of water and sewer rates to encourage conservation.
- Alter City and private sector landscaping procedures and requirements to reduce water usage.
- Reduce wasted water.
- Conduct a public awareness campaign.
- Integrate landscaping and irrigation standards review into the City Environmental Planning Commission, the County Planning Commission, and the Development Review Board processes.

RATIONALE: The City and County resolutions calling for the development of the GPPAP noted that the protection of the groundwater resource was an integral part of comprehensive water resources management.

The City Council authorized the City Public Works Department, which supplies water to about 88 percent of the County population, to develop a management strategy. The Public Works Department subsequently developed a work program to form a Water-Resources Management Plan (WRMP). The development and implementation of the Water Resources Management Plan will guide the conservation, use, protection, acquisition, development, and management of the region's water resources. Obtain the information required and develop the necessary technical tools and databases to quantitatively understand the quality and quantity of the region's water resources. Determine the true long-term cost of obtaining and protecting the drinking water supply. Refine these tools as needed to proactively manage, develop, and conserve the region's water resources.

The city and county will work with neighboring jurisdictions in a coordinated water-resources management effort. Develop a water and wastewater management strategy for all areas of the county.

RATIONALE: The GPPAP process must reflect the dynamic nature of environmental problems, federal and state regulations, and groundwater protection issues.

Establish review procedures to consistently update databases and add new information as it becomes available. Databases include: permits, facility monitoring, groundwater monitoring, releases, and enforcement actions. Establish review procedures to determine if the policy and technical tools developed for the GPPAP and the WRMP remain appropriate. When necessary, update the technical tools so that the goals of the programs will continue to be met. The Groundwater Protection Board shall report annually on the implementation of the GPPAP to the Board of County Commissioners and the City Council.

RATIONALE: Government must alert the public to possible hazards and proposed solutions to them. An informed public contributes to the solution of environmental problems.

The City and County shall inform the media and public about: (1) releases of hazardous materials or wastes that occur in Bernalillo County, (2) actions being taken to contain and/or clean up the releases, (3) the potential human health consequences of the releases, to the extent possible, and (4) the progress made in protecting public health and the quality of regional groundwater.

The city and county shall provide summary fact sheets listing the above for the Albuquerque groundwater basin at least annually.

RATIONALE: An informed public contributes to the solution of environmental problems. Government should facilitate the public's involvement. Recognition of public participation by individuals will encourage others to get involved. Clean, safe water is everyone's responsibility—individuals, government, business, and industry.

And what is needed is to do the following:

- Provide a program for public participation in the development of policies and regulations. Publicize opportunities to participate in this process.

- Provide, on a routine basis (perhaps annually or biannually), a fact sheet describing groundwater protection progress throughout the region.
- Solicit support and information useful to the GPPAP on a routine basis and develop public service announcements on public participation issues relative to the GPPAP.
- Enlist the support of professional and public interest groups in implementing the GPPAP.

Action Plan to Implement the Policy

To carry out the Groundwater Protection Policy, the city and county need to implement the following: Water and Wastewater Management activities, Threat Control, Interjurisdictional Coordination and Cooperation, a Water Resources Management Program, and Public Participation and Technical Assistance.

A Joint Powers Agreement will specify which city and county departments will be responsible for implementing groundwater protection measures involving Water and Wastewater Management activities.

In addition to city and county resolutions that adopt this GPPAP, its implementation will require developing a City/County Groundwater Protection Ordinance and a Joint Powers Agreement. Representatives of the County Environmental Health, Public Works, and Zoning/Building/Planning Departments and the City Environmental Health, Public Works, and Planning Departments shall be given the responsibility for drafting these documents.

City and county staff shall develop parallel Groundwater Protection Ordinances that do the following:

- Create an Albuquerque/Bernallio County Groundwater Protection Advisory Board.
- Establish the procedures by which the Groundwater Protection Advisory Board will recommend threat control regulations.

The Groundwater Protection Ordinances shall avoid duplication of existing ordinances and make reference to appropriate existing ordinances such as the County Liquid Waste Ordinance 88-1.

Activities to develop the Groundwater Protection Ordinances shall include:

- defining the elements of the ordinances
- providing for public participation
- drafting and recommending the ordinances to the Council and Commission for approval

The city and county shall develop and adopt a Joint Powers Agreement to define the roles and responsibilities (including funding arrangements) of appropriate city and county departments that will carry out the protection measures and enforce adopted regulations that are based on the GPPAP and the Groundwater Protection Ordinances. To reduce the need for new governmental agencies or new levels of government, implementation should draw on existing staff and programs to the extent possible.

The Groundwater Protection Advisory Board, in its advisory capacity, will oversee the program, develop and recommend regulations for adoption by City Council and County Commission, oversee implementation of the regulations, and review and update the GPPAP. It will also carry out the functions outlined in Section 3 of this policy in an advisory capacity.

City and county staff will manage the programs and activities required to implement regulations and protection measures. The City/County Joint Powers Agreement will need to specify the management structure that can be under the auspices of existing agencies or under a more centralized authority.

Within crucial areas and wellhead protection areas, certain activities will be prohibited or restricted as detailed in this policy. The regulations will define these areas, identify the restrictions and prohibitions, and appropriately amend the existing ordinances. Crucial areas in Bernalillo County have been defined, but this small-scale map and maps of wellhead protection areas need to be refined to parcel specificity and updated as new wells are added.

To facilitate the periodic updates that will be required and to make the maps accessible to users, a county-wide geographic information system is required. The City/County Joint Powers Agreement needs to define respective city and county roles in developing and maintaining the geographic information system.

In addition to recommending additions and revisions to local regulations and ordinances, the Groundwater Protection Advisory Board will see that city and county staff review and assess the GPPAP at least every five years. As the program is implemented, additional data generated as part of the GPPAP may change prior assumptions and interpretations, or state and federal regulatory changes may force certain components of the GPPAP to be modified. Groundwater Protection Advisory Board activities required to assess the need for changes in the Groundwater Protection Program include the following:

- evaluating the condition of regional groundwater resources
- evaluating the effectiveness of enforcement of existing regulations or new state and federal regulations
- reviewing the effectiveness of Memoranda of Understanding and Joint Powers Agreements related to the GPPAP
- establishing a review mechanism, which includes solicitation of public comment, to assess the need for updating the GPPAP

- producing an annual report summarizing the condition of the regional groundwater resources, reporting on the status and effectiveness of the GPPAP, and reflecting public input and all recommendations for changes in the GPPAP

A Memorandum of Understanding (MOU) between the City/County and the Federal government should be negotiated to formalize communications, encourage Federal facilities to voluntarily comply with this Groundwater Protection Policy and associated local regulations, and enable City/County scrutiny of federal facility compliance with applicable laws and regulations. Activities to institute the MOU include the following:

- identifying and negotiating the scope of the MOU
- working with appropriate liaison groups
- obtaining the necessary approvals

A Memorandum of Understanding or Joint Powers Agreement (JPA) between the city/county and the state is needed to assure the coordinated application of nonoverlapping state and local authorities, both of which are needed to achieve local groundwater quality protection. The MOU or JPA should call on the city and county to work with the state to ensure adequate enforcement of existing regulations, which only the state has the authority to enforce with Bernalillo County. The MOU or JPA should also provide for cooperative database exchanges and include the New Mexico Construction Industries Division.

Activities to develop the MOU include identifying and negotiating the scope of the MOU, including identifying the need and funding options for additional enforcement efforts. Obtaining additional state enforcement may involve the following:

- requesting the state to provide additional enforcement
- negotiating additional state enforcement through the JPA or MOU
- seeking additional funding for state enforcement efforts through the legislative process
- exploring the availability of grant funds for enforcement efforts
- funding for additional state enforcement effort through city and court sources

Obtaining the necessary approvals

The county shall enforce all elements of the County Liquid Waste Ordinance (88-1). In addition, the county should assess the need for interim modifications in the ordinance, such as requiring dwelling construction concurrent with installation of septic-tank systems and enhanced inspection requirements.

Continue and expand the monitoring program funded by the County's Environmental Services Gross Receipts Tax. Expand and fund the programs necessary to accomplish the following objectives:

- Assess groundwater degradation from existing on-site liquid waste disposal practices, particularly in the East Mountain, Sandia Heights, and North Albuquerque Acres areas.
- Assess the potential for natural processes, such as denitrification and dilution, to attenuate septic-tank contamination.
- Refine the planning tools used to develop the lot-size and density limitation guidelines.
- Develop criteria that can be used to determine lot-size guidelines for household liquid waste disposal alternatives.

It is anticipated that within two years, results from this monitoring program will be available to allow the Groundwater Protection Advisory Board to recommend specific liquid waste regulations and to develop and enforce a program, consistent with existing county rules, requiring hookup to existing sewers. The program should do the following:

- Identify existing on-site liquid waste disposal systems that need to be replaced by hookup to existing sewers.
- Include funding mechanisms to assist citizens that cannot otherwise afford to connect to existing sewers.
- Develop regulations or an ordinance requiring hookup to existing sewers and removal of on-site liquid waste disposal systems within the city limits if sewer services are with 200 feet of the property.

Demonstrating Alternative On-Site System and Other Liquid Waste Disposal Options

City and county staff should work with qualified professionals to design and implement an on-site liquid waste disposal system demonstration program. The program should include a public education component. Staff should work with appropriate state staff to identify changes to regulations (such as the Uniform Plumbing Code or State Liquid Waste Regulations) needed for alternative systems and effect the necessary changes.

The city and county should complete the wastewater service option studies for the Valley and East Mountain areas. Concurrent with the demonstration program and service option analyses, staff should develop performance criteria, operations and maintenance requirements, and enforcement procedures for on-site and cluster-scale liquid waste disposal systems. These programs should be designed and implemented concurrent with the two-year and ongoing monitoring effort.

Formulating County-Wide Wastewater Disposal Solutions

The results of the monitoring program, the alternative demonstration program, and the Valley and East Mountain service options analyses should be integrated to formulate county-wide wastewater disposal solutions. This effort should give appropriate consideration to alternative on-site systems in addition to centralized collection and treatment.

In addition, the existing facility master plans for water and wastewater need to be revised to reflect the other requirements of the GPPAP as well as new information on the groundwater resource and possible water-resource constraints identified under the Water Resources Management Plan and water and wastewater infrastructure.

City and county staff must prioritize the need for expanding the wastewater collection and treatment facilities. High priority should be given to extending sewers into areas where densities of existing septic-tank systems have caused or threaten groundwater contamination, especially where local water supply comes from private wells or within wellhead protection areas. The Groundwater Protection Advisory Board should recommend approval of the updated master plans to the City Council and County Commissions.

Extending Water and Sewer Services

The policy requires the continued prioritized and coordinated extension of water and sewer services where appropriate throughout the county. The updated master plans will specify the blueprint for this extension.

This includes the enforcement of existing regulations that require property owners to hook up to sewers if their property is within 200 feet of a sewer as new sewers are installed. (As used here, "sewers" refer to various levels of wastewater collection systems: urban, semi-urban, and community-level.) Appropriate consideration needs to be given to alternative on-site waste disposal systems. Variances to this requirement may be allowed where alternative systems are functioning effectively. Given the importance of this protection measure, the city and county should continue funding mechanisms to assist citizens that simply cannot otherwise afford to connect to available sewer lines.

Overseeing On-site Liquid Waste Disposal

Following the initial two-year phase of the monitoring program, the alternative demonstration/education program, and the development of county-wide wastewater disposal solutions, the Groundwater Protection Advisory Board will review this additional information, and recommend on-site wastewater disposal regulations. The Board will recommend regulations that specify the following:

- a certification procedure and program to certify installers and the personnel responsible for regulatory oversight
- performance-based standards for alternative on-site liquid waste disposal systems
- formal permit-application, design-review, and installation inspection procedures for new systems (requiring, for example, design and installation oversight by a Professional Engineer, registered in New Mexico, with expertise in wastewater, sanitary, or environmental engineering, for systems requiring site-specific design)
- mechanisms necessary to assure the proper operation, maintenance, and testing of treatment and disposal methods (for example, requiring a certified operator for treatment facilities, providing periodic inspection and oversight by a certified county or contracted operator, private testing of on-site discharges from alternative systems, or administrative procedures, such as vouchers or manifests for holding tank pumping, to verify the performance of nondischarging systems)
- that the locations of existing septic-tank systems not currently listed within the county database be determined and translated to a county-wide geographic information system to determine densities of existing conventional system and facilitate enforcement of existing regulations requiring hookup to available sewers
- development and implementation of an education program for the general public about the new requirement and options along with a technical assistance program for installers of on-site systems

Examining Sewer Exfiltration

Staff needs to establish and implement procedures to locate and measure rates of liquid waste exfiltration from sewer lines in Bernalillo County. Current efforts to repair or replace leaking sewer lines will be reprioritized as warranted and reported to Council and Commission annually.

Locating Abandoned and Improperly Constructed Wells

Staff needs to locate abandoned wells that may serve as conduits of contaminants. Wells still in use but which also may serve as conduits for contaminants (because of improper construction), should also be located. The determination as to whether a well is "abandoned" or not needs to consider the intent to use the well again.

Well Construction/Abandonment Permits and Standards

Staff needs to develop well construction and abandonment requirements for all public and private water-supply wells. Permits and adherence to construction and abandonment standards for groundwater monitoring wells will be required as will the licensing and testing of well drillers and drill contractors who desire to drill wells in Bernalillo County. The city and county shall implement regulations for owners to properly seal abandoned wells.

A program is required to identify, track, and evaluate potential threats to groundwater quality. This threat-Control program includes the following activities:

- identifying and regulating threats located within the wellhead protection areas of public water-supply wells
- managing the data generated by the program
- permitting and inspecting hazardous materials and waste storage facilities
- evaluating suspected groundwater contamination events
- monitoring groundwater quality and soil vapor at landfills
- supporting enforcement
- enforcing new crucial area and wellhead protection area overlay zoning
- identifying locations of french drains and detention basins
- continuing funding of the permanent household hazardous waste collection center
- determining best management practices and waste minimization techniques

Threats Within Wellhead Protection Areas

The state's Wellhead Protection Program requires a inventory of all potential sources of contamination within the wellhead protection areas around public water-supply wells. The city and county will inventory threats within the wellhead protection area of the public water-supply wells. These activities include the following:

- performing a field inventory and examining available data to identify existing threats
- taking measures to make threat information generated by federal, state, or local regulations available to public water supply operators
- requiring the necessary upgrades of underground storage tanks within wellhead protection areas
- providing data to allow monitoring plans for public water supply wells to be tailored to the risks posed by threat within their wellhead protection areas

- notifying threat owners and operators, and regulators of the higher degree of risk posed by the threats to a public drinking water supply well

Data Management and Mapping

The involved agencies need to actively share data generated by the new policies and pertinent programs now in place. Data management and mapping includes the following:

- assessing data and user needs
- collecting and integrating the databases—creating a county wide geographic information system
- maintaining/updating the database

Data components can include databases such as county on-site liquid waste disposal permits, well construction and abandonment permits, UST registrations, Superfund Amendments and Reauthorization Act (SARA) Title III notifiers, the Local Emergency Planning Committee, RCRA, NPDES, Groundwater Discharge Plan, groundwater monitoring data, sales of restricted-use pesticides, inventory-control data for pipelines, and so on. All of these databases need to be incorporated into a county-wide geographic information system.

In addition to these threat-related databases, the geographic information system needs to incorporate the data that define the crucial areas (depth to groundwater, recharge, aquifer media, soil types, topography or slope, impact of the vadose zone, and hydraulic conductivity, existing and planned public water-supply well locations and conditions, and 30-year capture zones) and wellhead protection areas.

A comprehensive, easily accessible, and regularly updated geographic information system that includes these data is absolutely necessary to effectively implement this GPPAP. The data should be accessible to all involved parties and levels of government: federal, state, county, and city.

City and county staff members will be required to distribute, obtain, format, and report data from various activities of the GPPAP that generate it. In addition to the data assembled to develop this policy, other potential data sources include results of well and drain sampling, the locations of potential threats to groundwater quality, groundwater remediation activities, and federal activities. Additional city, county, and support services may be required to maintain the system to relate the various databases to locations throughout the city and county.

Permitting and Inspecting

This activity applies to facilities that use, generate, store, treat, or dispose of hazardous materials or waste. The city and county have adopted the 1991 Uniform Fire Code. The city incorporated most amendments suggested in the HMWS policy. Bernalillo County should adopt the minor modifications suggested in the HMWS policy, and the city should evaluate additional modifications to make their respective Code more protective of groundwater quality. The city and county must permit and inspect HMWS facilities, with priority given to facilities within wellhead protection areas. Additional funding mechanisms may be required to assure that adequate resources are available to review permit applications. Identification of facilities should draw on the data management activity and the work of the Local Emergency Planning Committee. Once identified, the locations and quantities of all hazardous materials storage facilities need to be included in the SARA Title III (if appropriate) and Fire Code databases.

As part of the Fire Code inspection requirement, the city and county need to establish a program to identify the locations and activities of all small-quantity generators (including conditional exempt SQGS) of hazardous wastes and include them in the Fire Code database.

Identifying, Monitoring, and Evaluating Suspected Groundwater Contamination Events

The city and county need to identify, monitor, and evaluate sites where groundwater contamination is suspected or known to have occurred. This includes identifying the party or parties suspected of causing the contamination so that they can be made to remedy the pollution or so that cleanup costs may be borne by those responsible.

Monitoring Groundwater Quality and Soil Vapor at Landfills

The city and county need to expand on the present program to monitor groundwater quality and soil vapor at landfills by establishing prioritized groundwater and vadose-zone monitoring for landfills closed before April 14, 1989 (which are not subject to current New Mexico Solid Waste Management requirements). Prioritization depends on the type of landfill: municipal solid waste landfills will be monitored first, followed by landfills allegedly containing only construction/demolition debris, followed by dumps.

Supporting Enforcement

Through a Memorandum of Understanding, the city and county should work with the state to identify additional NMED staff needs and funding options for increased inspection and enforcement for Groundwater Discharge Plans and RCRA facilities.

Enforcing Crucial Area and Wellhead Protection Area Overlay Zoning

This aspect of the program involves enforcing the protection measures related to activities in crucial areas and wellhead protection areas. Certain activities will be prohibited or restricted in these areas; for example, new landfills and hazardous waste disposal facilities will be prohibited.

Identifying French Drains and Detention Basins

The Federal Clean Water Act regulations cover stormwater flows that may discharge to groundwater. These regulations apply in areas where there is a hydraulic connection between ground and surface waters. This effort involves identifying the location of french drains and detention basins and determining whether pollutants are likely to reach groundwater at the sites.

Continuing and Assessing Adequacy of the Permanent Household Hazardous Waste Collection Program

This effort involves continuing the household hazardous waste collection center established in July 1992 and assessing the need for expanding this service.

Determine Best Management Practices and Hazardous Waste Minimization Techniques

This effort involves the determination of appropriate best management practices and waste minimization techniques for a wide variety of waste-generating activities. The city and county will solicit the assistance of professional societies such as the New Mexico Hazardous Waste Management Society. In addition, the city and county may solicit the cooperation of the Waste Management Education and Research Consortium (which includes New Mexico State University, the University of New Mexico, the New Mexico Institute of Mining and Technology, Sandia National Laboratories, and Los Alamos National Laboratory) in this effort.

The implementation of a comprehensive groundwater protection program such as this will require the coordination and cooperation of many governmental jurisdictions. Specifically, Albuquerque and Bernallio County need to coordinate and cooperate with surrounding counties, communities, pueblos, and reservations to encourage their participation in protecting the regional groundwater resource.

Once this GPPAP is fully adopted, the city and county should involve adjacent jurisdictions to encourage them to adopt similar groundwater protection policies and measures.

Certain activities included in the Water-Resources Management Plan are critical to groundwater protection. These include the following:

- developing the computer models needed to assess how local pollution events threaten the regional water supply
- developing and implementing a regional groundwater monitoring program
- developing and implementing a water conservation program

Assessing the Impacts of Local Contamination

To evaluate the potential impacts of the identified groundwater contamination (for example, the widespread septic-tank contamination in the North and South Valleys) on the quality of the groundwater used for drinking water, the city needs to incorporate new information that has refined the conceptual understanding of the hydrogeologic conditions that control what happens to pollutants in groundwater. This new information needs to be assimilated into a computer model that will allow city staff to assess the potential impacts of relying exclusively on local groundwater for water supply.

Regional Groundwater Monitoring

Existing monitoring programs being conducted by the city and county will be integrated and expanded into a comprehensive regional program. A regional groundwater monitoring program is required to understand the natural variability in the aquifer and to determine where groundwater quality is being impacted. Strategically located monitoring wells will identify impacts associated with most of the groundwater threats. Where possible, data from existing public and private wells should be used, reducing costs.

Additionally, a surface-water drain monitoring program will help identify potential water-quality impacts caused by agricultural practices. Staff must determine the numbers, locations, and construction details of monitoring wells needed to characterize regional groundwater quality, in coordination with the U.S. Geological Survey, the NMED, and others, and choose the locations of drains to be sampled.

Water Conservation

A water conservation program is required to guide the use of the region's water resources. The final report of the Water Conservation Task Force identified short-term measures that, when implemented, will lead to wiser use of the resource. The Task Force also identified long-term measures to protect the permanent supply but recommended that the long-term measures not be implemented until the long-term supply has been quantified.

Consistent with the short-term recommendations of the Task Force, the low-flow plumbing requirements described in this policy have the secondary benefit of improving the functioning and reducing the costs of on-site liquid waste disposal systems.

Policy implementation requires an informed public and their active involvement. The city and county will develop a program to assure public participation in the regulatory process, educate the public, and provide technical assistance to the related community.

The public involvement and education program will include the following:

- establishing an outreach program to effectively educate schoolchildren
- establishing a public participation program to obtain public comment as regulations and ordinances are being developed to implement the GPPAP
- preparing informational handouts
- communicating and interacting with electronic and print media
- making material available by creating a centralize information clearinghouse
- disseminating the material through repositories such as libraries and field offices and public presentations

Technically qualified representatives from interested and affected agencies will provide input to the various work products. Elected and appointed officials will also be called upon to participate.

- teaching the regulated community about the new requirements
- soliciting the involvement of professional societies (such as the New Mexico Hazardous Waste Management Society) to develop best management practices, recycling, waste minimization strategies, and product substitution options
- providing necessary expert technical advice on how to comply with the new requirements

Conclusions

The review of numerous federal, state, and local laws, regulations, and ordinances shows that many of the regulations and ordinances deal only peripherally with hazardous materials and waste storage and siting and groundwater protection. A few provide comprehensive requirements covering some aspects for storage of hazardous materials, hazardous wastes, and other potential pollutants. However, even collectively, the reviewed regulations and ordinances do not include all components necessary for groundwater protection.

This suggests that even had the existing laws and regulations related to groundwater protection been in place before the existing contamination in Bernalillo County occurred, they would not have prevented many of these contamination cases. Albuquerque and Bernalillo County are not the first communities to recognize this, and many have implemented their own programs to fill the gaps in the existing regulatory work.

The Federal Resource Conservation and Recovery Act (RCRA) is the most detailed set of regulations covering "cradle-to-grave" management of hazardous wastes. RCRA also contains provisions for underground storage tanks, medical wastes, and solid wastes. The state is authorized to implement many of the RCRA provisions with U.S. Environmental Protection Agency (EPA) oversight.

The 1988 Uniform Fire Code (UFC), adopted by the City in 1990, provided comprehensive guidelines on storage, use, dispensing, and handling of hazardous materials. The code was designed to prevent accidents involving hazardous materials (which by code definition includes hazardous wastes) when implemented in accordance with its specifications. The improved 1991 UFC, adopted by the county and the city in 1993, contains improved measures. The version adopted by the city includes many of the amendments required by the Hazardous Materials and Waste Storage and Siting Policy— a principal component of the GPPAP.

Other regulations cover only narrow aspects of hazardous materials and waste storage and siting, treatment, disposal, and transportation. These include the reporting requirements under the Comprehensive Environmental Response, Compensation, and Liability Act (CERCLA) and the Superfund Amendments and Reauthorization Act (SARA), the Hazardous Materials Transportation Act, and provisions of the Atomic Energy Act (Radiation Protection Standards, etc.). Moreover, CERCLA focuses on remediation of contamination, not prevention of contamination. Even here, its effectiveness has been questioned.

A number of regulations and ordinances reviewed did not contain specific language relative to hazardous materials and waste siting and storage or groundwater protection. This includes the Metropolitan Environmental Health Advisory Board Ordinance, ordinances establishing the Air Quality Control Board, and the Public Nuisance Provision of the State Criminal Code. Others, such as the National Environmental Policy Act (NEPA) and the Sole-Source Aquifer provisions of the Safe Drinking Water Act, come into play only when Federally funded projects are involved.

State Hazardous Waste Act Regulations, Solid Waste Act Regulations, and Underground Storage Tank Regulations are similar to Federal RCRA Regulations, but lack of adequate resources almost certainly limits their effectiveness. The State Water Quality Control Commission (WQCC) Regulations provide a flexible framework, but government's resources to NMED have again been inadequate given the magnitude of the problem.

Albuquerque Public Works Department et al. "Groundwater Protection and Action Plan," As adopted by the Board of County Commissioners and City Council (1995).

American Society of Civil Engineers (ASCE). *Ground Water Management*. New York, American Society of Civil Engineers (1987).

Anonymous. "How Did We Get In This Mess," *The Bulletin of the Atomic Scientists*, pp. 65, May–June (1965).

Anonymous. "Rocky Mountain Arsenal: Landmark Case of Groundwater Polluted by Organic Chemicals, and Being Cleaned Up," *Civil Engineering*, 51:68–71 (1981).

Anonymous. "Savannah River Site Ground Water Cleanup Reaches 2 Billion Gallons," Editor, *Ground Water Monitoring Review*, 15(1):29–30 (1995).

Anonymous. "Environmental Cleanup; Progress in Resolving Long-Standing Issues at the Rocky Mountain Arsenal," National Technical Information Service, AD-A308 706/1NEG, 22 pp. (1996).

Anzzolin, A. R., Siedlecki, M. and J. Lloyd. "The Challenge of Ground Water Quality Monitoring," *Ground Water Monitoring & Remediation*, 19(2):57–60 (1999).

Archey, C. and C. Mawson. "Municipal Watershed Management, A Unique Opportunity in Massachusetts," *Journal NEWWA* (June 1984).

Arnade, L. J. "Seasonal Correlation of Well Contamination and Septic Tank Distance," *Ground Water*, 37(6):920–923 (1999).

Atomic Energy Commission. "Effluent and Environmental Monitoring and Reporting," AEC Manual, Chapter 0513. Atomic Energy Commission, Washington, D.C. (1973).

Austin, T. "Federal Cleanups: Good News from Bad," *Civil Engineering*, March, 48 pp. (1994).

Austin, T. "Partnering with the Enemy." *Civil Engineering*, March, 40 pp. (1995).

Aziz, M. et al. "Laboratory Studies for Prediction of Radionuclide Migration in Groundwater," *Radioactive Waste Management and Environmental Restoration*, 18:243–256 (1994).

Baca, E. "On the Misuse of the Simplest Transport Model," *Ground Water*, 37(4):483 (1999).

Bacon, M. J. and W. A. Oleckno. "Groundwater Contamination: A National Problem with Implications for State and Local Environmental Health Personnel," *Journal of Environmental Health*, 48(3):116–121 (1985).

Bagtzoglou, A. C. et al. "Application of Particle Methods to Reliable Identification of Groundwater Pollution Sources," *Water Resources Management*, 6:15–23 (1992).

Bagtzoglou, A. C. et al. "Groundwater Quality Management of a Low Inertia Basin: Application to the San Mateo Basin, California," *Water Resources Management*, 7:189–205 (1993).

Baik, M. H. et al. "Effect of Chelating Agents on the Migration of Radionuclides," *Transactions of the American Nuclear Society*, 64(10):160 (1991).

Barber, M. J. "The Federal Regulation of Groundwater," *Environmental Permitting*, Winter 2(1):103–110 (1992).

Barcelona, M. J. "Beyond BTEX," *Ground Water Monitoring & Remediation*, (19(1):4–6 (1999).

Barney, G. S. et al. "Removal of Plutonium from Low-Level Process Wastewaters by Absorption," in Environmental Remediation: Removing Organic and Metal Ion Pollutants, Vandegrift, G. F. Reed, D. T., and I. R. Tasker, editors, American Chemical Society, Washington, D.C. (1991).

Barrett, K. R. "Ecological Engineering in Water Resources: The Benefits of Collaborating with Nature," *IWRA, Water International*, 24(3):182–188 (1999).

Batchelor, B. "A Framework for Risk Assessment of Disposal of Contaminated Materials Treated by Solidification/Stabilization," *Environmental Engineering Science*, 14(1):3–13 (1997).

Berg, R. C., Curry, B. B. and R. Olshansky. "Tools For Groundwater Protection Planning: An Example from McHenry County, Illinois, USA," *Environmental Management*, 23(3):321 (1999).

Biswas, K. A. "Environmental Impact Assessment for Groundwater Management," *Water Resources Development*, 8(2):113–117 (1992).

Bjorklund, L. W. and B. W. Maxwell. "Availability of Groundwater in the Albuquerque Area, Bernalillo and Sandoval Counties, New Mexico." New Mexico State Engineer Technical Report 21, 117 pp. (1961).

Blair, S. and W. W. Wood. "A Civil Action—What Will Be the Legacy of Wells G and H?" *Ground Water*, Vol. 37(2):161 (1999).

Borton, R. L. "Bibliography of Groundwater Studies in New Mexico, 1873–1977." New Mexico State Engineer Special Publication, 121 pp. (1978).

Borton, R. L. "Bibliography of Groundwater Studies in New Mexico, 1848–1979." New Mexico State Engineer Special Publication, 46 pp. (1980).

Borton, R. L. "Bibliography of Groundwater Studies in New Mexico, 1903–1982, a supplement to Bibliography of Groundwater Studies in New Mexico, 1873–1977." New Mexico State Engineer Special Publication, 84 pp. (1983).

Bredehoeft, J. "A New Paradigm for Cleanup," *Ground Water*, 34(4):577 (1996).

Brusseau, M. L. "Transport of Reactive Contaminants in Heterogeneous Porous Media," *Reviews of Geophysics*, 32(3):285–313 (1994).

Bugai, D. A. et al. "Risks from Radionuclide Migration to Groundwater in the Chernobyl 30-KM," *Health Physics*, 71(1):9–18 (1996).

Bullard, C. W. et al. "Managing the Uncertainties of Low-Level Radioactive Waste Disposal," *Journal of the Air & Waste Management Association*, 48(8):701–710 (1998).

Burger, J. "How Should Success be Measured in Ecological Risk Assessment? The Importance of Predictive Accuracy." *Journal of Toxicology and Environmental Health*, 42:367–376 (1994).

Burke, J. J., Sauveplane, C. and M. Moench. "Groundwater Management and Socioeconomic Responses," *Natural Resources Forum*, 23(4):303 (1999).

Butler, G. C. *Principles of Ecotoxicology*. New York:Wiley (1978).

Campbell, A. and T. Z. C. "Less Regulation More Enterprise," *The Bulletin of the Atomic Scientists*, May–June, pp. 45–46 (1995).

Canter, L. W. and R. C. Knox. *Ground Water Pollution Control*. Chelsea, MI:Lewis Publishers, Inc., 525 pp. (1986).

Canter, L. W. and K. M. Maness. "Groundwater Contaminants and Their Sources—A Review of State Reports," *Intern. J. Environmental Studies*, 47:1–17 (1995).

Chapin, R. E. et al. "Toxicology Studies of a Chemical Mixture of 25 Groundwater Contaminants (III. Male Reproduction Study in B6C3F₁ Mice)." *Fundamental and Applied Toxicology*, 13:388–398 (1989).

Chiang, P. D., Petkovsky, P. H. and P. M. McAllister. "A Risk-Based Approach for Managing Hazardous Waste," *Ground Water Monitoring Reviews*, 15(1):79–89 (Winter 1995).

Christakos, G. and D. T. Hristopulos. "Stochastic Indicators for Waste Site Characterizaton," *Water Resources Research*, 32(8):2563–2578 (August 1996).

Christensen, C. W. et al. "Soil Adsorption of Radioactive Wastes at Los Alamos," *Sewage and Industrial Wastes*, 30(12):1478–1489 (1958).

Chung, Y. J. and K. J. Lee. "Applying the Filtration Equation to Radioactive Colloid Transport," *Transactions of the American Nuclear Society*, 64(1):159 (1991).

Colten, C. E. "Groundwater and the Law: Records and Recollections," *The Public Historian*, 20(2):25 (1999).

Cornaby, B. W. et al. "Application of Environmental Risk Techniques to Uncontrolled Hazardous Waste Sites," *Proc. of the Nat. Conf. on Management of Uncontrolled Hazardous Waste Sites, 1982*, Hazardous Materials Control Research Inst., Silver Spring, MD, pp. 390–395 (1982).

Cressman, K. R. "Cost Components of Remedial Investigation/Feasibility Studies," *Cost Engineering*, 33(7):25–29 (1991).

Crittenden, J. C. et al. "Sun Fuels Groundwater Remediation," *Water Environment and Technology*, 7(2):15–16 (1995).

Crowley, K. D. "Nuclear Waste Disposal: The Technical Challenges," *Physics Today*, 50(6):32–39 (1997).

Datskou, I. and K. North. "Risks due to Groundwater Contamination at a Plutonium Processing Facility," *Water and Soil Pollution*, 90(1–2):133–141 (July 1996).

De La Cruz, S. and E. Pena. "Method to Improve Water Resources Management in Groundwater Pumping Areas and a Case Study," *Water Resources Development*, 10(3):329–337 (1994).

DeSena, M. "TOPAZ Uses Innovative Strategies to Provide Landscape Drainage Data for Hydrological Models," *Water Environment & Technology*, 11(7):22 (1999).

Devarakonda, M. and M. Seiler. "Radioactive Wastes," *Water Environment Research* 67(4):585–596 (1995).

Eklund, W. W. "Federal Facilities and Federal Groundwater Law," National Technical Information Service, AD-A318 869/5NEG, 76 pp. (1996).

Elliot, C. N., Dunbar, M. J. and M. C. Acreman. "A Habitat Assessment Approach to the Management of Groundwater Dominated Rivers," *Hydrological Processes*, 13(3):459 (1999).

Farrara, R. A. et al. *Ground Water Contamination from Hazardous Wastes*. Princeton University Water Resources Program, Englewood Cliffs, NJ:Prentice Hall, Inc. (1984).

Federal Register. Hazardous Waste Management System: Identification and Listing of Hazardous Waste; Proposed Rule. Washington, D.C., U.S. EPA (1992). Also see, *Federal Register*, Vol. 57, No. 130, Notices 29871 (1992).

Feldman, D. L. and R. A. Hanahan. "Public Perceptions of a Radioactively Contaminated Site: Concerns, Remediation Preferences, and Desired Involvement," *Environmental Health Perspectives*, 104(12):1344–1352 (1996).

Fine II, R. L. "Remediation of Contaminated Soil and Groundwater Using Air Stripping and Soil Venting Technologies," *Colorado Engineering*, 9(3):41–43 (1991).

Gass, T. E. "Ground Water in the News," *Ground Water*, 23:148–149 (1985).

Gehringer, P. et al. "Remediation of Groundwater Polluted with Chlorinated Ethylenes by Ozone-Electron Beam Irradiation Treatment," *App. Radiat. Isot.*, 43(9):1107–1115 (1992).

Germolec, D. R. et al. "Toxicology Studies of a Chemical Mixture of 25 Groundwater Contaminants (II. Immunosuppression in B6C3F Mice)," *Fundamental and Applied Toxicology*, 13:377–387 (1989).

Gibbons, R. D., Dolan, D. G., May, H., O'Leary, K. and R. O'Hara. "Statistical Comparison of Leachate from Hazardous, Codisposal, and Municipal Solid Waste Landfills," *Ground Water Monitoring & Remediation*, 19(4):57–72 (1999).

Gershey, E. L. et al. *Low-Level Radioactive Waste: From Cradle to Grave*. New York, NY.:Van Nostrand Reinhold Publishers (1990).

Gerty, M. et al. *"History and Geophysical Description of Hazardous Waste Disposal Area A Technical Area 21."* LA-11591-MS (1989).

Gertz, C. P. and P. L. Cloke. "Site Characterization at the Potential High-Level Radioactive Waste Repository Site at Yucca Mountain, Nevada." Environmental Sciences General, *Transactions of the American Nuclear Society*, 69:58 (1993).

Goldblum, D. K. et al. "Use of Risk Assessment Groundwater Model in Installation Restoration Program (IRP) Site Decisions," *Environmental Progress*, 11 (2):91–97 (1992).

Gorelick, S. M. et al. *Groundwater Contamination: Optimal Capture and Containment*. Boca Raton, FL: Lewis Publishers (1993).

Gray, R. *Environmental Monitoring, Restoration, and Assessment: What Have We Learned?* Twenty-Eighth Hanford Symposium on Health and the Environment. R. Gray, ed., Pacific Northwest Laboratory, Richland, WA (1990).

Gupta, D. A. and P. R. Onta. "Groundwater Management Models for Asian Developing Countries," *Water Resources Development*, 10(4):457–473 (1994).

Haas, C. N. "Editorial: The Risk of Over-Reliance on Risk Assessment," *Water Environment Research*, 67(1):1 (1996).

Hakonson, T. E. et al. *"Ecological Investigation of Radioactive Materials in Waste Discharge Areas at Los Alamos (for the Period July 1, 1972 through March 31, 1973)."* LA-5282-MS (1973).

Hale, W. E. et al. "Characteristics of the Water Supply in New Mexico," Tech. Rep. 31, New Mexico State Engr. and U.S. Geological Survey, Santa Fe, NM (1965).

Hallenbeck, W. H. "Risk Analysis of Exposure to Radium-226/228 in Groundwater," *The Environmental Professional*, 11:171–177 (1989).

Hamed, M. and P. Bedient. "On the Performance of Computational Methods for the Assessment of Risk from Groundwater Contamination," *Ground Water*, 35(4):638–646 (1997).

Haque, R. et al. *Dynamic, Exposure and Hazard Assessment of Toxic Chemicals*. Ann Arbor, MI:Ann Arbor Science, Pub., Inc. (1980).

Harris, B. B. "Reducing the Risk of Groundwater Contamination by Improving Livestock Holding Pen Management," Texas Agricultural Extension Service, Texas A & M University System. B-6031 (1997a).

Harris, B. B. "Reducing the Risk of Groundwater Contamination by Improving Wellhead Management and Condition," Texas Agricultural Extension Service, Texas A & M University System. B-6024. (1997b).

Harris, B. B. "Reducing the Risk of Groundwater Contamination by Improving Hazardous Waste Management," Texas Agricultural Extension Service, Texas A & M University System. B-6028 (1997c).

Hall, D. H. "Ground Contamination, Impacts on Groundwater, and Uncertainty," *The Nuclear Engineer*, 40(3):97 (1999).

Harris, B. B. "Reducing the Risk of Groundwater Contamination by Improving Fertilizer Storage and Handling," Texas Agricultural Extension Service, Texas A & M University System. B-6026 (1997d).

Harris, B. B. "Reducing the Risk of Groundwater Contamination by Improving Petroleum Product Storage," Texas Agricultural Extension Service, Texas A & M University System. B-6027 (1997e).

Harris, B. B. "Reducing the Risk of Groundwater Contamination by Improving Milking Center Wastewater Treatment," Texas Agricultural Extension Service, Texas A & M University System. B-6032 (1997f).

Harris, B. B. "Reducing the Risk of Groundwater Contamination by Improving Livestock Manure Storage and Treatment Facilities," Texas

Agricultural Extension Service, Texas A & M University System. B-6030 (1997g).

Hartley, W. R., Englande Jr., A. J. and D. J. Harrington. "Health Risk Assessment of Groundwater Contaminated with Methyl Tertiary Butyl Ether (MTBE)," *Water Science and Technology*, 39(10/11):305 (1999).

Hawley, J. W. and C. S. Haase. "Hydrological Framework of the Northern Albuquerque Basin," Socorro, New Mexico Bureau of Mines and Mineral Resources Open-File Report 387, pp. IX-7 (1992).

Heindel, J. et al. "Assessment of the Reproduction and Developmental Toxicity of Pesticide/Fertilizer Mixtures Based on Confirmed Pesticide Contamination in California and Iowa Groundwater," *Fundamental and Applied Toxicology*, 22:605-621 (1994).

Heindel, J. et al. "Assessment of the Reproductive Toxicity of a Complex Mixture of 25 Groundwater Contaminants in Mice and Rats," *Fundamental and Applied Toxicology*, 25:9-19 (1995).

Higley, K. A. and Geiger, R. A. "*Environmental Monitoring at U.S. Department of Energy Facilities*," in *Environmental Monitoring, Restoration, and Assessment*, R. H. Gray, ed., U.S. Department of Energy (DOE) Washington, D.C. and Pacific Northwest Laboratory, Richland, WA (1990).

Hinds, J. J., Shemin, G. and C. Fridrich. "Numerical Modeling of Perched Water Under Yucca Mountain, Nevada," *Ground Water*, 37(4):498-504 (1999).

Hofsetter, K. J. "Continuous Monitoring for Tritium in Aqueous Effluents at SRS Using Solid Scintillators," *Transactions of the American Nuclear Society*, 29:26 (1993).

Hong, H. L. et al. "Residial Damage to Hematopoietic System in Mice Exposed to a Mixture of Groundwater Contaminants," *Toxicology Letters*, 57:101-111 (1991).

Horseley, S. W. "California's New Ground Water Management Law," *Ground Water Monitoring Reviews*, 14(4):114-115 (1995a).

Horseley, S. W. "Comprehensive State Ground Water Protection Programs: A Preliminary Examination of Two States," *Ground Water Monitoring Review*, 14(1):71-72 (Winter 1995b).

Hoskins, B. et al. "Variation in the Use of Risk-Based Groundwater Cleanup Levels at Petroleum Release Sites in the United States," *Human and Ecological Risk Assessment*, 3(4):521-535 (1997).

Hurst, B. H. "Removing Groundwater Contaminants Through Irrigation," *Irrigation Journal*, 45(50):24-25 (1995).

Huyakorn, P. et al. "An Improved Sharp-Interface Model for Assessing NAPL Contamination and Remediation of Groundwater Systems," *Journal of Contaminant Hydrology*, 16:203-234 (1994).

Hwang, P. L. et al. "Bimodal Filtration Coefficient for Radiocolloid Migration in Porous Media," *Transactions of the American Nuclear Society*, 64(10):160 (1991).

Icenhour, A. S. et al. "Low-Level Radioactive Waste Performance Assessments: Source Term Modeling," *Transactions of the American Nuclear Society*, 72(1):57 (1995).

Illman, D. "Hanford Tank Farm Safety, Monitors Found Lacking," *C&EN*, March 1, pp. 22 (1993).

Institution of Civil Engineers. *Nuclear Contamination of Water Resources*. Telford House, 1 Heron Quay, London E149XF. Thomas Telford Ltd. Publishers (1990).

International Atomic Energy Agency. *Disposal of Radioactive Wastes*. Proceedings of the Scientific Conference on the Disposal of Radioactive Wastes. Iaea, Vienna:STI/PUB/18, Austria (1959).

International Technology Corporation. *Surface Gravity Survey for Fault Delineation and Hydrogeologic Characterization*, Unpublished Report for Sandia National Laboratory Environmental Restoration Division (1992).

IWRA, Water International. "Special Section on Water Resources and the Internet," *IWRA, Water International*, 24(2):126-175 (1999).

Jacob, T. L. et al. "2nd Moment Method for Evaluating Human Health Risks from Groundwater Contamination by Trichloroethylene," *Environmental Health Perspectives*, 104(8):866-870 (August 1996).

Jones, J. R. "The Clean Water Act: Groundwater Regulation and the National Pollutant Discharge Elimination System," *Dickinson Journal Of Environmental Law & Policy*, 8(1):93 (1999).

Josephson, J. "Groundwater Strategies," *Environmental Science and Technology*, 14(9):1031-1032 (1980).

Jury, W. A., and K. Roth, "Transfer Functions and Solute Movement Through Theory and Applications," Birkhauser, Verlag, Basel, 226 pp. (1990).

Kaplan, E. and W. F. McTernan. "Overview of the Risk Assessment Process in Relation to Groundwater Contamination," *The Environmental Professional*, 15:334-340 (1993).

Keller, J. F. "*Regulatory Requirements for Groundwater Monitoring Networks at Hazardous-Waste Sites*," in *Environmental Monitoring, Restoration, and Assessment: What Have We Learned?* R. H. Gray, ed., U.S. Department of Energy, Batelle, Pacific Northwest Laboratory, Richland, WA (1990).

Kelley, V. C. "Geology of Albuquerque Basin, New Mexico." Socorro, New Mexico Bureau of Mines and Mineral Resources Memoir 33, 60 pp. (1977).

Kernodle, J. M. et al. "Three-Dimensional Model Simulation of Transient Groundwater Flow in the Albuquerque-Belen Basin, New Mexico." U.S. Geological Survey Water-Resources Investigations Report 90-4037, 35 pp. (1987).

Kernodle, J. M. and W. B. Scott. "Three-Dimensional Model Simulation of Steady-State Groundwater Flow in the Albuquerque-Belen Basin, New Mexico." U.S. Geological Survey Water-Resources Investigations Report 84-4353, 58 pp. (1986).

Knox, C. et al. *Subsurface Transport and Fate Processes*. Boca Raton, FL: Lewis Publishers (1993).

Korte, N. E. et al. "The Inadequacy of Commonly Used Risk Assessment Guidance for Determining Whether Solvent-Contaminated Soils Can Affect Groundwater at Arid Sites," *J. Environ. Sci. Health*, A27(8):2251-2261 (1992).

Kufs, C. et al. "Rating the Hazard Potential of Waste Disposal Facilities," *Proceedings of the National Conference on Management of Uncontrolled Hazardous Wastes Sites*, Hazardous Materials Control Research Inst., Silver Spring, MD, pp. 30-41 (1980).

Kuo Chin-Hwa, et al. "Design of Optimal Pump-and-Treat Strategies for Contaminated Groundwater Remediation Using the Simulated Annealing Algorithm," *Advances in Water Resources*, 15:95-105 (1992).

Lantzy, R. J. et al. "Use of Geographical Maps to Manage Risk from Groundwater Contamination," *Journal of Hazardous Materials*, 61(1-3):319-328 (1998).

Lee, K. K., Kim, K. R. and S. H. Cho. "Evaluation of Groundwater Contamination from Glass Fiber Dumping at Gozan-Dong, Incheon, Korea," *Environmental Pollution*, 104(3):459 (1999).

Lee, M. "The Low-Down on Groundwater," *Environmental and Planning Law Journal*, 16(3):265 (1999).

Lesage, S. and R. Jackson. *Groundwater Contamination and Analysis at Hazardous Waste Sites*. New York, NY:Marcel Dekker, Inc. Publ. (1992).

Leusink, A. "The Planning Process for Groundwater Resources Management," *Water Resources Development*, 8(2):98-102 (1992).

Long, J. "National Performance Review Spurs Clinton's New Round of Budget Cuts," *C&EN*, Jan. 2, 17 pp. (1995).

Los Alamos National Laboratory. "*Environmental Restoration and Waste Management Five-Year Plan; Site-Specific Plan*," Health, Safety, and Environment Division, Los Alamos, NM (1990).

Los Alamos National Laboratory. "*Work Plan for Operational Unit—Technical Area 21*," Environmental Restoration Program (EM-13; Reading Room), Los Alamos, NM (1992).

Los Alamos National Laboratory. "*General Employee Radiological Training*," ES&H Course 8530, Los Alamos, NM (1993).

Los Alamos National Laboratory. "*Groundwater Protection Management Program Plan*." Water Quality and Hydrology Group (ESH-18) (1990, 1995).

Los Alamos National Laboratory. "*Environmental Surveillance at Los Alamos*," Environmental Protection Group, LA-11306-ENV, LA-12000-ENV, LA-122271- MS, and related series, Los Alamos, NM (1987-1997).

MacDonald, J. A. "Cleaning Up the Nuclear Weapons Complex," *Environmental Science & Technology*, 33(15):314 (1999).

Macdonald, J. A. and M. C. Kavanaugh. "Restoring Contaminated Groundwater an Achievable Goal?" *Environ. Sci. Technol.*, 28(8):362A (1994).

Maxwell, R. M., Permulder, S. D. and W. E. Kastenberg. "On the Development of a New Methodology for Groundwater Driven Health Assessment," *Water Resources Research*, 34(4):833 (1999a).

Maxwell, R. M., Kastenberg, W. E. and Y. Rubin. "A Methodology to Integrate Site Characterization Information Into Groundwater Driven Health Risk Assessment," *Water Resources Research*, 35(9):2841 (1999b).

Mays, C. W. et al. "Cancer Risk from the Lifetime Intake of Ra and U Isotopes," *Health Physics*, 44:635–648 (1985).

McCabe, W. J. et al. "History of the Sole Source Aquifer Program: A Community-Based Approach for Protecting Aquifers Used for Drinking Water Supply," *Ground Water Monitoring and Remediation*, 17(3):78–86 (1997).

McCord, J. T. et al. *"Detailed Mapping and Preliminary Geostatistical Analysis of Alluvial Fan Deposits Exposed in Six Miles of Trench*, Sandia National Laboratories, SAND93-0680 (1993).

McKee, J. E. et al. "Gasoline in Groundwater," *Journal Water Pollution Control Federation*, 44:293–302 (1972).

McKone, T. E. and K. T. Bogen. "Predicting the Uncertainties in Risk Assessment," *Environ. Sci. Technol.*, 25(10):1674–1681 (1991).

Missimer, T. M. "The Search for Groundwater Contamination: Discovery, Verification, and Remediation," *Environmental Permitting*, 2(1):91–102 (Winter 1992).

Monogham, G. W. and G. J. Larson. "A Computerized Ground-Water Resources Information System," *Ground Water*, 23:233–239 (1985).

Moorehouse, J. L. "Groundwater Protection: A Contrast in State Style," *Water Engineering and Management*, pp. 23–25 (March 1985).

Moriarity, F. *Ecotoxicology: The Study of Pollutants in Ecosystems*. New York, London: Academic Press (1983).

Morrison, A. "If Your City's Well Water Has Chemical Pollutants, Then What?" *Civil Engineering*, 51(9):65–67 (1981).

Muntzing, L. M. and J. C. Person. "Environmental Remediation and Waste Management in the United States," *Proceedings 9th Pacific Basin Nuclear Conference*, Sydney, Australia, 1–6 May (1994).

National Academy of Sciences (NAS). *Report to the U.S. Atomic Energy Commission: Disposal of Radioactive Waste on Land*. Washington, D.C.:National Academy of Sciences (1957).

National Research Council. *Drinking Water and Health, Vol(s). 1–5*. Washington, D.C.:National Academy Press (1977–1983).

National Research Council. *Pesticides and Groundwater Quality*, Board of Agriculture, Washington, D.C.:National Academy Press (1986a).

National Research Council. *Groundwater Quality Protection, State and Local Strategies*. Washington, D.C.:National Academy Press (1986b).

Naturman, L. "DOE's Budget: Is It Getting the Message?, Editorial, *Energy*, 20(3):6 (1995).

Neel, D. and J. P. McCord. *"Summary of May–November 1992 Field Operations South Fence Road Project."* Unpublished Report for Sandia National Laboratory Environmental Restoration Division (1993).

Nelson, K. A. and R. C. Janke. "Establishing a Comprehensive Risk Assessment Document of Fernald," *Transactions of the American Nuclear Society*, 72(1):59 (1995).

Nie, N. H. *SPSS-X, UsersGuide*. New York, NY:McGraw-Hill Book Co. (1983).

Nuttall, H. E. and R. Kale. "Remediation of Toxic Particles from Groundwater," *Journal of Hazardous Materials*, 37:41–48 (1994).

Nyhan, J. W. et al. "Distribution of Plutonium and Americium Beneath a 33-yr-old Liquid Waste Disposal Site," *Journal Environmental Quality*, 14(40):501–509 (1985).

Nyler, E. K., Carman, E. P. and R. M. Flynn. "Other Types of Contamination," *Ground Water Monitoring & Remediation*, 19(2):61–64 (1999).

Office of Technology Assessment. *Protecting the Nations Groundwater from Contamination*, U.S. Congress, Office of Technology Assessment, OTA-0-233, Washington, D.C.:U.S. Government Printing Office (1984).

Office of Technology Assessment. *An Evaluation of Options for Managing Greater-than-Class-C Low-Level Radioactive Waste*, Washington, D. C.: U.S. Government Printing Office (1988).

Office of Technology Assessment. *Partnerships Under Pressure: Managing Commercial Low-Level Radioactive Waste*. OTA-0-426 Washington D.C.:U. S. Government Printing Office (1989).

Olsen, R. L. and M. C. Kavanaugh. "Can Groundwater Restoration be Achieved?" *Water Environment & Technology*, 5(3):42–47 (1993).

Olsthoorn, T. N. "The Power of the Electronic Worksheet: Modeling Without Special Programs," *Ground Water*, 34:381–390 (1985).

O'Neill, R. V. et al. "Ecosystem Risk Analysis: A New Methodology," *Environmental Toxicology and Chemistry*, 1:167–177 (1982).

Paasivirta, J. *Chemical Ecotoxicology*. Chelsea, MI:Lewis Publ. Inc. (1991).

Penrose, W. R., et al. "Mobility of Plutonium and Americium through a Shallow Aquifer in a Semiarid Region," *Environ. Sci. Technol.*, 24(2):228–234 (1990a).

Penrose, W. R. et al. "Mortandad Canyon Studies," *Environ. Sci. Technol.*, 24:228 (1990b).

Petts, G. E., Bickerton, M. A. and D. Evans. "Flow Management To Sustain Groundwater-Dominated Stream Ecosystems," *Hydrological Processes*, 13(3):497 (1999).

Pillay, K.K. "Environmental Safety and Health Vulnerabilities of Plutonium at the Los Alamos National Laboratory," *Transactions of the American Nuclear Society*, 72(1):56 (1995).

Piontek, K. "Science for Non-Scientists: Current Trends in Groundwater Remediation," Journal of Environmental Law & Practice, 6(3):58 (1999).

Pohll, G., Hassan, A. E., Chapman, J. B., Papelis, C. and R. Andricevic. "Modeling Ground Water Flow and Radioactive Transport in a Fractured Aquifer," *Ground Water*, 37(5):770–784 (1999).

Psilovikos, A. A. "Optimization Models in Groundwater Management, Based on Linear and Mixed Integer Programming: An Application to a Greek Hydrological Basin," *Physics and Chemistry of Earth*, 24(1/2):139 (1999).

Rail, C. D. "Groundwater Monitoring Within an Aquifer—A Protocol," *Journal of Environmental Health*, 48(3):128–132 (1985a).

Rail, C. D. *Plague Ecotoxicology: Including Historical Aspects of the Disease in the Americas and the Eastern Hemisphere*. Springfield, IL:Charles C. Thomas, Publ. (1985b).

Rail, C. D. *Summarization of Water Quality Data (Bernalillo County 1960–1976) the Inorganic Ions*. Environmental Services Division, Environmental Health Dept., City of Albuquerque, Volumes I and II, unpublished manuscripts, 700 pp. (1986).

Rail, C. D. *Groundwater Contamination: Sources, Control, and Preventive Measures*. 1st edition, Lancaster, PA.:Technomic Publishing Co., Inc. (1989).

Rail, C. D. *"Joint Environmental, Safety, and Health Plan - TA-21 - Buildings 3 & 4 (South) Decommissioning, the Original Uranium-Plutonium Facility at Los Alamos, NM*, unpublished manuscript, JCI Environmental, Safety, and Health, Los Alamos, NM (1992).

Ramade, F. *Ecotoxicology*. 2nd edition, Paris:Masson (1979).

Rayson, G. D. et al. "Recovery of Toxic Heavy Metals From Contaminated Groundwaters," *Radioactive Waste Management and Environmental Restoration*, 18:99–108 (1994).

Reddy, K. R., et al. "A Review of *In-Situ* Air Sparging for the Remediation of VOC-Contaminated Saturated Soils and Groundwater," *Hazardous Waste & Hazardous Materials*, 12(2):97–118 (1995).

Reeder, H. O. et al. "Quantitative Analysis of Water Resources in the Albuquerque Area, New Mexico—Computed Effects on the Rio Grande of Pumpage Groundwater, 1960–2000." New Mexico State Engineer Technical Report 33, 34 pp. (1967).

Rezendes, V. "Reinventing the Energy Department: The World is not the Same as it Was When the Agency Was Created in 1977," *Energy*, 20(3): 4–6 (1995).

Richte, E. and J. Safi. "Pesticide Use, Exposure, and Risk: A Joint Israeli-Palestinian Perspective," *Environmental Research*, 73(1–2):211–218 (1997).

Rogers, M. A. "History and Environmental Setting of LASL Near-Surface Land Disposal Facilities for Radioactive Wastes (Areas A, B, C, D, E, F, G, and T)," Los Alamos National Laboratory, LA-6848-MS Vol. I (1977a).

Rogers, M. A. "History and Environmental Setting of LASL Near-Surface Land Disposal Facilities for Radioactive Wastes (Areas A, B, C, D, E, F, G, and T)," Los Alamos National Laboratory, LA-6848-MS Vol. II (1977b).

Rogers, D. B. and B. M. Gallaher. "The Unsaturated Hydraulic Characteristics of the Bandalier Tuff," Los Alamos National Laboratory, LA-12968-MS (1995).

Rothstein, L. "Nothing Clean about Cleanup," *The Bulletin of the Atomic Scientists*, May/June (1995).

Rowe, W. D. "Superfund and Groundwater Remediation: Another Perspective," *Environ. Sci. Technol.*, 25(3):370–371 (1991).

Rushton, K. R. "Groundwater Aspects: Losses are Inevitable but Re-Use Is Possible?" *Agricultural Water Management*, 40(1):111 (1999).

Sandhu, S. "Trace Element Distribution in Various Phases of Aquatic Systems of the Savannah River Plant," *in Environmental Remediation: Removing Organic and Metal Ion Pollutants*, Vandegrift, G. F., D. T. Reed, and I. R. Tasker, editors, American Chemical Society, Washington, D.C. (1991).

Sandia National Laboratory. *"Site-Wide Hydrogeologic Characterization Project."* Calendar Year Annual Report. Environmental Restoration Program (1992).

SAS Institute, Inc. *SAS Users Guide*. Raleigh, NC:Spark Press (1979).

Schintu, M., Koussih, L. and J. M. Robert. "Monitoring of Labile Zinc in Cultures of Skeletonema Costatum Using a Groundwater Salt," *Ecotoxicology And Environmental Safety*, 42(3):207 (1999).

Schuller, T. A. et al. "Groundwater Modeling for an NPL Risk Assessment," *Environmental Toxicology and Chemistry*, 11:1355–1362 (1991).

Shanklin D. E. et al. "Micro-Purge Low-Flow Sampling of Uranium-Contaminated Ground Water at the Fernald Environmental Management Project," *Ground Water Monitoring Reports*, 15(3):169–175 (1995).

Shen, Y. "In-Vitro Cytotoxicity of BTEX Metabolites in Hela-Cells," *Archives of Environmental Contamination and Toxicology*, 34(3): 229–234 (1998).

Sherwood, D. R. et al. *"Identification of Contaminants of Concern in Hanford Groundwaters"* in *Environmental Monitoring, Restoration, and Assessment: What Have We Learned*, Twenty-Eighth Hanford Symposium on Health and the Environment, R. H. Gray, ed., U. S. Department of Energy and Batelle, Pacific Norhtwest Laboratory, Richland, WA (1990).

Shimp, J. F. et al. "Beneficial Effects of Plants in the Remediation of Soil and Groundwater Contaminated with Organic Materials," *Critical Reviews in Environmental Science and Technology*, 23(1):41–77 (1993).

Shirley, P. A. "Use of STORET as a Data Base for Ground-Water Quality Management," *Proc. Sixth National Ground Water Quality Symposium, NWWA, Worthington, OH* (1982).

Shukla, S. et al. "A Risk-Based Approach For Selecting Priority Pesticides for Groundwater Monitoring Programs," *Transactions of the ASAE*, 39(4):1379–1390 (1996).

Simmons, J. E. et al. "Toxicology Studies of a Chemical Mixture of 25 Groundwater Contaminants: Hepatic and Renal Assessment, Response to Carbon Tetrachloride Challenge, and Influence of Treatment-Induced Water Restriction," *Journal of Toxicology and Environmental Health*, 43:305–325 (1994).

Slough, W. et al. "Margins of Uncertainty in Ecotoxicological Hazard Assessment," *Environmental Toxicology and Chemistry*, 5:841–852 (1986).

Steel, R. G. and J. H. Torrie. *Principles and Procedures of Statistics*. New York, NY:McGraw-Hill Book Co. (1960).

Stone, W. J. and N. H. Mizell. "Availability of Geophysical Data for the Eastern Half of the U.S. Geological Survey's Southwestern Alluvial Basin Regional Aquifer-System Study," Socorro, New Mexico Bureau of Mines and Mineral Resources Open-File Report 109, 80 pp. (1979).

Swenson, E. "Public Trust Doctrine and Groundwater Rights," *University of Miami Law Review*, 53(2):363 (1999).

Sun, M. and C. Zheng. "Long-Term Groundwater Management by a MODFLOW Based Dynamic Optimization Tool," *Journal of the American Water Resources Association*, 35(1):99 (1999).

Teng, S. H. and C. H. Lee. "Improved Technique for Estimating Parameters of Diffusion Experiments," *Transactions of the American Nuclear Society*, 64(10):163 (1991).

Thomson, B. M. "Radioactive Wastes," *Research Journal WPCF*, 63(4): 510–518 (1991).

Thomson, B. M. "Radioactive Wastes," *Water Environment Research*, 64(4):479–472 (1992).

Thorn, C. R. et al. "Geohydrologic Framework and Hydrologic Conditions in the Albuquerque Basin, Central New Mexico." U.S. Geological Survey Water-Resources Investigations Report 93-4149, 106 pp. (1993).

Toran, L. "Radionuclide Contamination in Groundwater: Is There a Problem?" in *Groundwater Contamination and Control*, U. Zoller, ed., Marcel Dekker, Inc. (1993).

Trelease, F. J. *Water Law: Resource Use and Environmental Protection*. St Paul, MN:West Pub. Co. (1974).

Truhaut, R. "Ecotoxicology—A New Branch of Toxicology," in *Ecological Toxicology Research*, A. D. McIntyre and C. F. Mills, eds., *Proc. NATO Science Comm. Conf Mt. Gabriel, Quebec, May 6–10, 1974*. 323 pp., New York:Plenum Press (1975).

Truhaut, R. "Ecotoxicology: Objectives, Principles, and Perspectives," *Ecotoxicology and Environmental Safety*, 1:151 (1977).

Tuinhoff, A. "Organization and Management of Groundwater Planning: Implementation and Feedback," *Water Resources Development*, 8(2):118–125 (1992).

Tung, Y. K. and G. E. Koltermann. "Some Computational Experiences Using Embedded Techniques for Ground-Water Management," *Ground Water*, 23:455–464 (1985).

Tykva, R. and J. Sabol. *Low-Level Environmental Radioactivity*. Lancaster, PA: Technomic Publishing Co., Inc. (1995).

U.S. Army Corps of Engineers, "Special Flood Hazard Information, Tijeras Arroyo and Arroyo de Coyote, Kirtland AFB," New Mexico, U.S. Army Corps. of Engineers, Albuquerque District, Albuquerque, NM (1979).

U.S. Department of Energy. "The Development and Production of Nuclear Weapons: A Summary Including Organizations, Procedures, and Interfaces." Informational Purposes Only), Washington, D.C. (1984).

U.S. Department of Energy. "Data Base for 1988: Spent Fuel and Radioactive Waste Inventories, Projections, and Characteristics," DOE/RW-0006, Rev. 4, Washington, D. C. (1988a).

U.S. Department of Energy. "Environmental Survey Preliminary Report Los Alamos National Laboratory, Los Alamos, NM," Environment, Safety and Health Office of Environmental Audit, DOE/EH/OEV-12-P (1988b).

U.S. Department of Energy. "Environmental Restoration and Waste Management (Five Year Plan; Fiscal Years 1992–1996, Executive Summary)," DOE/S-0077P or NTIS-PR-360 (1990a).

U.S. Department of Energy. "Closure of Hazardous and Mixed Radioactive Waste Management Units at DOE Facilities," Environmental Guidance. Office of Environmental Guidance. RCRA/CERCLA Division EH-23 (1990b).

U.S. Department of Energy. "Environmental Restoration and Waste Management Five-Year Plan; FY 1993–1997 Plan Guidance," Albuquerque Operations Office, Environmental Management Staff, Albuquerque, NM (1990c).

U.S. Department of Energy. "Applicability of Land Disposal Restrictions to RCRA and CERCLA Groundwater Treatment Reinjections," Memorandum, March 8 (1990d).

U.S. Department of Energy. "Transporting Radioactive Materials: Answers to Your Questions," Environmental Restoration and Waste Management. DOE/EM-0097, April (1993a).

U.S. Department of Energy. "U.S. Department of Energy Interim Mixed Waste Inventory Report: Waste Streams, Treatment Capacities and Technologies," DOE/NBM-1100 (1993b).

U.S. Department of Energy. "Department of Energy 1977–1994," Human Resources and Administration, Energy History Series, November (1994a).

U.S. Department of Energy. "The Manhattan Project: Making the Atomic Bomb," Human Resources and Administration, Energy History Series, September (1994b).

U.S. Department of Energy. "Environmental Management: Fact Sheets," Office of Environmental Management, August (1994c).

U.S. Department of Energy. "Clinton Administration Releases Domestic Natural Gas and Oil Plan," *The Landman,* 39(1):25–28 (1994d).

U.S. Department of Energy. "Implementation Plan for the Programmatic Environmental Impact Statement for the Department of Energy UMTRA Ground Water Project," DOE/AL62350-26 (1994e).

U.S. Department of Energy. "Closing the Circle on the Splitting of the Atom." The Environmental Legacy of Nuclear Weapons Production in the United States and What the Department of Energy is Doing About It. Office of Environmental Management, January (1995a).

U.S. Department of Energy. "U. S. Department of Energy Environmental Justice Strategy, Executive Order 12898," U. S. DOE, April (1995b).

U.S. Department of Energy. "Estimating the Cold War Mortgage: The 1995 Baseline Environmental Management Report," Vol. I, March (1995c).

U.S. Department of Energy. "Estimating the Cold War Mortgage: The 1995 Baseline Environmental Management Report," Vol. II, March (1995d).

U.S. Department of Energy. "Contaminated Plumes Containment and Remediation Focus Area," Technology Summary, Office of Environment Management Technology Development, DOE/EM-0248, June (1995e).

U.S. Department of Energy. "Landfill Stabilization Focus Area," Technology Summary, Office of Environment Management Technology Development, DOE/EM-0251, June (1995f).

U.S. Department of Energy. "Decontamination and Decommissioning Focus Area," Technology Summary, Office of Environment Management Technology Development, DOE/EM-0253, June (1995g).

U.S. Department of Energy. "Characterization, Monitoring, and Sensor Technology Crosscutting Program," Technology Summary, Office of Environment Management Technology Development, DOE/EM-0254, June (1995h).

U.S. Department of Energy. "Mixed Waste Characterization, Treatment, and Disposal Focus Area," Technology Summary, Office of Environment Management Technology Development, DOE/EM-0252, June (1995i).

U.S. Department of Energy. "Radioactive Tank Waste Remediation Focus Area," Technology Summary, Office of Environment Management Technology Development, DOE/EM-0255, June (1995j).

U.S. Department of Energy. "Robotics Technology Crosscutting Program," Technology Summary, Office of Environment Management Technology Development, DOE/EM-0250, June (1995k).

U.S. Department of Energy. "Efficient Separations and Processing Crosscutting Program," Technology Summary, Office of Environment Management Technology Development, DOE/EM-0249, June (1995l).

U.S. Department of Energy. "Draft: Programmatic Environmental Impact Statement for the Uranium Mill Tailings Remedial Action Ground Water Project," DOE/EIS-0198 (1995m).

U.S. Department of Energy. "Spent Nuclear Fuel Management," Draft Environmental Impact Statement, DOE/EIS-0279D, Savannah River Operations Office, Aiken, S.C. (1998).

U.S. Department of Energy. "Environmental Assessment for the Proposed Construction and Operation of the Non-Proliferation and International Security Center," DOE-EA-1238, Los Alamos National Laboratory, Los Alamos, N.M., U.S. DOE Los Alamos Area Office (1999).

U.S. Department of the Interior. Geothermal Leasing Program. NTIS Accession No. PB 203 102-D, 156 pp., Washington, D.C. (1971).

U.S. Environmental Protection Agency. "Quality Criteria for Water," prepublication copy (1976a).

U.S. Environmental Protection Agency. "National Interim Primary Drinking Water Regulations," EPA-570/9-76-003. Office of Water Supply, Washington, D.C. (1976b).

U.S. Environmental Protection Agency. "Monitoring Groundwater Quality: Economic Framework and Principles," Environmental Monitoring and Support Laboratory. Office of Res. and Dev. EPA-600/4-76/045, Las Vegas, NV (1976c).

U.S. Environmental Protection Agency. "Water-Related Environmental Fate of 120 Priority Pollutants," Vols. I and II. EPA 440/4-79-029a (1979).

U.S. Environmental Protection Agency. "Rapid Assessment of Potential Ground-Water Contamination Under Emergency Response Conditions,"

Office of Health and Environmental Assessment, EPA-600/8-83-030, Washington, D.C. (1983).

U.S. Environmental Protection Agency. "Best Management Practices for Agricultural Nonpoint Source Control," IV. Pesticides, Office of Research and Development, Washington, D.C. ES-NWQEP-84/02 (1984a).

U.S. Environmental Protection Agency, Committee on the Challenges of Modern Society (NATO/CCMS): Drinking Water Microbiology, *NATO/CCMS Drinking Water Series,* EPA 570/9-84-006 (1984b).

U.S. Environmental Protection Agency. "Ground-Water Protection Strategy," Office of Ground Water Protection, Washington, D.C. (1984c).

U.S. Environmental Protection Agency. *Septage Treatment and Disposal Handbook.* EPA 625/6-84-009, Cincinnati, OH (1984d).

U.S. Environmental Protection Agency. *Practical Guide to Groundwater Sampling.* Robert S. Kerr Environmental Research Laboratory, EPA/600/2-85/104, Ada, OK (1985).

U.S. Environmental Protection Agency. *Pesticides in Ground Water: Background Document.* Office of Ground Water Protection (WH-550G), Washington, D.C. (1986a).

U.S. Environmental Protection Agency. Summary of State Reports on Releases from Underground Storage Tanks. Office of Underground Storage Tanks, EPA 600/M-86/020 (1986b).

U.S. Environmental Protection Agency. *"Guidelines for Groundwater Classification under the EPA Groundwater Protection Strategy,"* Washington D.C. (1986c).

U.S. Environmental Protection Agency. *Proposed Regulations for Underground Storage Tanks: What's in the Pipeline?* Office of Underground Storage Tanks, Washington, D.C. (1987a).

U.S. Environmental Protection Agency. Nitrate/Nitrite, and N-nitroso Compounds, Washington, D.C. USEPA. Office of Drinking Water (1987b).

U.S. Environmental Protection Agency. *"Uncontrolled Hazardous Waste Site Ranking System."* A Users Manual, HW-10, 40 CFR Part 300, 300.86, appendix A, pp. 55–84 (1988).

U.S. Environmental Protection Agency. "Transport and Fate of Contaminants in the Subsurface," Technology Transfer, EPA/625/4-89/019, September (1989a).

U.S. Environmental Protection Agency. "Injection Well Mechanical Integrity." Office of Research and Development. Washington, D.C. EPA/625/9-89/007 (1989b).

U.S. Environmental Protection Agency. "Environmental Indicators of Water Quality in the United States," Office of Water. Washington, D.C. EPA 841-F-96-002 (1996a).

U.S. Environmental Protection Agency. "Groundwater Disinfection Rule: Workshop on Predicting Microbial Contamination of Groundwater Systems, July 10–11, 1996: Proceedings Report," U.S. EPA, Office of Groundwater and Drinking Water (1996b).

U.S. Environmental Protection Agency. "Is Someone Contaminating Your Drinking Water?" EPA (800/K-96/900) (1996c).

U.S. Geological Survey. "Ground-Water Geochemistry of the Albuquerque-Belen Basin, Central New Mexico," *Water-Resources Investigations Report 86-4094* (1988).

U.S. Geological Survey. "Simulation of Groundwater Flow in the Albuquerque Basin, Central New Mexico, 1901–1994. With Projections to 2020," U.S.G.S. Water-Resources Investigations Report 94-4251 (1995).

Valenti, M. "Taming Hanford's Most Troublesome Nuclear Waste Tank," *Mechanical Engineering,* pp. 68–72 (November 1993).

Vandermeulen, J. H. and S. E. Hrudey. *Oil in Freshwater: Chemistry, Biology and Countermeasure Technology.* Proceedings of a Symposium on Freshwater Oil Pollution. Alberta, Canada: Pergamon Press (1987).

Van der Molen, W. H. "Technical Aspects of Groundwater Management," *Water Resources Development,* 8(2):103–112 (1992).

Vandijk, H. F. and F. A. Dehaan. "Risks of Pesticides to Groundwater Ecosystem," *Human and Ecological Risk Assessment,* 3(2):151–155 (May 1997).

Vogel, J. "Old Landfill Problems Require New Solution," *Public Works,* 130(9):32 (1999).

Walker, B. "The Present Role of the Local Health Departments in Environmental Toxicology," *Journal of Environmental Health,* 48(3):133–137 (1985).

Williams, D. "DOE Shows of Budget Scars; Late Breaking News," *Pollution Engineering,* 27(8):3, (1995).

Wise, H. F. "Policy Implications of Urban Land Practices for Groundwater Quality," *Water and Sewage Works,* 84–85 (1977).

Wright, A. F. "Bibliography of the Geology and Hydrology of the Albuquerque Greater Urban Area, Bernalillo and Parts of Sandoval, Santa Fe, Socorro, Torrance, and Valencia Counties, New Mexico." U.S. Geological Survey Bulletin 1458, 31 pp. (1978).

Yang, R. H. and E. J. Rauckman. "Toxicological Studies of Chemical Mixtures of Environmental Concern at the National Toxicology Program: Health Effects of Groundwater Contaminants," *Toxicology,* 47:15–34 (1987).

Yang, R. S. et al. "Toxicological Studies of a Chemical Mixture of 25 Groundwater Contaminants," *Fundamentals and Applied Toxicology,* 13:366–376 (1989).

Yosie, T. F. "EPAs Risk Assessment Culture," *Environmental Science and Technology,* 21(6):526–531 (1987).

URL Internet Hyperlink Reference Numbers

Internet Hyperlinks [Universal Resource Locators (URLs)][1] and Reference Numbers Related to *Groundwater Contamination*

[1] GROUNDWATER ONLINE (NATIONAL GROUNDWATER ASSOCIATION)[2,3]
   http://www.ngwa.org
   http://www.ngwa.org/gwonline/index.html
   http://www.ngwa.org/about/index.html
   http://www.ngwa.org/publication/pubmenu.html
   http://www.ngwa.org/publication/bookrev.html

[2] NATIONAL GROUNDWATER ASSOCIATION LINKS
   http://www.ngwa.org/links/links.html
   http://www.groundwatersystems.com/

[3] THE WATER LIBRARIAN'S HOME PAGE
   http://www.wco.com/~rteeter/waterlib.html [Home]
   http://www.tec.org/tec/terms2.html [Encyclopedia of Water Terms]
   http://www.wqa.org/WQIS/Glossary/GlossHome.html [Glossary of Water Terms]
   http://lcweb.loc.gov/homepage/lchp.html [Library of Congress Home]
   http://www.melvyl.ucop.edu/ [University of California—Melvyl System]

[4] USGS WRSIC Research Abstracts—1967 to October 1993
   http://www2.uwin.siu.edu/databases/wrsic [Universities Water Information Network]

[5] USGS Selected Water Resources Abstracts—USGS Authors, Reports only, 1977–Present
   http://water.usgs.gov/public/swra/index.html [Selected Abstracts]
   http://water.usgs.gov/public/swra/help.html [Instructions for Searching]

http://www2.uwin.siu.edu/databases/wrsic/search.html [Keyword Search]

[6] PACIFIC NORTHWEST LABORATORY AND HYDROLOGY WEB
   http//www.pnl.gov/ [Home]
   http://terrassa.pnl.gov:2080/Hydrology/about.html
   http://www.pnl.gov/links.html [Links]
   http://www.pnl.gov/science.html [Science and Technology]
   http://www.pnl.gov/ecology/Index.html [Ecology Group]

[7] U.S. ENVIRONMENTAL PROTECTION AGENCY'S OFFICE OF WATER
   http://www.epa.gov/ow [Home]
   http://www.epa.gov/owow [Office of Wetlands, Oceans, and Watersheds]
   http://www.epa.gov/ow [Links]
   http://www.epa.gov/OW/programs.html [Water Programs]
   http://www.epa.gov/OGWDW/ [Groundwater and Drinking Water]

[8] U.S. GEOLOGICAL SURVEY WATER RESOURCES DIVISION
   http://h2o.usgs.gov/index.html [Water Resources of the United States]
   http://water.usgs.gov/techr.html [Technical Resources]
   http://water.usgs.gov/programs.html [State and Regional Programs]
   http://water.usgs.gov/pandp.html [Publications and Products]

[9] AMERICAN WATER RESOURCES ASSOCIATION
   http://www.awra.org
   http://www.awra.org/jawra/index.html
   http://www.awra.org/connections/assoc.html
   http://www.awra.org/proceedings/proceedings.html
   http://www.awra.org/proceedings/paper.html

[10] AMERICAN WATER WORKS ASSOCIATION
   http://www.awwa.org

[11] CANADIAN WATER RESOURCES ASSOCIATION
   http://www.cwra.org/cwra

[12] GROUNDWATER AND THE INTERNET
   http://gwrp.cciw.ca/internet

[1] **Examples of URLs:**
http://www.w3.org/default.html; http://www.acme.co.uk:8080/images/map.gif; http://wombat.doc.ic.ac.uk/?Uniform+Resource+Locator; ftp://wuarchive.wustl.edu/mirrors/msdos/graphics/gifkit.zip; ftp://spy:secret@ftp.acme.com/pub/topsecret/weapon.tgz; mailto:Santocrail@aol.com; news:alt.hypertext; telnet://dra.com
[2] Some *URLs* might have changed or the system may be down when accessed. If this is the case, use a Search Engine such as http://www.metacrawler.com and search for the general subject as presented in this *URL* listing. Other Search Engines can also be evaluated at: http://www.Albany.net/allinone/alllwww.html. At the time of the writing of this manuscript, all of the WWW pages were accessible.
[3] Ideally, these URLs should be accessed from a diskette or hard drive that is connected to the Internet WWW via the use of a word processing program such as Microsoft® Word with the *Find* function that can search for the [x] or {x}.

http://gw2.cciw.ca/internet/online.html
http://gw2.cciw.ca/internet/servers.html
http://gw2.cciw.ca/internet/software.html

[13] HYDROGEOLOGIST'S HOME PAGE—Data Sources, Software, Web Links

http://www.ems.psu.edu/Hydrogeologist
http://www.ems.psu.edu/Hydrogeologist/org_chrt.htm
http://www.ems.psu.edu/Hydrogeologist/search.htm
http://www.ems.psu.edu/Hydrogeologist/pubs.htm
http://www.ems.psu.edu/Hydrogeologist/gengeo.htm

[14] HYDROLOGY PAGE from the U.S. Geological Survey

http://www.usgs.gov/network/science/earth/water.html
http://wwwrvares.er.usgs.gov/nawqa/index.html
http://www.mines.edu/research/igwmc/
http://www.hwr.arizona.edu/hydro_link.html
http://www.hwr.arizona.edu/globe/h2oissues.html
http://hydrolab.arsusda.gov/

[15] SEEPAGE/GROUNDWATER MODELING SOFTWARE, Links to Websites

http://www.et.byu.edu/~asce-gw

[16] U.S. WATER NEWS ONLINE

http://www.uswaternews.com
http://www.uswaternews.com/arc96-97.html
http://www.uswaternews.com/links.html
http://www.uswaternews.com/news.html

[17] WATER ON-LINE, Putting California Water Information on the Net

http://ceres.ca.gov/theme/water_resources.html
http://wwwdwr.water.ca.gov/dir-CA_water_infoR2/GroundwaterR2.html
http://wwwdpla.water.ca.gov/cgi-bin/supply/gw/main.pl
http://wwwdwr.water.ca.gov/
http://ceres.ca.gov/
http://ceres.ca.gov/watershed/
http://ceres.ca.gov/wetlands/
http://ceres.ca.gov/topic/env_law/newsletters/1995water.html

[18] WATER WORLD

http://waternet.com

[19] U.S. ENVIRONMENTAL PROTECTION AGENCY HOME PAGE

http://www.epa.gov
http://www.epa.gov/epahome/research.htm
http://www.epa.gov/epahome/Programs.html
http://www.epa.gov/epahome/locate3.htm
http://www.epa.gov/epahome/program2.htm
http://www.epa.gov/epahome/general.htm
http://www.epa.gov/epahome/locate3.htm#ow
http://www.epa.gov/OST/
http://www.epa.gov/OST/pubs/

[20] Federal Government Agencies VIRTUAL LIBRARY

http://www.lib.lsu.edu/gov/fedgov.html

[21] FEDERAL LEGISLATION from Thomas (Library of Congress)

http://thomas.loc.gov
http://thomas.loc.gov/home/thomas2.html
http://thomas.loc.gov/r105/r105.html

[22] GOVERNMENT INFORMATION from Fedworld

http://www.fedworld.gov
http://www.fedworld.gov/detail.htm#general
http://www.fedworld.gov/detail.htm#search

[23] U.S. GEOLOGICAL SURVEY WATER RESOURCES INFORMATION

http://h2o.usgs.gov
http://h2o.usgs.gov/index.html
http://water.usgs.gov/techr.html
http://water.usgs.gov/programs.html
http://water.usgs.gov/pandp.html
http://wwworegon.wr.usgs.gov/pubs_dir/twri-list.html
http://toxics.usgs.gov/toxics/pubs/pubs.shtml
http://www.nap.edu/readingroom/reader.cgi?auth=free&label=ul.book.NI000131
http://www.usgs.gov/pubprod/

[24] U.S. Geological Survey's BAY AND DELTA Page

http://bard.wr.usgs.gov/Access/Access-sfb.html

[25] DESERT RESEARCH INSTITUTE

http://www.dri.edu/Library
http://www.dri.edu/Library/pubs/
http://www.dri.edu/Library/pubs/bulletin/
http://www.dri.edu/Library/epubs/
http://nevada.usgs.gov/biblio/bibsearch.html
http://www.dri.edu/Library/other/

[26] U.S. Environmental Protection AGENCY LIBRARIES

http://ww.epa.gov/natlibra/index.html
http://www.epa.gov/natlibra/overback.htm
http://www.epa.gov/natlibra/policy.htm
http://www.epa.gov/natlibra/ols.htm
http://www.epa.gov/natlibra/liblists.html
http://www.epa.gov/natlibra/moreinfo.htm

[27] U.S. Geological Survey LIBRARY

http://library.usgs.gov
http://library.usgs.gov/svcspol.html#Begin
http://library.usgs.gov/specoll.html#Begin
http://ngmdb.usgs.gov/
http://library.usgs.gov/onlinext.html#Begin

[28] Developing a CORE WATER COLLECTION, a List of Useful Resources

http://www.wco.com/~rteeter/watrcore.html
http://www.epa.gov/natlibra/core/water.htm
http://www.englib.cornell.edu/eld/publications.html
http://www.nal.usda.gov/wqic/biblios.html
http://www.nal.usda.gov/wqic/
http://www.nal.usda.gov/wqic/#1
http://www.nal.usda.gov/wqic/aboutwq.html

[29] AMERICAN WATER RESOURCES ASSOCIATION (AWRA)

http://www.unin.siiu.edu/~awra

[30] American Water Works Association (AWWA) PUBLICATIONS

http://www.awwa.org/asp/pubs/asp

[31] Geraghty and Miller's Water Information Center

http://www.gmgw.com/

[32] U.S. Environmental Protection Agency OFFICE OF WATER PUBLICATIONS

http://www.epa.gov/OW/pubs.html
http://www.epa.gov/ow/programs.html
http://www.epa.gov/OGWDW/
http://www.epa.gov/OGWDW/orgcht3.html
http://www.epa.gov/OGWDW/regs.html
http://www.epa.gov/ogwdw/swp/gwpgrt.html
http://www.epa.gov/OGWDW/rdp.html
http://www.epa.gov/OGWDW/ncod/ncod.html

http://www.epa.gov/OGWDW/standard/occsel.html

http://www.epa.gov/OGWDW/sdwa/contamin.html

[33] SEARCH U.S. Geological FORMAL REPORTS—Bulletins, Professional Papers, Circulars, Water Supply

http://greenwood.cr.usgs.gov/formal/reports.html

http://greenwood.cr.usgs.gov/bulletin.html

http://greenwood.cr.usgs.gov/propaper.html

http://greenwood.cr.usgs.gov/circular.html

http://greenwood.cr.usgs.gov/wsp.html

http://greenwood.cr.usgs.gov/thmaps.html

[34] WATER RESOURCES PUBLICATIONS

http://www.waterplus.com/srp/index.html

[35] NATIONAL GROUNDWATER ASSOCIATION LINKS

http://www.ngwa.org/links/links.html

http://www.ngwa.org/links/links.html#Othersites

http://www.ngwa.org/publication/pubmenu.html

http://www.groundwatersystems.com/

http://www.ngwa.org/publication/gwmrinfo.html

http://www.ngwa.org/publication/wwjinfo.html

[36] Groundwater REMEDIATION TECHNOLOGIES Analysis Center

http://www.gwrtac.org

http://www.gwrtac.org/html/about.html

http://www.gwrtac.org/html/techdocs.html

http://www.gwrtac.org/html/tech_status.html

[37] CENTER FOR GROUNDWATER STUDIES (AUSTRALIA)

http://www.cls.csiro.au/CGS

[38] POLLUTION ONLINE

http://www.pollutiononline.com

http://news.pollutiononline.com/month_in_review.html

http://news.pollutiononline.com/wisewire/wisewire_intro.html

[39] WATER WISER—A Cooperative Project of the AWWA, the U.S. EPA, and the U.S. Bureau of Reclamation

http://www.waterwiser.org

http://www.waterwiser.org/books.html

http://www.waterwiser.org/wwlinks.html

http://www.waterwiser.org/forums/main.cfm?CFID=17880&CFTOKEN=5193&CFApp=44&

[40] BRITISH COLUMBIA GROUNDWATER ASSOCIATION

http://www.drilshop.com/bcgwa

[41] CALIFORNIA GROUNDWATER ASSOCIATION

http://www.drilshop.com/bcgwa

[42] NATIONAL GROUNDWATER ASSOCIATION FORUM

http://www.ngwa.org/suresite/forum.html

http://www.ngwa.org/gwonline/index.html

[43] PRINCETON Groundwater

http://www.princeton-groundwater.com

http://www.princeton-groundwater.com/brochdr.htm

http://www.princeton-groundwater.com/training.htm

http://www.flowpath.com/

http://www.princeton-groundwater.com/othersit.htm

[44] WATER SCIENCE FOR SCHOOLS

http://wwwga.usgs.gov/edu

http://wwwga.usgs.gov/edu/mwater.html

http://wwwga.usgs.gov/edu/mearth.html

http://wwwga.usgs.gov/edu/wateruse.html

http://wwwga.usgs.gov/edu/specials.html

http://wwwga.usgs.gov/edu/msac.html

http://wwwga.usgs.gov/edu/dictionary.html

http://wwwga.usgs.gov/edu/links.html

http://water.usgs.gov/education.html

http://water.usgs.gov/public/education.html

[45] U.S. DEPARTMENT OF ENERGY

http://www.em.doe.gov [Environmental Management Home Page]

http://www.doe.gov/ [DOE Home Page]

http://ende.lbl.gov/EE.html [Environmental Energy Technologies Division]

http://www.lanl.gov/Internal/projects/IPO/DTIN/open/labtitl.html [DOE Labs and Facility Services]

http://doe-is.llnl.gov/ [DOE Information Security Home Page]

http://www.bnl.gov/bnl.html [Brookhaven National Laboratory]

http://www.ohre.doe.gov/ [DOE Human Radiation Experiments]

http://terrassa.pnl.gov:2080/DFE/ [Office of Pollution Prevention]

http://www.tis.eh.doe.gov/websites/websites.html [Websites]

http://gpo.osti.gov:901/dds/easy.html [Easy Search]

[46] USGS GROUNDWATER ATLAS of the United States Index

http://wwwcapp.er.usgs.gov/publicdoes/gwa

[47] U.S. EPA OFFICE OF RESEARCH AND DEVELOPMENT, Including National Center for Environmental Research and Quality Assurance

http://www.epa.gov [EPA Home]

http://www.epa.gov/OWOW/watershed/tools/model.html [Modeling Tools]

http://www.epa.gov/ORD/WebPubs/stratplan/ [Update to Office Research and Development Plan]

http://www.camerata.net/eed/reports.htm [EPA Benefit Cost Analysis]

http://www.epa.ohio.gov/opp/tanbook/fppgbgn.html [EPA Ohio]

http://www.epa.gov/ORD/WebPubs/final/ [Research Plans and Strategies]

http://www.epa.gov/docs/ORD/ [ORD Science Network]

http://www.epa.gov/ORD/ [Science Network]

http://www.epa.gov/attic/index.html

[48] GROUNDWATER MONITORING AND REMEDIATION

http://www.ngwa.org/publication/pubmenu.html [National Groundwater Association Publications]

http://www.library.wisc.edu/li . . . s/Water_Resources/wrrsjml.htm [Library University Wisconsin]

http://www.rti.org/units/ese/cemqa/geosci/bat.html [BAT Groundwater Monitoring System]

http://www.ntis.gov/fcpc/cpn6938.htm [Groundwater Monitoring Training CD-ROM]

http://www.epa.gov/epaoswer/non-hw/muncpl/gwm.htm [Groundwater Monitoring]

http://www.dep.state.pa.us/dep . . . SrceProt/GrdMonitor/tblcnt.htm

http://www.emt.com/ [Environmental Monitoring and Technologies, Inc]

http://www.scvwd.dst.ca.us/wtrqual/wqgwqm.htm [Water Quality Groundwater Monitoring Network]

http://www.rcgrd.uvm.edu/ [Research Center for Groundwater Remediation Design]

http://www.sci-sms.com/gw-monitoring.htm [Groundwater Monitoring Software]

http://www.hydromodels.com/awdb.htm [Groundwater Database Software] http://www.groundsearch.co.nz/Groundwater.htm [Subsurface Imaging Services]

[49] Underground Tank Technology Update—Newsletter

http://epdwww.engr.wisc.edu/uttu

[50] Waterloo Hydrogeologic—Developers of Groundwater Software

http://www.flowpath.com

http://www.doe.ca/water/en/nature/grdwtr/e_gdwtr.htm [Groundwater - Canada]

[51] USGS Water Resources of the United States

http://water.usgs.gov

[52] USGS Groundwater Information Pages

http://water.usgs.gov/ogw [USGS Groundwater Information Pages]

http://webserver.cr.usgs.gov/ [Colorado Project]

http://water.usgs.gov/ [Water Resources of the United States]

http://h2o.usgs.gov [Water Resources of the United States]

http://www.usgs.gov/network/science/earth/usgs.html [Index of USGS Web Servers]

http://www-nmd.usgs.gov/www/gnis/gnisform.html [Geographic Names]

http://info.er.usgs.gov/research/gis/title.html [Geographic Information Systems]

http://water.wr.usgs.gov/ [USGS Headquarters]

http://water.wr.usgs.gov/mine/ [Mine Drainage Interest Group]

http://www.usgs.gov/network/index.html/ [Internet Resources]

http://svr1dutslc.wr.usgs.gov/ [USGS—UTAH]

[53] USGS WATER RESOURCES DIVISION—NEW MEXICO

http://wwwdnmalb.cr.usgs.gov [New Mexico District]

http://water.usgs.gov/public/pubs/FS/FS-031-96/ [Programs in New Mexico]

http://www.riogrande.org/putting/organize.htm [Links to Border-Related and Environmental]

http://geology.cr.usgs.gov/states/NM.html [USGS Information about New Mexico]

http://www.cabq.gov/resources/links.html [City of Albuquerque Water Conservation Links]

[54] USGS Local Office for Water Resources

http://water.usgs.gov/public/srd002.html

[55] USGS MIDDLE RIO GRANDE BASIN STUDY

http://rmmcweb.cr.usgs.gov/public/mrgb/home.html [Home]

http://rmmcweb.cr.usgs.gov/public/mrgb/mrgb_extent.html [Extent of Basin]

http://rmmcweb.cr.usgs.gov/public/mrgb/mrgb_projects.html [Agencies Involved]

http://rmmcweb.cr.usgs.gov/public/mrgb/mrgb_references.html [References]

[56] ACCESS EPA

http://earth.epa.gov/Access

http://www.epa.gov/oms/olstelnt.htm [Telnet Access to the EPA Online System]

http://www.epa.gov/epahome/finding.html [EPA Other Resources]

http://epawww.ciesin.org/national/epahome/epahome.html

http://es.epa.gov/ncerqa/rfa/empact.html [International Earth Science Network

http://www.epa.gov/epahome/search.html [Search the EPA Internet]

http://www.epa.gov/records/tools/info_acc/index.htm [Records Network Access]

http://www.epa.gov/records/policy/schedule/sched/608a.htm [Records Control Schedule]

http://nsdi.epa.gov/epahome/search.hints.html [Public Access Server Search Hints]

http://www.epa.gov/reinvent/notebook/elpa.htm [Electronic Public Access]

http://www.epa.gov/epahome/about.html [About the EPA Public Access Server]

http://epainotes1.rtpnc.epa.gov:7777/ [Internet Support]

http://www.epa.gov/enviro/index_java.html [Envirofacts Warehouse]

http://atsdr1.atsdr.cdc.gov:8080/superfnd.html [Superfund Related]

http://es.epa.gov/oeca/idea/epa.html [Getting Started]

[57] U.S. EPA GOPHER Server

http:gopher://gopher.epa.gov/

http://www.envirosw.com/agcyfed.html [EPA WWW Server]

http://www.einet.net/GJ/federal.html [Federal Agencies and Related Gopher Sites]

http://www.epa.gov/oppe/custrep/contacts.html [Additional Information]

http://www.swapca.org/resources.html [Environmental Links]

http://www.englib.cornell.edu/eld/listserv/9508/9508.26.html [EPA Chemical Information]

[58] GROUNDWATER.COM LINKS DIRECTORY

http://www.groundwater.com/links3.html

http://www.groundwater.com/links1.html

http://www.groundwater.com/links2.html

[59] SAGE—Solvent Alternatives Guide (U.S. EPA)

http://clean.rti.org

http://clean.rti.org/other.htm

http://clean.rti.org/tools.htm

http://clean.rti.org/links.htm

[60] AMES LABORATORY Environmental Technology Development (ETD)

http://www.edt.ameslab.gov

[61] U.S. DOE, OFFICE OF ENVIRONMENTAL MANAGEMENT (EM)

http://www.em.doe.gov [Environmental Management]

http://www.doe.gov [Home Page]

http://eande.lbl.gov/EE.html [Environmental Energy Technologies Divisions]

http://www.lanl.gov/Internal/projects/IPO/DTIN/open/labtitl.html [DOE Facility Servers]

http://doe-is.llnl.gov/ [Information Security]

http://www.em.doe.gov/index.html [Environmental Management Web]

http://www.bnl.gov/bnl.html [Brookhaven National Laboratory]

http://www.ohre.doe.gov/ [Human Radiation Experiments]

http://terrassa.pnl.gov:2080/DFE/ [Pollution Prevention by Design]

[62] EPIC (Energy Pollution Prevention Information Clearinghouse)

http://146.138.5.107/EPIC.EXE?EPIC

[63] Office of INDUSTRIAL TECHNOLOGIES (OIT; U.S. Department of Energy)

http://www.nrel.gov/uit/oit.html

http://www.oit.doe.gov/ [Office of Industrial Technologies]

http://www.oit.doe.gov/News/wisconsin.html [News]

http://www.oit.doe.gov/Access/locator/ipl.html [Industrial Projects Locator]

http://www.oit.doe.gov/Links/links.html [Related Industry Links]

[64] U.S. Department of Energy's TECHNICAL INFORMATION SERVICES (TIS)

http://www.tis.eh.doe.gov [TIS for Environmental, Health, and Safety Professionals] http://tis-nt.eh.doe.gov/oepa/policy.html [DOE Environmental Policy and Guidance]

http://www.hyperk.com/

http://tis-nt.eh.doe.gov/nepa/docs/docs.htm [DOE NEPA Analysis]

http://tis-nt.eh.doe.gov/extreg/ [External Regulation of DOE Safety]

http://tis.eh.doe.gov/gils/gils_tis.html

http://tis.eh.doe.gov/fire/fire_handbook.html [Fire Protection Handbook]

http://www.tis.eh.doe.gov/whatsnew/whatsnew.html [What's New]

http://www.tis.eh.doe.gov/javamenu/ [TIS Applet Menu]

http://www.tis.eh.doe.gov/library/library.html [Digital Library]

http://tis-hq.eh.doe.gov/web/chem_safety/ [DOE Chemical Safety Program]

http://tis-nt.eh.doe.gov/oepa/guidance/risk.htm [DOE Guidance—Risk Assessment]

http://www.tis.eh.doe.gov/websites.html

[65] National Institute of Standards and Technology (NIST)—Manufacturing Extension Program

http://www.nist.gov

http://www.nist.gov/public_affairs/general2.htm

http://www.nist.gov/public_affairs/pubs.htm

http://www.nist.gov/public_affairs/siteindex.htm

http://www.nist.gov/weblinks.htm

http://www.boulder.nist.gov/timefreq/javaclck.htm [Atomic Clock Time—Java Enabled]

[66] Cornell University Archive of Federal Regulations

http://www.law.cornell.edu/topics/environmental.html

http://www.law.cornell.edu/topics/state_statutes.html#water

http://www.law.cornell.edu/statutes.html#state

http://www.law.cornell.edu/states/nm.html#codes

http://www.clay.net/statag.html

http://www.nmenv.state.nm.us/

[67] CODE OF FEDERAL REGULATIONS

http://www.pls.com:8001/his/cfr.html

http://law.house.gov/cfr.htm

http://law.house.gov/usc.htm

http://www.cpe.com/

http://www.gpo.ucop.edu/search/cfr.html

http://www.epa.gov/Rules.html

http://www.dot.gov/general/orders.html

http://www.envirotech.org/info/cfr.html

http://www.environmental-law.com/

http://fatty.law.cornell.edu/regs.html

[68] FEDERAL REGISTER—DAILY TABLE OF CONTENTS

http://earth1.epa.gov/EPAFR-CONTENTS

http://www.epa.gov/fedrgstr/EPAFR-CONTENTS/1998/

http://www.epa.gov/fedrgstr/EPAFR-CONTENTS/1997/

http://www.epa.gov/fedrgstr/EPAFR-CONTENTS/1996/

http://www.epa.gov/fedrgstr/EPAFR-CONTENTS/1998/November/

[69] FEDERAL REGISTER—EPA ENVIRONMENTAL SUBSET

{1} http://www.epa.gov/Rules.html [Regulations and Rules]

{2} http://www.epa.gov/epahome/rules.html#proposed [Proposed]

{3} http://www.epa.gov/epahome/rules.html#legislation [Current]

{4} http://www.epa.gov/epahome/rules.html#laws [U.S. Code Database]

{5} http://www.epa.gov/epahome/rules.html [Laws and Regulations]

{6} http://www.epa.gov/epahome/rules.html#codified [Codified Regulations]

{7} http://www.epa.gov/epacfr40/ [Title 40]

{8} http://www.access.gpo.gov/nara/cfr/index.html [Federal Regulations]

{9} http://www.access.gpo.gov/nara/cfr/cfr-table-search.html [Code of Federal Regulations]

{10} http://www.access.gpo.gov/nara/cfr/cfr-retrieve.html#page1 [Retrieve CFR by Citation]

{11} http://www.access.gpo.gov/nara/cfr/cfr-table-search.html#page1 [Search or Browse]

[70] FedWorld

http://www.fedworld.gov

[71] National Technology Transfer Center (NTTC)

http://iridium.nttc.edu/gov_res.html [Search]

http://iridium.nttc.edu/nttc.html [Home Page]

http://www.nttc.edu/homepage/terms.html [Copyright Terms]

http://www.procurenet.com/nttc/ntcnwsfl.htm [Newsflash]

http://www.nttc.edu/brs/brs.html [Search NTTC Developed Databases]

http://www.nttc.edu/training/guide/secb02g.html [Information Guide]

http://www.nttc.edu/byrd/history.html [History of NTTC]

http://www.nasatech.com/advertisers_feb97/nttc.html [General Information]

[72] Oak Ridge National Laboratory Environmental Sciences Division (ESD)

http://www.esd.ornl.gov [Oak Ridge Environmental Services Division]

http://www.ornl.gov/ceea [Center for Energy and Environmental Analysis]

http://www.ornl.gov/ORNL/Energy_Eff/Energy_Eff.html [Energy Efficiency and Renewable Energy Program]

http://www.ornl.gov/ORNLFacilities.html

http://www-eosdis.ornl.gov/ [Biogeochemical Analysis]

http://www.ncsa.uiuc.edu/General/GIBN/oak.ridge.lab.html [Computational Studies on the Fundamental Properties of Water]

http://www.tanks.org/ [Underground Tank Storage Remediation—Hanford]

[73] SANDIA NATIONAL LABORATORY

http://www.sandia.gov [Home]

http://www.em.doe.gov/idb94/sec8419.html [Links]

http://www.sandia.gov/geothermal/staff/draymo.htm [Geothermal Research]

http://larosio.upc.es/DOC-HTML/doc/htmlref/sandia/index.html [Search Sandia and Links]

http://www.sandia.gov/geothermal/staff/rdjacob.htm [Geothermal Research Department]

http://www.cmc.sandia.gov/about/visit/visit4.htm [Visiting]

http://www.sandia.gov/hist_sum.html [History]

http://www.sandia.gov/library/lib_hmpg.html [Technical Library]

[74] U.S. ARMY CORPS OF ENGINEERS

http://lms61.mvs.usace.army.mil/ [Water Control Management]

http://www.usace.army.mil/ [Home Page]

http://www.wes.army.mil/ [Waterways Experiment Station]

[75] Australian Environment and Water Related

http://kaos.erin.gov.au [Environment Australia Online]

http://www.agso.gov.au/

http://www.us.net/adept/links.html [Links]

http://www.acenz.com/ [Clean Air and Water, New Zealand]

http://www.wrc.wa.gov.au/public/waterwise/
contamination.html [Groundwater Contamination—Perth]

http://online.anu.edu.au/pad/ANURep/V29-3/tourism.html
[Water Shortage]

http://www.agso.gov.au/informa . . . on/ausgeonet/1997/
pitjant.html

[76] WATER WISER (Water Efficiency Clearinghouse)

http://www.waterwiser.org

http://www.waterwiser.org/books.html

http://www.waterwiser.org/wwlinks.html

http://www.waterwiser.org/wwlinks.html#related

http://www.awwa.org/forums/main.cfm?CFID=
48909&CFTOKEN=13274&CFApp=57&

http://www.waterwiser.org/forums/index.cfm?cfapp=44

[77] International Association of Environmental Hydrology (IAEH)

http://www.hydroweb.com

http://www.hydroweb.com/ehr.html

http://www.hydroweb.com/iaeh.html

http://www.hydroweb.com/iaehlink.html

[78] JOURNAL OF ENVIRONMENTAL HYDROLOGY

http://www.hydroweb.com/jeh.html

http://www.hydroweb.com/jehtit98.html

http://www.hydroweb.com/jehtit97.html

http://www.hydroweb.com/jehtit96.html

http://www.hydroweb.com/jeh_3_2.html

[79] Environmental HYDROLOGY REPORT

http://www.hydroweb.com/her.html

[80] TEXAS WATERNET

http://twri.tamu.edu

http://twri.tamu.edu/subjindex/techreps/

http://twri.tamu.edu/reports/

http://twri.tamu.edu/twriorg/front.html

http://twri.tamu.edu/wrlinks/

[81] GROUNDWATER.COM LINKS DIRECTORY

http://www.groundwater.com/links1.html

[82] WATERMODELING.org

http://watermodeling.org

http://watermodeling.org/html/resources.html

http://watermodeling.org/html/references.html

http://watermodeling.org/html/reviews.html

http://watermodeling.org/html/events.html

http://gwrp.cciw.ca/internet/online.html

[83] UNIVERSITY OF NEW MEXICO—Department of Earth
& Planetary Sciences

http://eps.unm.edu

http://eps.unm.edu/eps_info.htm

http:ftp:eps.unm.edu/pub/internet/research.htm

[84] U.S. EPA Groundwater FORUM HOMEPAGE

http://www.epa.gov/r10earth/offices/oea/gwf/gwfmain/htf

http://www.lkmichiganforum.org/ [Lake Michigan Forum]

http://www.epa.gov/oam/forum/f3qameta.htm [Region 3 Forum]

http://www.csun.edu/~vchsc006/tom.html [Risk Contamination
Forum]

[85] POLLUTION ONLINE and Related Groundwater Pollution

{1} http://www.pollutiononline.com

{2} http://gwrp.cciw.ca/internet/online.html [Online
Resources for Groundwater Studies]

{3} http://home.pacbell.net/gfredlee/index.html [Landfill Impact Publications]

{4} http://www.cedar.univie.ac.at/ . . . h/enveng-1/96apr/
msg00153.html

{5} http://www.princeton-groundwater.com/ [Princeton
Groundwater]

{6} http://www.hydromodels.com/awplfr.htm
[Groundwater Forecasting System]

{7} http://www.springer-ny.com/cat . . . ep95np/DATA/
0-387-94212-2.html

{8} http://www.ess.co.at/GAIA/models/gwp.htm
[Groundwater Pollution Models]

{9} http://www.loc.gov/lexico/liv/g/Groundwater_
pollution.html [Groundwater Pollution]

{10} http://www.umkc.edu/umkc/catalog/htmlc/engineer/ce/
c447.html [Groundwater Pollution and Modeling]

{11} http://www.tec.org/almanac/map.toc.html [Map Section,
Table of Contents]

{12} http://twri.tamu.edu/twripubs/NewWaves/v3n1/
report-6.html [Groundwater and Real Estate Values]

{13} http://area.ba.cnr.it/~cerimp01/TAORM.HTM
[Groundwater Pollution and Overdevelopment]

{14} http://isvapcs1.isva.dtu.dk/grc/1994/proj-2.htm [Oil and
Creosote Related Pollution]

{15} http://www.bgs.ac.uk/bgs/w3/hydro/Rept_SP.htm
[British Geological Survey—Selected Publications]

{16} http://users.aol.com/gfredlee/landfill.htm [Landfills and
Groundwater Quality]

{17} http://www.bgs.ac.uk/bgs/w3/hydro/PollProb.htm
[Groundwater Pollution a Problem in Store, British
Geological Survey]

{18} http://www.scisoftware.com/inhyd14.htm
[Groundwater Software]

{19} http://www.bgs.ac.uk/bgs/w3/hydro/Pollute.htm
[Nitrate Movement]

{20} http://www.doe.ca/water/en/manage/poll/e_tanks.htm
[Underground Storage Tanks and Piping]

{21} http://www.epa.gov.tw/english.new/bwqp.htm [Bureau of
Water Quality Protection]

[86] PUBLIC WORKS ONLINE and Related Groundwater Concerns

http://www.publicworks.com

http://www.cityofla.org/SAN/swmd/index.htm [Stormwater]

http://www.pwmag.com/ [Public Works Magazine Online]

http://www.engineersonline.com/ [Engineers Online]

http://www.fwrj.com/ [Florida Water Resources Journal]

[87] Groundwater for Windows and Related

http://www.geocities.com/EUREKA/8409

[88] U.S. DOE 1996 Baseline Environmental Management Report and
Related

http://www.em.doe.gov/bemr96

http://www.em.doe.gov/bemr96/execsum.html

http://www.em.doe.gov/bemr96/chp1_v11.html

http://www.em.doe.gov/bemr96/chp5_11.html

http://www.em.doe.gov/bemr96/chp8_v11.html

http://www.em.doe.gov/bemr96/appfv11.html

http://www.em.doe.gov/bemr96/glossary.html

http://www.em.doe.gov/bemr96/nm.html

http://www.em.doe.gov/bemr96/aloo.html

http://www.em.doe.gov/bemr96/amla.html

http://www.em.doe.gov/bemr96/itri.html

http://www.em.doe.gov/bemr96/lanl.html

http://www.em.doe.gov/bemr96/prgb.html

http://www.em.doe.gov/bemr96/snln.html

http://www.em.doe.gov/bemr96/ship.html

http://www.em.doe.gov/bemr96/svss.html
http://www.em.doe.gov/bemr96/wipp.html
http://www.epa.gov/OWOW/indic/I.html
http://www.epa.gov/R5Super/kisawyer.htm
http://www.epa.gov/radiation/mixed-waste/mw_pg3.htm
http://www.rockwell.com/About/Env/enviro_data.html
http://www.epa.gov/radiation/mixed-waste/mw_pg3.htm
http://ombwatch.org/www/ombw/regs/omb-rpt.html
http://www.dtic.mil/envirodod/brac/cerfa.html
http://www.envintl.com/projects.htm

[89] BERNALILLO COUNTY ENVIRONMENTAL HEALTH DEPARTMENT

http://www/Bernco.gov/eh [Environmental Health]
http://www/bernco.gov/ [County Government]
http://www.bernco.gov/pi/index.html [Public Information]
http://www.cabq.gov/rgvls/ [Rio Grande Library System]
http://www.bernco.gov/pw/faq.html [Public Works]
http://www.bernco.gov/news/pr050698.html [County Government]

[90] HYDROWEB

http://www.hydroweb.com
http://www.hydroweb.com/cdrom98.html
http://www.hydroweb.com/member.html

[91] THE UNIVERSITIES WATER INFORMATION NETWORK

http://www.uwin.siu.edu
http://www.uwin.siu.edu/welcome/index.html
http://www.uwin.siu.edu/ucowr/
http://www.uwin.siu.edu/news/index.html
http://www2.uwin.siu.edu/IWRN/orgs/
http://www2.uwin.siu.edu/databases/wrsic/index.html
http://www.uwin.siu.edu/NIWR/index.html
http://www.uwin.siu.edu/pick/index.html
http://www.uwin.siu.edu/tocnoframes.html

[92] Groundwater Remediation Technologies Analysis Center

http://www.gwrtac.org
http://www.gwrtac.org/html/about.html
http://www.gwrtac.org/pdf/gw10-6re.pdf
http://www.gwrtac.org/html/links.html

[93] Water Information Center/Geraghty-Miller

http://www.gmgw.com

[94] U.S. Occupational Safety and Health Administration Computerized Information System

http://www.osha.gov

[95] EPA CHEMICAL FACTS SHEETS AND RELATED

{1} http://www.epa.gov/enviro/html/emci/chemref/index.html [Chemical Reference WWW]
{2} http://ecologia.nier.org/english/level1/substance.html [Toxic Substances]
{3} http://www.envirolink.org/issues/pollution-map/index.html [Toxics Databases]
{4} http://www.pp.okstate.edu/ehs/LINKS/MSDS.HTM [Material Safety Data Sheets]
{5} http://www.cedar.univie.ac.at/ . . . /infoterra/96jan/msg00082.html
{6} http://www.brynmawr.edu/Admins/OOES/net.html [Online MSDS]
{7} http://www.berea.edu/chemmist/links/links.html [Chemist Links]
{8} http://www.ntis.gov/fcpc/cpn7016.htm [Health Effects Notebook]
{9} http://www.chemical.net/html/resources.html [Chemical Related Sources and Links]

{10} http://www.epa.gov/epahome/r2k.htm [Right to Know]
{11} http://ace.orst.edu/info/nain/aichem.htm [Antimicrobial Chemicals]
{12} http://atsdr1.atsdr.cdc.gov:8080/toxfaq.html [Agency for Toxic Substances and Disease Registry]
{13} http://www.epa.gov/opptintr/cie/factshee.htm [Chemical Fact Sheets in the Environment]
{14} http://www.epa.gov/ERNS/headline/headline.htm [Headline Spills]
{15} http://www.potomac.net/users/rbhuis/toxinfo.htm [TOXINFO—Toxicology Information]
{16} http://www.best.com/~akkana/RFG/health.txt [Health Effects—Gasoline]
{17} http://ecologia.nier.org/english/level1/substance.html [Toxic Substance Information]
{18} http://ingis.acn.purdue.edu:9999/ctic/pestfact.html [Pesticide Facts]

[96] LAWRENCE LIVERMORE NATIONAL LABORATORY

http://www.llnl.gov [Home Page]
http://www-ep.es.llnl.gov/ [Earth and Environmental Sciences]
http://www-ep.es.llnl.gov/www-ep/esd.html [Geosciences Home Page]
http://www.ukrweekly.com/Archive/1997/089710.shtml [Robots Cleanup Chornobyl]
http://www-ep.es.llnl.gov/www-ep/atm.html [Atmospheric Research]
http://www.llnl.gov/sci_educ/ [Science and Math Education]
http://www.nrl.navy.mil/clementine/clementine.html [Water on the Moon]

[97] LOS ALAMOS NATIONAL LABORATORY

http://www.lanl.gov [Home Page]
http://lib-www.lanl.gov/ [Research Library]
http://www-emtd.lanl.gov/TD/Technology.html [Environmental Problem Solving]
http://www.em.doe.gov/rtc1993/fslanl.html [General Information]
http://mwanal.lanl.gov/CST/imagemap/periodic/periodic.html [Periodic Table of the Elements]
http://www.em.doe.gov/ffaa/lan1ffca.html [Compliance Order]
http://www.education.lanl.gov/ [Education at LANL]
http://stb.lanl.gov/stb_edu_header.html [Science Education Programs]

[98] U.S. EPA OSWER RCRA Solid Waste

http://earth1.epa.gov/oswrcra

[99] CENTER FOR GROUNDWATER RESEARCH AND RELATED

{1} http://www.ese.ogi.edu/ese_docs/cgr.html
{2} http://cmr.sph.unc.edu/ [Groundwater Research Group]
{3} http://www.nwl.ac.uk/gwf/ [UK Groundwater Forum]
{4} http://isvapcs1.isva.dtu.dk/grc/1994/ANNUAL1.HTM [Annual Report]
{5} http://gw2.cciw.ca/nhri/nhrigw.html [Canadian Groundwater Issues]
{6} http://www.webdirectory.com/Water_Resources/Groundwater/ [Water Resources: Groundwater]
{7} http://www.isc.tamu.edtu/PICS/Princeton.html [Groundwater Research at Princeton]
{8} http://www.isva.dtu.dk/grc/ [Groundwater Research Center—Denmark]
{9} http://www.library.wisc.edu/li . . . raries/Water_Resources/UWS.htm
{10} http://www.api.org/ehs/sgresbul.htm [Groundwater Research Bulletins]

{11} http://civil.queensu.ca/environ/groundwater/main.htm [DNAPLs in Groundwater Research Group]

{12} http://cbsc.org/nb/bis/5135.html [New Brunswick Groundwater Study Group]

{13} http://www.mem.dk/geus/dvk/gvc-uk.htm [Danish Water Resources Committee]

{14} http://www.eerc.und.nodak.edu/960828/grnd.htm [Key Groundwater Activities]

{15} http://www.bgs.ac.uk/bgs/w3/hydro/HG_Prof.htm [Hydrology Group Profile—British Geological Survey]

{16} http://www.science.uwaterloo.c . . . ch_groups/ucsgrp/behavior.html

[100] Karst Waters Institute Inc.

http://www.uakron.edu/geology/karstwaters/kwi.html [Karst Water Institute]

http://terrassa.pnl.gov:2080/hydrology/research.html [Research Organizations—Hydrology]

http://www.dyetracing.com/ [Home of Karst on the Web]

http://www.geo.unizh.ch/~heller/SSS/BBS/90/23CN.html [Karst—China's People Republic]

http://www.halcyon.com/samara/nssccms/events.html

http://www.uwin.siu.edu/announce/event/1997/event1030.html [Karst Water Symposium]

http://www.ihe.nl/hy/staff/nen.htm [Hydrology People]

http://www.wittenberg.edu/acad . . . s/geol/progers/geol220/porter/

[101] Los Alamos National Laboratory/EES-5, Geoanalysis

http:ees-www.lanl.gov

http:ees-www.lanl.gov/capa.html

http:ees-www.lanl.gov/program.html

http://ees5db.1anl.gov/Publications/

http://ees-www.lanl.gov/org.html

[102] University Hygienic Laboratory/University of Iowa

http://www.uhl.uiowa.edu

http:www.uhl.uiowa.edu/Publications/index.html

[103] HOW GROUNDWATER IS CONTAMINATED

{1} http://www.gwconsortium.org/GWCON4.html [Chemicals Drawn in Public Water Supply]

{2} http://www.papers24-7.com/ecology_papers.htm [Term Papers in Related Environmental Subjects]

{3} http://www.traverse.com/groundwater/facts.htm [Groundwater Facts and Trivia]

{4} http://www.anl.gov/LabDB/Current/Ext/H601-text.002.html [NAPL Groundwater Contamination]

{5} http://www.ag.ohio-state.edu/~ohioline/b820/b820_10.html [Bulletin Pesticides and Groundwater Contamination]

{6} http://www.oag.state.ny.us/environment/reports/golf95.html [Toxic Fairways—Golf]

{7} http://www.flint.umich.edu/Dep . . . gional Groundwater/rgchome.html

{8} http://gwrp.cciw.ca/gwrp/ [Groundwater Remediation Project]

{9} http://www.rkkengineers.com/GASOLINE.HTM [Gasoline Groundwater Contamination]

{10} http://gw2.cciw.ca/gwrp/abstracts/crowe-044.html [Pesticide Groundwater Contamination in Canada]

{11} http://gwrp.cciw.ca/canada/ [Canadian Groundwater Issues]

{12} http://waterhome.tamu.edu/texa . . . yst/farmworkbooks/fastbl1.html

{13} http://atsdr1.atsdr.cdc.gov.80 . . . HAC/PHA/memphisdep/ddm_p2.html

{14} http://www.rcgrd.uvm.edu/ [Research Center for Groundwater Remediation Design]

{15} http://gwrp.cciw.ca/gwrp.conferences/pest_conf.html [Special Session on Groundwater Contamination by Pesticides]

{16} http://www.pubaf.bnl.gov/pr/bnldoeepapr040397.html [Groundwater Contamination at Brookhaven]

{17} http://www.epa.gov/superfund/o . . . products/nplsites/0301029n.htm

{18} http://www.ec.gc.ca/water/en/manage/poll/e_howgrd.htm [How We Contaminate Water]

{19} http://www.frtr.gov/abstracts/00000022.html [Soil Vapor Extraction]

{20} http://nesen.unl.edu/csd/illustrations/ec11/ec11text.html [Fundamentals of Groundwater Contamination]

{21} http://www.dtic.mil/envirodod/ . . . erpreport95/vol_2/nara093.html

{22} http://www.cowi.dk/div3_2_7.htm [Soil and Groundwater Contamination]

[104] Map of States with Approved Comprehensive State Groundwater Protection Programs and Related

http://www.epa.gov/OGWDW/csgwppnp.html [Maps of States]

http://www.us.net/adept/links.html [Links to Hydrology Related]

http://www.uwin.siu.edu/announce/event/1997/event0903b.html [Groundwater Tools]

http://www.tnrcc.state.tx.us/w . . . gpc/meetings/Fy95-1st_min.html

http://www.uaex.edu/publications/pub/fsa2039.htm [The Nature of Water]

http://owr.ehnr.state.nc.us/ref01/00066.htm [Federal Policies and Programs to Protect Groundwater]

http://www.uswaternews.com/links.html [Links to Other Water-Related Sites on Web]

[105] Groundwater Protection Council Online

http://gwpc.site.net

http://gwpc.site.net/Sourcewater/

http://gwpc.site.net/legislat.htm

http://gwpc.site.net/State%20Directors.htm

http://gwpc.site.net/FedStSites.htm

[106] U.S. EPA OFFICE OF WATER PUBLICATIONS

http://www.epa.gov/OGWDW/Pubs/index.html [Groundwater and Drinking Water Publications]

http://www.epa.gov/owm/ssodesc.htm [Sanitary Sewer Overflows]

http://www.lifewater.ca/links.htm [Interesting Water Links]

http://www.cpa.gov/region06/6en/w/sso/ssodesc.htm [EPA Sanitary Sewer Overflows]

http://www.anr.ces.purdue.edu/anr/whpa/whparesc.htm [Wellhead Protection Resources—Purdue]

[107] National Academy of Sciences Report on Radon in Drinking Water

http://www.epa.gov/OGWDW/radon/nas.html

http://www.nap.edu/readingroom/enter2.cgi?0309062926.html

http://www.epa.gov/OGWDW/radon/nasdw.html

http://www.epa.gov/OGWDW/radon/approach.html

http://www2.nas.edu/whatsnew/2936.html

http://www.epa.gov/OGWDW/standard/pp/radonpp.html

[108] Revised Source Water Protection Pages and Related

http://www.epa.gov/OGWDW/pmtect.html [Protecting Water Sources]

http://www.tnrcc.state.tx.us/w . . . ter/quality/gw/tgpc/index.html

http://www.tnrcc.state.tx.us/water/quality/gw/gwprotct.html
[Groundwater Protection]

http:httttp://www.tnrcc.state.tx.us/water/quality/gw/index.html
[Groundwater Assessment]

http://www.gov.ab.ca/env/water/WMRC/ground.html
[Groundwater]

http://www.deq.state.mi.us/ogp/ [Wellhead Protection Unit]

http://www.unites.uqam.ca/idea . . . ta/Papers/
fthguelph1996-4.html

http://www.bgs.ac.uk/bgs/w3/hydro/Manage.htm
[Groundwater Management and Protection]

http://www.epa.gov/rgytgrnj/sp . . . cinit/p2/conference/
weber2.htm

http://www.acnatsci.org/erd/ea/gwnews2.html
[The Challenge of Protecting Water]

[109] U.S. EPA Review of Monitoring Requirements

http://www.epa.gov/OGWDW/pws/cmr-fr.html

http://www.epa.gov/fedrgstr/EPA-WATER/1998/July/
Day-30/o-w20414.htm

http://www.epa.gov/fedrgstr/EPA-WATER/1997/July/
Day-03/w17210.htm

http://www.epa.gov/fedrgstr/EPA-WATER/1997/July/
Day-03/s-w17210.htm

http://www.epa.gov/docs/fedrgstr/EPA-
WATER/1997/July/Day-03/water.htm

[110] U.S. EPA Water Conservation Guidelines and Related

http://www.epa.gov/OGWDW/epa.gov/own/genwave.htm

http://www.cabq.gov/resources/ [City of Albuquerque]

http://www.swcs.org/ [Soil and Water Conservation Society]

http:http//www.watershedthesystem.com/ [Watershed—
The System]

http://www.issaquah.org/COMORG/gwac/gwac.htm
[Groundwater Advisory Committee]

http://www.geocities.com/capecanaveral/lab/8375/proj1.html
[The Significance of Water]

http://www.sawwa.org/index.html [Sacramento Water Works
Association]

http://www.twca.org/ [Texas Water Conservation Association]

http://www.cabq.gov/resources/outdoor.html [City of
Albuquerque Water Conservation]

http://www.ccme.ca/wpgwater/ [Winnipeg's Water
Conservation Program]

http://ogee.hydlab.do.usbr.gov/ [Water Research
Laboratory—Bureau of Reclamation]

http://wrri.nmsu.edu/wrdis/conserve/conserve.html
[New Mexico Water Conservation Programs]

http://www.dakotaswcd.org/ [Dakota County Soil and Water]

http://www.cnr.colostate.edu/CWK/conslink.htm [Links to
Other Water Conservation Sites]

http://www.unitedwater.org/ [United Water Conservation
District]

http://www.tempe.gov/water/water1.htm [Water
Conservation Tips]

[111] WATERBORNE DISEASES and Related

http://www.epa.gov/OGWDW/standard/wbornew.html

http://www3.uchc.edu/~wdc/ [Waterborne Disease Center]

http://www.hlth.gov.bc.ca/library/statsmas/in2033.html
[Waterborne Diseases in Canada]

http://www.mbnet.mb.ca/wpgwate . . .
om/sprynet/geraldf/XCBASIC.HTM

http:/www.uwin.siu.edu/announ . . . s/1996/waterborne-
disease.html

http://www.pure-water.com/headline.html

http://www.nvpdc.state.va.us/4MileRun/4mab-fc.htm
[Coliform Related]

http://bugs.uah.ualberta.ca/webbug/envbug/faecal.htm
[Coliform Related]

http://bugs.uah.ualberta.ca/webbug/envbug/faecal.htm
[Coliform Related—USGS]

http://www.cockatiels.org/ecoli.html [Coliform Related]

http://www.ae.iastate.edu/HTMDOCS/ae3060.htm [Coliform
Bacteria]

http://www.citiwater.com/ [Coliform Related]

http://www.citiwater.com/definition.htm [Coliform Related]

http://wilkes1.wilkes.edu/~eqc/coliform.htm [Coliform
Bacteria]

http://www.iwr.msu.edu/edmodule/water/fc.htm [Fecal
Coliform]

http://hermes.ecn.purdue.edu/cgi/convertwq?6187 [Coliform
Bacteria as Indicator]

http://btwebsh.macarthur.uws.edu.au/new_site/depart/
biology/ck/micro21.htm [Microbiology]

http://www.cdc.gov/travel/camerica.htm [Health Information to
Travelers—Mexico]

http://www.asmusa.org/pasrc/h20con.htm [Safe Drinking Water
Act Related]

http://www.cdc.gov/travel/temsam.htm [Health Information to
Travelers—South America]

http://www.eurekalert.org/releases/psu-clchan.html [Climate
Change and Waterborne Diseases]

http://www.mediconsult.com/tra . . .
el/shareware/safrica/food.html

http://www.cdc.gov/travel/nafrica.htm [Health Information to
Travelers—Africa]

http://www.theservicecenter.net/aquadoc/report.htm [Ground-
water Quality Report]

http://enrp.tamu.edu/hot/crypto/ [Microbial Contamination of
Drinking Water]

http://www.giardia.com/ [Giardia Related]

http://martin.parasitology.mcgill.ca/jimspage/
GIARDIA.HTM [Giardia Related]

http://www.naturesstandard.com/ [Giardia]

http://martin.parasitology.mcgill.ca/jimspage/biol/giardia.htm
[Giardia Related]

http://www.state.me.us/dhs/eng/water/giarcryp.htm [Giardia
and Cryptosporodium]

http://www.utoronto.ca/env/lib_hold/db2/files/7788_TE.htm
[Virus Related]

http://www.webdirectory.com/Pollution/
Water_Pollution/Products_and_Services/ [Virus Related]

http://www.phlsnorth.co.uk/weqa/index.html [Virus Related]

http://www.atlas.co.uk/listons/dwi/summary/wqhvirus/
dwi0743.htm [Virus Related]

http://www.ce.vt.edu/enviro2/gwprimer/virus/virus.html [Viral
Contamination]

[112] U.S. EPA Review of Monitoring Requirements for Chemical
Contaminants in Drinking Water

http://www.epa.gov/OGWDW/pws/cmr-fr.html

[113] U.S. EPA Background Materials for Groundwater Rule Stakeholder
Meeting, May, 1998

http://www.epa.gov/OGWDW/standard/gwr.html#back

http://www.epa.gov/OGWDW/ndwacsum.html

http://www.epa.gov/OGWDW/mdbp/mdbp.html

http://www.epa.gov/OGWDW/standard/bmp.html

http://www.epa.gov/OGWDW/standard/baseline.html

http://www.epa.gov/OGWDW/standard/monitor.html

http://www.epa.gov/OGWDW/standard/montbl1.html

http://www.epa.gov/OGWDW/standard/montbl2.html

http://www.epa.gov/OGWDW/standard/occur.html

http://www.epa.gov/OGWDW/standard/outbreak.html
http://www.epa.gov/OGWDW/standard/phs.html
http://www.epa.gov/OGWDW/standard/statereq.html

[114] U.S. EPA New Project Plan for Occurrence and Contaminant Selection

http://www.epa.gov/OGWDW/standard/pp/cclpp.html
http://www.epa.gov/OGWDW/ccl/cclfs.html
http://www.epa.gov/safewater/standards.html

[115] U.S. EPA Fact Sheet: Drinking Water Contaminant Regulations

http://www.epa.gov/OGWDW/epa.gov/safewater/source/therule.html

[116] U.S. EPA Summary of SOLE SOURCE AQUIFER DESIGNATIONS

http://www.epa.gov/OGWDW/swp/sumssa.html
http://www.epa.gov/OGWDW/swp/ssa.html
http://www.epa.gov/region02/water/petition.htm
http://www.epa.gov/region02/water/petition.htm

[117] U.S. EPA Drinking Water Contaminant Candidate List

http://www.epa.gov/OGWDW/ccl/cclfs.html
http://www.epa.gov/OGWDW/ccl/ccl_fr.html [Federal Register Reading]
http://www.epa.gov/OGWDW/ccl/ccl_fr.pdf [pdf Format]

[118] U.S. EPA State Methods for Delineating Source Water Protection Areas for Surface Water Supplied Sources of Drinking Water

http://www.epa.gov/OGWDW/swp/delin.html
http://www.epa.gov/OGWDW/ccl/ccl_fr.html
http://www.epa.gov/docs/fedrgstr/EPA-TRI/1995/June/Day-16/pr-15.html
http://www.epa.gov/fedrgstr/EPA-WATER/1997/October/Day-06/w26433.htm

[119] U.S. EPA State Source Water Assessment and Protection Progam

http://www.epa.gov/OGWDW/swp/swappg.html
http://www.epa.gov/OGWDW/source/swpguid.html
http://www.epa.gov/OGWDW/swp/swapes.html
http://www.epa.gov/OGWDW/swp/fs-swpg.html
http://www.epa.gov/OGWDW/dwsrf/ffland.html
http://www.epa.gov/OGWDW/source/contacts.html
http://www.epa.gov/OGWDW/swp/response.html

[120] U.S. EPA Office of Groundwater and Drinking Water Asks for Public Input on DEVELOPMENT OF GUIDANCE FOR STATE SOURCE WATER ASSESSMENT AND PROTECTION PROGRAMS

http://www.epa.gov/OGWDW/disc_frn.htm
http://www.epa.gov/OGWDW/discguid.htm
http://www.epa.gov/OGWDW/discguid.htm#I
http://www.epa.gov/OGWDW/discguid.htm#II
http://www.epa.gov/OGWDW/discguid.htm#III
http://www.epa.gov/OGWDW/discguid.htm#IV
http://www.epa.gov/OGWDW/discguid.htm#AppD

[121] U.S. EPA GROUNDWATER PROTECTION

http://www.epa.gov/OGWDW/Pubs/gwprotct.html
http://www.epa.gov/OGWDW/Pubs/12ground.html
http://www.epa.gov/OGWDW/Pubs/06ground.html
http://www.epa.gov/OGWDW/Pubs/11ground.html

[122] U.S. EPA Kid's Stuff (and for Teachers too)

http://www.epa.gov/OGWDW/kids
http://www.epa.gov/OGWDW/kids/art/index.htm
http://www.epa.gov/OGWDW/kids/tuar.html
http://www.epa.gov/OGWDW/kids/aquifer.html
http://www.epa.gov/OGWDW/kids/cycle.html

http://www.groundwater.org/kids/kids.htm [Non EPA]
http://www.epa.gov/OGWDW/kids/where.pdf
http://www.epa.gov/OGWDW/kids/non-pt.pdf

[123] U.S. EPA Underground INJECTION WELLS

http://www.epa.gov/OGWDW/uic.html [Underground Injection Control]
http://danpatch.ecn.purdue.edu . . . mstead/inject/src/varclass.htm
http://www.law.cornell.edu/uscode/42/300h-3.html [Interim Regulations]
http://www.state.sd.us/state/e . . . ve/denr/enviro/underinject.htm
http://www.eq.state.ut.us/eqwq/r317-007.txt [Water Quality]
http://waterquality.deq.state . . . us/wq/groundwa/EPAProposal.htm
http://www.epa.gov/OGWDW/regs/classv.html [Groundwater—Underground Injection Control]
http://dnr.state.il.us/ildnr/offices/mines/uic.html [Underground Injection Control Program]
http://www.emnrd.state.nm.us/OCD/aboutocd.htm [Oil Conservation Division]
http://www.emnrd.state.nm.us/O . . ./Environ/Handbook/undergro.htm
http://www.georgianet.org/dnr/ . . ./branches/geosurv/uiccovlt.htm

[124] U.S. EPA Regional Water Offices

http://www.epa.gov/OGWDW/links-r.html
http://www.epa.gov/region01/eco/drinkwater/
http://www.epa.gov/region02/water.htm
http://www.epa.gov/reg3wapd/
http://www.epa.gov/region4/waterpgs/wtr.html
http://www.epa.gov/r5water/sdw/
http://www.epa.gov/earth1r6/6wq/6wq.htm
http://www.epa.gov/region07/programs/wwpd/wwp.html
http://www.epa.gov/unix0008/
http://www.epa.gov/region09/water/
http://epainotes1.rtpnc.epa.gov:7777/r10/water.NSF/Office+of+Water/EPA+Region+10+Office+of+Water

[125] U.S. DOE Independent Oversight

http://tis-hq-eh.doe.gov/oversight

[126] U.S. EPA Guidance for Future State GROUNDWATER PROTECTION GRANTS

http://www.epa.gov/ogwdw/swp/gwpgrt.html

[127] Protecting the Nation's Groundwater: EPA's Strategy for the 1990s

http://www.epa.gov/ogwdw/pubs/11ground.html

[128] U.S. EPA Office of WETLANDS, OCEANS, AND WATERSHEDS

http://www.epa.gov/owow
http://www.epa.gov/OWOW/monitoring/
http://www.epa.gov/OWOW/monitoring/techmon.html
http://www.epa.gov/OWOW/monitoring/wqreport.html
http://www.epa.gov/OWOW/watershed/
http://www.epa.gov/OWOW/watershed/public.htm
http://www.epa.gov/OWOW/watershed/database.htm
http://www.epa.gov/OWOW/highlight.html
http://www.epa.gov./OWOW/tours/
http://www.epa.gov/OWOW/tours/wtrshed.html
http://www.epa.gov/OWOW/tours/wetlands.html
http://www.epa.gov/OWOW/tours/links.html

[129] U.S. EPA NONPOINT SOURCE POLLUTION Control Programs

http://www.epa.gov/owow/nps

http://www.epa.gov/OWOW/NPS/qa.html

http://www.epa.gov/OWOW/NPS/npsie.html

http://www.epa.gov/OWOW/NPS/elistudy/

http://www.epa.gov/OWOW/info/NewsNotes/

http://www.epa.gov/OWOW/NPS/nps3.html

http://www.epa.gov/OWOW/NPS/prevent.html

[130] GROUNDWATER AND SOIL CONTAMINATION DATABASE Plus Related

http://cd-rom-guide.com/cdprod1/cdhrec/009/098.shtml

http://www.infonordic.se/GROUND.html [Groundwater and Soil Contamination Database]

http://www.agiweb.org/agi/pubs/newpub.html [Glossary, Dictionary, and Database]

http://publish.uwrl.usu.edu/faculty/kemblowski.html [Utah Water Research Laboratory]

http://syssrv9vh1.nre1.gov/Access/locator/3881.htm [DOE Remediation of Contaminated Groundwater]

[131] U.S. DOE Research and Development Summaries

http://www.doe.gov/rnd/quick.html [Search]

http://www.osti.gov/

http://www.doe.gov/rnd/mainhelp.html

[132] Site Remediation Technologies: In si tu Bioremediation of Organic Contaminants

http://www.doe.gov/rnd/data/28073.html

http://www.inel.gov/technology_transfer/fact-htm/fact268.html

http://www.bioactive.com/

http://www.gwrtac.org/pdf/gw10-6re.pdf

http://bordeaux.uwaterloo.ca/bio1447/groundwater/remediation_menu.html

http:httttp://bordeaux.uwaterloo.ca/bio1447/groundwater/bioremediation_in_situ groundwater.html

http://bordeaux.uwaterloo.ca/bio1447/bioremediation excel/index.html

http://www.hanford.gov/eis/twrseis/rod/twrsrod.htm

http://www.epa.gov/radiation/mixed-waste/library/ref159.htm

http://dticam.dtic.mil/sbir/index/sba249.html

http://rap.nas.edu/lab/PA/22430510.html

http://iridium.nttc.edu/env/In_Situ_Rem/In_Situ_TOC.html

http://iridium.nttc.edu/env/In_Situ_Rem/In_Situ_overview.html

http://www.em.doe.gov/define/surface/inorganc.html

http://isvapcs1.isva.dtu.dk/grc/1994/PROJ-1.HTM

[133] State and National Energy and Environmental Risk Analysis System for Underground Injection Control

http://www.doe.gov/rnd/data/30855.html

http://oil/bpo.gov/Data/10220.html

http://www.grac.org/fall96/solutions.htm

http://www.state.nh.us/des/gwcatlog.htm

http://www.state.nh.us/des/gwcatlog.htm#Groundwater

http://www.state.nh.us/des/gwcatlog.htm#Groundwater Protection

http://www.tnrcc.state.tx.us/water/quality/gw/tgpc/meetings/Fy97-4th_min.html

http://www.cpn.org/sections/tools/manuals/groundwater3.html

[134] Advanced Subsurface Water Containment Systems Design and Performance Evaluation

http://www.doe.gov/rnd/data/45820.html

http://www.magnet.state.ma.us/dep/sero/mmr/files/contain.htm

http://www.migov.state.mi.us/rules/96/t/067

http://www.mbakercorp.com/environmental/gwmodel.htm

http://www.brad.ac.uk/acad/civeng/gwbfnet.html

[135] GROUNDWATER DIGEST

http://www.groundwater.com

http://www.groundwater.com/gwlist.html

http://www.groundwater.com/articles.html

http://www.groundwater.com/search.html

[136] Groundwater in the Great Plains Now Online

http://twri.tamu.edu/twripubs/ggp

http://twri.tamu.edu/

http://twri.tamu.edu/watertalk/archive/1997-Sep/Sep-30.1.html

[137] Canadian Groundwater Issues

http://gwrp.cciw.ca/canada

http://gw2.cciw.ca/cgwd.html

[138] CALIFORNIA WATER LINKS

http://www.wrd.org/links.htm

http://wwwdwr.water.ca.gov/

http://www.swrcb.ca.gov/

[139] Other World Wide Web Servers Related to Groundwater

http://gw2.cciw.ca/internet/servers.html [WWW Servers]

http://www.nwl.ac.uk/gwf/ [UK Groundwater Forum]

http://dino.wiz.uni-kassel.de/ecobas.html [University of Kassel—Ecological Modeling]

http://www.inforamp.net/~cgs/ [Canadian Geotechnical Society]

[140] Water Resources Authority, Groundwater Pollution Risk Mapping

http://www.wra-ja.org/riskmap.htm

http://www.wra-ja.org/glossary.htm#aquifer

http://www.wra-ja.org/ref.htm#fetter

http://www.wra-ja.org/profile.htm#activities

[141] EDUCATIONAL MATERIALS for the Protection of Groundwater

http://dpr.clemson.edu/publications/gwater.html

http://www.glc.org/docs/gwstrat/gwstrat.txt [Strategy]

http://www.gwsoftware.com/envgwd.htm [Software]

http://water.usgs.gov/public/education.html [USGS]

http://www.uwsp.edu/acad/uwexcoop/gndwater/Index.htm [Central Wisconsin Center]

http://danpatch.ecn.purdue.edu/~epados/ground/src/you3a.htm [Groundwater Guardian Program]

http://ohioline.ag.ohio-state.edu/b820/index.html [Bulletin Pesticides and Groundwater Contamination]

http://www.glc.org/docs/gwstrat/gwstrat.txt [Education Strategy]

http://www.extension.umn.edu/Documents/D/D/DD5867.html [What is Groundwater?]

http://gwrp.cciw.ca/education/ [Groundwater Education]

[142] GROUNDWATER PROTECTION COUNCIL

http://gwpc.site.net/mainsite.htm [Mainsite]

http://gwpc.site.net/factshee.htm [Underground Injection Fact Sheet]

http://gwpc.site.net/classii.htm [Class II Injection Wells]

http://gwpc.site.net/legislat.htm [National Legislation Links]

http://gwpc.site.net/best.htm [Best Management Practices]

http://gwpc.site.net/best.htm#Farming Practices for Ground Water Protection [Farming Practices for Ground Water Protection]

[143] Welcome to the GROUNDWATER FOUNDATION

http://www.groundwater.org

http://www.groundwater.org/guard/gg_index.htm

http://www.groundwater.org/catalog/catalog.htm

[144] Groundwater Message Board

http://www.groundwater.org/site/wwwboard/wwwboard.html

[145] GROUNDWATER MANAGEMENT, Site Remediation, and Related
  {1} http://www.clw.csiro.au/research/groundwater
  {2} http://www.agwa.org/ [Association of Groundwater Agencies]
  {3} http://www.flint.umich.edu/Dep . . . gional Groundwater/rgchome.html
  {4} http://www.foxcanyongma.org/ [Fox Canyon Water Management]
  {5} http://www.tnrcc.state.tx.us/water/quality/gw/index.html [Groundwater Assessment]
  {6} http://integratedwater.com/services/gm.html [Groundwater Management—California]
  {7} http://www.hayboo.com/leading/presslev4.htm [Water Resources Management Rules]
  {8} http://watershed.lake-coe.k12.ca.us/lakeinfo/gwater.html [Local Management]
  {9} http://www.globaltechs.com/ [Site Remediation]
  {10} http://bordeaux.uwaterloo.ca/b . . . undwater/remediation_menu.html
  {11} http://www.falcongdp.com/toc.htm [Mining Remediation]
  {12} http://gwrp.cciw.ca/ [Groundwater Remediation Project]
  {13} http://www.skwsystem.com/GOTEBORG/CHALMERS.htm [Soil Remediation—Creosote and Heavy Metals]
  {14} http://www.gwrtac.org/ [Groundwater Remediation Analysis]
  {15} http://www.opengroup.com/open/tabooks/156/156670281X.shtml [Fundamentals of Hazardous Waste Site Remediation—Book]

[146] Welcome to Tex A Syst
  http://waterhome.tamu.edu
  http://waterhome.tamu.edu/projects/index.html
  http://waterhome.tamu.edu/news/index.html
  http://waterhome.tamu.edu/presentation/index.html
  http://waterhome.brc.tamus.edu/tsswcb/index.html
  http://waterhome.brc.tamus.edu/tsswcb/index.html
  http://waterhome.brc.tamus.edu/tsswcb/index.html

[147] University Water Information Network New Books
  http://www.uwin.siu.edu/announce/newbooks
  http://www.uwin.siu.edu/welcome/index.html
  http://www.uwin.siu.edu/index.html

[148] NORTH CAROLINA Division of Water Quality and Waste Management
  http://gw.ehnr.state.nc.us/INDEXOLD.HTM
  http://gw.ehnr.state.nc.us/CONTENTS.HTM
  http://gw.ehnr.state.nc.us/PUBLIC.HTM
  http://gw.ehnr.state.nc.us/LINKS.HTM

[149] Policy and Practice (100,000 Map Series)
  http://www.soton.ac.uk/~bopcas/data/fullrecs/7075.htm

[150] THE UNIVERSITY WATER INFORMATION NETWORK
  http://www.uwin.siu.edu/index.html
  http://www.uwin.siu.edu/welcome/index.html
  http://www.uwin.siu.edu/news/index.html
  http://www.uwin.siu.edu/topics/WaterPolicy/waterpolicy.html
  http://www2.uwin.siu.edu/IWRN/orgs/
  http://www.uwin.siu.edu/NIWR/index.html
  http://www2.uwin.siu.edu/databases/wrsic/index.html
  http://www.uwin.siu.edu/pick/index.html
  http://www2.uwin.siu.edu/WaterSites/index.html [Wetlist]

[151] The Hamilton to Baltimore Groundwater Consortium
  http://www.gwconsortium.org/index.html
  http://www.gwconsortium.org/Contam.html
  http://www.gwconsortium.org/Consortium.html
  http://www.gwconsortium.org/Sites.html

[152] Related Internet Sites for Groundwater
  http://www.engineering.usu.edu/bie/gw.html
  gopher://igwmc.mines.colorado.edu:3851/

[153] REGIONAL GROUNDWATER CENTER (The University of Michigan)
  http://www.flint.umich.edu/Departments/RegionalGroundwater/rgchome.html
  http://www.flint.umich.edu/Departments/RegionalGroundwater/workshop.htm
  http://www.flint.umich.edu/Departments/RegionalGroundwater/whatsnew.htm

[154] Waterloo Centre for Groundwater Research Groundwater Notes
  http://darcy.uwaterloo.ca/gwn/march96/index.html
  http://darcy.uwaterloo.ca/gwn/march96/gw-can.html
  http://darcy.uwaterloo.ca/gwn/march96/ontario.html
  http://darcy.uwaterloo.ca/gwn/march96/westt.html
  http://darcy.uwaterloo.ca/gwn/march96/reprints.html
  http://darcy.uwaterloo.ca/gwn/index.html

[155] Groundwater Remediation Project National Water Research Institute
  http://gwrp.cciw.ca
  http://gwrp.cciw.ca/gwrp/
  http://gwrp.cciw.ca/canada/
  http://gwrp.cciw.ca/internet/
  http://gwrp.cciw.ca/education/

[156] ON-LINE RESOURCES for Groundwater Studies
  http://gwrp.cciw.ca/internet/online.html

[157] Gasoline Service Station Soil/Groundwater Contamination Characterization and Remediation
  http://www.rkkengineers.com/GASOLINE.HTM

[158] Earth Sciences 144—OVERVIEW of Groundwater Contamination
  http://wwwcatsic.ucsc.edu/~eart144/overview.html

[159] *Cryptosporidium* and Related
  http://www.mayohealth.org/mayo/9606/htm/cryptosp.htm
  http://www.nal.usda.gov/wqic/Bibliographies/eb9612.html
  http://www.nal.usda.gov/wqic/crypto.html
  http://www.cdc.gov/ncidod/diseases/crypto/crypto.htm
  http://www.nalusda.gov/wqic/cornell.html
  http://www.nalusda.gov/wqic/cryptfac.html
  http://www.ksu.edu/parasitology/
  http://www.phlsnorth.co.uk/weqa/index.html
  http://www.phlsnorth.co.uk/weqa/index.html
  http://www.epa.gov/OGWDW/crypto.html

[160] City of Albuquerque Water Conservation
  http://www.cabq.gov/resources/index.html

[161] UK Groundwater Forum
  http://www.nwl.ac.uk/gwf
  http://www.nwl.ac.uk/gwf/gwfnews1.htm
  http://www.nwl.ac.uk/gwf/gwfdbas2.htm
  http://www.nwl.ac.uk/gwf/about01.htm
  http://www.nwl.ac.uk/gwf/gwfbook1.htm

[162] Learn about Groundwater
  http://www.groundwater.org/learn/learn.htm
  http://www.groundwater.org/learn/hydro_cy.htm

http://www.groundwater.org/learn/contam.htm
http://www.groundwater.org/learn/protect.htm
http://www.groundwater.org/learn/wells.htm

[163] GEOGRAPHIC INFORMATION SYSTEM Support for GIS for Groundwater

http://www.flint.umich.edu/departments/regionalgroundwater/rgc-pg17.htm

[164] U.S. DOE Richland Operations (RL) Hanford

http://www.hanford.gov
http://www.doe.gov/people/peopae.htm
http://www.hanford.gov/misc_info/trilinks.htm

[165] U.S. DOE - Hanford Groundwater/Vadose Zone Integration Project

http://www.bhi-erc.com/vadose/vadose.htm
http://www.bhi-erc.com/vadose/docs.htm
http://www.bhi-erc.com/vadose/links.htm

[166] NATIONAL ENVIRONMENTAL HEALTH ASSOCIATION

http://www.NEHA.org
http://www.neha.org/neha.html
http://www.neha.org/radonpage.html
http://www.neha.org/pubs.html
http://www.neha.org/pubs.html#ww
http://www.csn.net/~beckyr/
http://www.neha.org/links.html#ww

[167] The Death of Common Sense (Phillip K. Howard)

http://www.amazon.com
http://www.barnesandnoble.com

[168] FEDERAL WEB Locator

http://www.law.vill.edu/fed_agency/fedwebloc.html
http://www.vcilp.org/Fed-Agency/fedwebloc.html#latest
http://www.vcilp.org/Fed-Agency/fedwebloc.html#toc

[169] TECHNOMIC PUBLISHING COMPANY, INC.

http://www.techpub.com
http://www.techpub.com/enews/
http://www.techpub.com/tech/default.asp
http://www.techpub.com/Content.asp?Nav=Contact
http://www.techpub.com/enews/default.htm#Instructions_for_viewing

[170] Conservation Foundation

http://www.metacrawler.com

[171] NATIONAL RESEARCH COUNCIL

http://www.nas.edu
http://www.nas.edu/about/
http://www.nas.edu/publications/
http://www4.nas.edu/cp.nsf
http://www4.nas.edu/webcr.nsf/(MeetByDocID)/AF9A49524EBC1D84852566C1005BC8BD?OpenDocument
http://www4.nas.edu/webcr.nsf/(MeetByDocID)/5C61BA818AFC691C852566B30077912A?OpenDocument
http://www.nas.edu/browse.html

[172] DARCY'S Law

http://res.agrica/CANIS/GLOSSARY/darcys_law.html

[173] Law MASS ACTION

http://www.chem.uidaho.edu/~honors/massact.html

[174] NERST Equation

http://www.camchem.rutgers.edu/chemicals/chem.html

[175] UNESCO

http://www.unesco.org
http://www.unesco.org/general/eng/programmers/index.html
http://www.unesco.org/general/eng/publish/index.html
http://www.unesco.org/general/eng/TOC.html

[176] English COMMON LAW

http://www.ionet.net/~okclaw/common.html

[177] COUNCIL ON ENVIRONMENTAL QUALITY

http://www.whitehouse.gov/CEQ/
http://www.whitehouse.gov/CEQ/About.html

[178] OFFICE OF TECHNOLOGY ASSESSMENT Online

http://www.wws.princeton.edu/~ota/

[179] ROCKY MOUNTAIN ARSENAL

http://www.defenselink.mil/new . . . /Jun1996/b061196
http://www.e-21.qpg.com/
http://www.defenselink.mil/news/Jun1996/b061196_bt354-96.html
http://www.gao.gov/AIndexFY97/abstracts/n197033.htm
http://www.pmrma-www.army.mil/htdocs/misc/about.html
http://www.em.doe.gov/plutrod/

[180] THE BIBLE

http://www.gospelcom.net/bible

[181] WATERBORNE DISEASE Center

http://www3.uchc.edu/nwdc

[182] Aquatic ECOSYSTEMS

http://www.zoo.Toronto.edu/zooweb/aeg
http://www.zoo.toronto.edu/zooweb/aeg/landwat.htm

[183] CHEMICAL SPECIES

http://www.lerc.nasa.gov/www/chemsensors
http://www.ito.umnw.ethz.ch/SoilProt/staff/gfeller/csgloss.html
http://www.chem.msu.su/~rudnyi/tdlib/
http://webbook.nist.gov/chemistry/name-ser.htm

[184] DISSOLVED IONS in Water

http://www.culligan.ca/ions.html
http://webbook.nist.gov/chemistry/name-ser.htm

[185] GEOCHEMICAL CYCLE

http://www.agu.org/pubs/toc/gb/gb_10_3.html
http://www.agu.org/pubs/toc/gb/gb_10_3.html
http://sflwww.er.usgs.gov/projects/evergl_merc/factsheets/fs-166-96/food_chain.html
http://www.agro.wau.nl/ssg/Library/huisman.htm

[186] CHEMICAL COMPOSITION of Natural Water

http://gopher.well.sf.ca.us:70 . . . Environment/Barr/Barr.Appendix
http://webbook.nist.gov/chemistry/name-ser.htm
http://webbook.nist.gov/chemistry/
http://bordeaux.uwaterloo.ca/bio1447new/epadocs/naturalmenu.htm
http://denr1.igis.uiuc.edu/isgsroot/isgshome/isgshome.html
http://www.dnr.state.wi.us/org/water/wm/nps/

[187] HYDROLOGIC Methods

http://www.prenhall.com/ptrobooks/esm_013227924.html
http://ewre-www.cv.ic.ac.uk/urban.html
http://www.sigann.com/
http://hoth.gcn.ou.edu/~jahern/enviro/course_description.html
http://www.gsf.de/UNEP/finweri.html
http://www.univ.kiev.ua/GEG/GEG_SC.HTML
http://rap.nas.edu/lab/NOAA/26980021.html
http://www.cs.indiana.edu/hyplan/vmenkov/HSV/tk980113.html

[188] GEOCHEMICAL Investigations

http://moontan.marine.usf.edu/gloria.htlm
http://www-odp.tamu.edu/publications/prosp/172_prs/172specsamp.html

http://www.core.hu/geochem/
http://www.uni-wuerzburg.de/mineralogie/index.html
http://www.glg.msu.edu/micropostdoc.html
http://pubs.usgs.gov/factsheet/fs50-97/
http://www.le.ac.uk/geology/map2/pander/wresearch.html

[189] NATURAL WATER Chemistry

http://www.nrri.umn.edu/nrri/cwe.html
http://webbook.nist.gov/chemistry/name-ser.htm
http://webbook.nist.gov/chemistry/

[190] TOTAL DISSOLVED SOLIDS in Water

http://www.culligncom/tdsinfo.htm
http://webbook.nist.gov/chemistry/
http://www.culligan.ca/tds.html
http://www.greatwaterco.com/tds.htm
http://weather-mirror.nmsu.edu/Teaching_Material/
    soil456/ECtotal.htm
http://www.academyofgolf.com/pglg/tds.htm
http://www.ci.shreveport.la.us/stormwtr/tds.htm
http://www.reskem.com/equipment/tds.html

[191] SODIUM in Water

http://www.wqa.org/Glossary/sodium.html
http://www.cwqa.com/
http://webbook.nist.gov/chemistry/name-ser.htm
http://webbook.nist.gov/chemistry/
http://encarta.msn.com/index/conciseindex/02/002ab000.htm
http://www.srhip.on.ca/bgoshu/Water/WaterSodiumFS.html
http://www.alfaenv.com/dsodium.htm

[192] CALCIUM in Water

http://www.execpc.com/~magnesum/calcium.html
http://www.wqa.org/Glossary/calcium.html
http://webbook.nist.gov/chemistry/name-ser.htm
http://webbook.nist.gov/chemistry/
http://www.calciuminfo.com/
http://members.aol.com/thera1234/caldigest.html
http://www.bc-dairy-foundation.org/bcdf/calcium.htm
http://www.asmusa.org/press/genmic16.htm
http://www.galicia.simplenet.com/calmagplus.htm
http://comptons2.aol.com/encyclopedia/ARTICLES/
    00778_A.html

[193] MAGNESIUM in Water

http://www.wqa.org/Glossary/magnesium.html
http://www.execpc.com/~magnesum/
http://www.execpc.com/~magnesum/rylandr.html
http://www.execpc.com/~magnesum/calcium.html
http://www.execpc.com/~magnesum/anderson.html
http://www.execpc.com/~magnesum/
    fdaweek.html#California
http://www.execpc.com/~magnesum/freeweb.html
http://webbook.nist.gov/chemistry/name-ser.htm
http://webbook.nist.gov/chemistry/
http://www.execpc.com/~magnesum/
http://www.menopause-online.com/magnesium.htm
http://bvsd.k12.co.us/cent/Newspaper/may96/
    Tom_Blackburn.html
http://www.babycenter.com/refcap/659.html
http://www.shef.ac.uk/~chem/web-elements/Mg.html
http://minerals.er.usgs.gov/minerals/pubs/commodity/
    magnesium/

[194] HARDNESS in Water

http://www.wqa.org/Glossary/hardness.html

http://webbook.nist.gov/chemistry/name-ser.htm
http://webbook.nist.gov/chemistry/
http://www.wqa.org/WQIS/Glossary/Hardness.html
http://water.nr.state.ky.us/ww/ramp/rmhard.htm
http://www.siouxlan.com/water/faq.html

[195] IRON in Water

http://www.wqa.org/Glossary/iron.html
http://webbook.nist.gov/chemistry/name-ser.htm
http://webbook.nist.gov/chemistry/
http://www.wqa.org/Consumer/the-stainers.html
http://www.tpssite.com.au/water/stain.htm
http://www.bobvila.com/cleanwater.html
http://www.culligan.ca/fe_state.html
http://www.ae.iastate.edu/HTMDOCS/ae3059.htm

[196] MANGANESE in Water

http://www.wqa.org/Glossary/manganese.html
http://webbook.nist.gov/chemistry/name-ser.htm
http://webbook.nist.gov/chemistry/
http://wilkes1.wilkes.edu/~eqc/iron1.htm
http://www2.inetdirect.net/~ecoindy/chems/mang.html
http://patent.womplex.ibm.com/details?patent_
    number=5443729
http://www.sfes.com/iron.htm
http://www.ianr.unl.edu/PUBS/water/g1280.htm
http://water.nr.state.ky.us/ww/ramp/rmmag.htm
http://www.mj.domhost.com/mjwater/ironmang.htm
http://www.cyberone.com.au/~enviro/manganese.htm
http://www.clo2.com/reading/drinking/system.html
http://www.lamotte.com/instruc/7518.htm
http://hermes.ecn.purdue.edu/server/water/bib/
    Drinking_Water_Quality/Contaminants.html

[197] SULFATE Bacteria in Water

http://www.wqa.org/Glossary/Thiobacillus/html
http://webbook.nist.gov/chemistry/name-ser.htm
http://webbook.nist.gov/chemistry/
http://www.wqa.org/Glossary/sulfate-bacteria.html
http://www.wqa.org/Glossary/S-List.html

[198] CHLORIDE in Water

http://water.nr.state.ky.us.ww/ramp/rmel.htm
http://webbook.nist.gov/chemistry/name-ser.htm
http://webbook.nist.gov/chemistry/
http://www.healthmatrix.com/Minerals/Min_all/chloride.htm
http://www.state.ky.us/nrepc/water/wcpcl.htm
http://www.herbaldave.com/Minerals/Chlorine.htm
http://water.wr.usgs.gov/fact/b07/low.html

[199] FLUORIDE in Water

http://www.wqa.org/Glossary/fluoride.html
http://webbook.nist.gov/chemistry/name-ser.htm
http://webbook.nist.gov/chemistry/
http://home.cdsnet.net/~fluoride/info.htm
http://www.shef.ac.uk/chemistry/web-elements/nofr-
    uses/F.html
http://www.eastonutilities.com/water/wflourid.html
http://hammock.ifas.ufl.edu/txt/fairs/1569
http://www.watercheck.com/docs/fulltext.htm
http://www.zerowasteamerica.org/Flouride.htm
http://www.healthy.net/othersites/citizens/fluoride.htm
http://www.healthy.net/library/articles/schacter/fluoride.n.htm

[200] U.S. PUBLIC HEALTH SERVICE

http://phs.os.dhhs.gov/phs/phs.html

[201] NITRATES in Water
http://www.wqa.org/Glossary/nitrates.html
http://gilligan.esu7.k12.ne.us . . .
    b/Lakeview/science/nitrate.htm
http://csdm.k12.mi.us/pages/bios/nitrate.html
http://www.idahonews.com/020598/THE_WEST/13402.htm
http://www.wqa.org/Consumer/nitrates.html
http://www.montana.edu/wwwpb/home/nitrates.html
http://www.uswaternews.com/archive/96/quality/nitrate.html
http://muextension.missouri.edu/xplor/waterq/wq0256.htm
http://webbook.nist.gov/chemistry/name-ser.htm
http://webbook.nist.gov/chemistry/
http://www.healthy.net/library/articles/schacter/fluoride.n.htm
    [Methemoglobinemia]
http://www.wizard.com/NHL/faq/nitrate.htm
http://cancernet.nci.nih.gov/clinpdq/risk/Nitrate_in_Drinking_
    Water_Associated_With_Increased_Risk_for_NHL.html
http://www.epa.gov/OWOW/indic/fs11.html
http://www.hhs.gov/news/press/1996pres/960906.html
http://wwwrcolka.cr.usgs.gov/midconherb/isoprop.final.html
http://www.srhip.on.ca/bgoshu/Water/WaterNitratesFS.html
http://hammock.ifas.ufl.edu/txt/fairs/2632

[202] PHOSPHATE in Water
http://www.wqa.org/Glossary/phosphate.html
http://www.northstarnet.org/prkhome/living/phosp.html
http://portia.advanced.org/3336/act.3poll.html
http://www.pacific.ccg-gcc.gc . . . . EPAGES/OFFBOAT/
    PAE/keeping.htm
http://webbook.nist.gov/chemistry/name-ser.htm
http://webbook.nist.gov/chemistry/

[203] pH and Water
http://www.wqa.org/Glossary/pH.html
http://webbook.nist.gov/chemistry/name-ser.htm
http://webbook.nist.gov/chemistry/
http://www.eosc.osshe.edu/peers/lessons/water/waterpH.html
http://wilkes1.wilkes.edu/~eqc/ph.htm
http://wwwga.usgs.gov/edu/phdiagram.html
http://www.thekrib.com/Chemistry/
http://www.srl.rmit.edu.au/condin/pH.htm
http://jersey.uoregon.edu/~djohnson/wswq/pH.html
http://www.velda.nl/talen/uk/water.html
http://w4u.eexi.gr/~andreask/ph.htm
http://www.crop.cri.nz/curresea/soil/vegph.htm

[204] ALKALINITY and Water
http://www.wqa.org/Glossary/alkalinity.html
http://webbook.nist.gov/chemistry/name-ser.htm
http://webbook.nist.gov/chemisty/
http://water.nr.state.ky.us/ww/ramp/rmalk.htm
http://www.ces.msstate.edu/bad_url.cgi?file=/pubs/is1334.htm
http://www.cleaningstuff.com/glossary.htm
http://bellnet.tamu.edu/res_grid/elementry/WaterTesting.htm

[205] ACIDITY in Water
http://www.wqa.org/Glossary/acidity.html
http://webbook.nist.gov/chemistry/name-ser.htm
http://webbook.nist.gov/chemistry

[206] AMERICAN SOCIETY FOR TESTING MATERIALS
http://www.astm.org

[207] TRACE METALS in Water
http://www.auburn.edu/~pritcme/thesis.html
http://www.chemtronics.com.au/hmetals.htm

http://www.centrelab.com/trace.htm
http://vm.cfsan.fda.gov/~lrd/pestadd.html
http://rap.nas.edu/lab/USGS/90244302.html
http://rap.nas.edu/lab/USGS/90309605.html
http://www.floridaplants.com/CR/natural.htm
http://webbook.nist.gov/chemistry/name-ser.htm
http://webbook.nist.gov/chemistry/

[208] AMERICAN PUBLIC HEALTH ASSOCIATION
http://www.apha.org

[209] ANALYTICAL CHEMISTRY
http://www.anachem.umu.se/jumpstation.htm
http://webbook.nist.gov/chemistry/
http://webbook.nist.gov/chemistry/

[210] ORGANICS in Water
http://www.wqa.org/Glossary/organics/html
http://www.talloaks.com/techinfo/organics.html
http://www.pathfinder.com/mone . . .
    st/press/BU/1998Sep15/924.html
http://www.midwestlabs.com/certification.html
http://h2osparc.wq.ncsu.edu/lake/bass/orgsel.html
http://www.dhs.cahwnet.gov/ps/ . . . chemicals/MTBE/
    mtbesummary.htm
http://www.barringer-labs.com/
http://webbook.nist.gov/chemistry/name-ser.htm
http://webbook.nist.gov/chemistry/
http://www.isas-dortmund.de/pgroups/pgroup1/
http://www.agnic.org/agdb/ishow.html
http://ucaswww.mcm.uc.edu/geology/maynard/organic/

[211] WATER SCIENCE AND TECHNOLOGY
http://www.iawq.org.uk

[212] TRIHALOMETHANES in Water
http://www.wqa.org/Glossary/trihalomethanes/html
http://www.ulaval.ca/vrr/rech/Proj/54359.html
http://www.mec.cuny.edu/biology/cstep98lc.html
http://webbook.nist.gov/chemistry/name-ser.htm
http://webbook.nist.gov/chemistry/
http://www.dwp.ci.la.ca.us/water/quality/wq_tthm.htm
http://members.aol.com/lcjts/water_problems.html
http://tango.cheec.uiowa.edu/seed/fy93/93c.html
http://www.epa.gov.tw/english.now/analysis/policye87/
    1302034e.htm

[213] MICROBIOLOGY of Water
http://www.phlsnorth.co.uk/weqa/index.html
http://www.millipore.com/analy . . .
    ical/technote/micro/index.html
http://www.europe.apnet.com/textbook/lbs/new9596/
    envmic.htm
http://commtechlab.msu.edu/sites/dlc-me/zoo/
http://www.groundwatersystems.com/
http://www.iso.ch/cate/0710020.html

[214] MICROBIAL ECOLOGY
http://www.egr.msu.edu/DER/labs/cme.html
http://www.ifas.ufl.edu/~dmsa/course/overview.htm
http://isar.mpi-bremen.de/fog/fog.html
http://www.ls.huji.ac.il/~MicEco/home.htm
http://www.msue.msu.edu/msue/iac/iac1046.html
http://commtechlab.msu.edu/sites/dlc-me/
http://microbes.org/
http://www.cme.msu.edu/cme/
http://www.kent.edu/biology/courses/40363.htm

http://www.edv.agrar.tu-muenchen.de/micbio/ecolo.htm

http://wwwsoc.nacsis.ac.jp/jsme2/

[215] MICROORGANISMS in Water

http://www.wqa.org/Glossary/microorganisms.html

http://www.science.uwaterloo.ca/biology/mayfield.html

http://www.cnr.colostate.edu/~bobw/sam_o_1/water/water35.htm

http://markun.cs.shinshu-u.ac.jp/learn/kenbikyou/picture/bisei/bisei-e.htm

[216] ENDOTOXINS in Water

http://www.wqa.org/Glossary/endotoxins.html

http://www.md.huji.ac.il/md/mi . . . ology/bact330/lectureendo.html

http://www.healthworks.co.uk/h . . . blisher/churchill/church6.html

[217] SEPTIC TANKS and Leachfields

http://www.propump.com/septic101.html

http://klingon.util.utexas.edu/Septic_Tanks/Septic_FixQ.html

http://www.tanks-a-lot.com/

http://danpatch.ecn.purdue.edu . . . /farmstead/onsite/approach.htm

http://nhresnet.sr.unh.edu/gra . . . /planning/guide/docs/hdoc2.txt

http://www.calepa.cahwnet.gov/epadocs/invenpub.txt

http://www.inspect-ny.com/septbook.htm

http://www.zeitec.com/ssr/flier.html

http://www.fuzzylu.com/greencenter/q17/septic.htm

http://www.kistner.com/uus-aeration.html

http://www.state.ga.us/legis/1997_98/leg/fulltext/hb1113.htm

http://www.standards.com.au/Ne . . . ash/1998/19980429/19980429.htm

[218] SEPTAGE and Water

http://www.mahoning-health.org/septage.htm

http://www.dep.state.pa.us/dep . . . e/wm/MRW/Docs/septage_haul.htm

http://www.psma.net/

http://www.hillmurray.com/projects/solution/sol-bb.htm

http://www.bright.net/~wchd/wc_biosolids_recycling.html

http://www.mahoning-health.org/septage.htm

http://www.uwin.siu.edu/announce/newbooks/1997/book0725.html

http://www.deq.state.or.us/od/news97/septage.htm

http://twri.tamu.edu/twripubs/Insights/v2n2/article-3.html

http://www.dep.state.pa.us/dep . . . e/wm/MRW/Docs/septage_haul.htm

http://www.psma.net/

http://www.mahoning-health.org/septage.htm

[219] PRIVIES and Water

http://www.eq.state.ut.us/eqwq/r317-560.txt

http://www.plcmc.lib.nc.us/branch/main/carolina/plum/outdoor.htm

http://curtain.dyndns.com:6970/779

http://www.cloudnet.com/~renfest/privies.htm

http://www.mia.mb.ca/resources/govtfacts/govtfact25.html

http://pen2.ci.santa-monica.ca.us/city/municode/art05/5.08.060.html

[220] LANDFILLS and Groundwater

http://www.Kirkworks.com/grrnlee.htm

http://members.aol.com/gfredlee.gfl.htm

http://home.pacbell.net/gfredlee/index.html

http://www.eieio.org/

http://www.actionenv.com/

http://www.epa.state.oh.us/

http://www.water-ed.org/briefing.html

http://www.pagnet.org/waterqual.html

http://www.dep.state.pa.us/

http://www.imt.dtu.dk/research/grounpro.htm

[221] LEACHATE in Groundwater

http://www.swana.org/ground.htm

http://www.cedar.univie.ac.at/ . . . h/enveng-1/96feb/msg00037.html

http://www.essential.org/listproc/dioxin-l/msg00410.html

http://twri.tamu.edu/reports/1991/153.html

http://gwrp.cciw.ca/education/landfill/landfill.html

http://www.epa.gov/superfund/o . . . products/nplsites/0504952n.htm

http://gw2.cciw.ca/gwrp/abstracts/lesage-006.html

[222] MUNICIPAL SEWAGE DISPOSAL SYSTEMS and Groundwater

http://www.env.gov.bc.ca/epd/epdpa/mwr/agfasf.html

http://www.co.washtenaw.mi.us/depts/eis/eisehfee.htm

http://www.jeffcity.com/cityclerk/data/chap29.htm

http://www.law.indiana.edu/codes/in/36/ch-36-9-24.html

http://www.pugetsound.org/sewage/report/regca.html

http://www.epa.gov/cgi-bin/claritgw?op-Display&document=clserv:epa-cinn:0399;&rank=4&template=epa

[223] DETERGENTS in Water

http://www1.huji.ac.il/www_teva_law/d3.html

http://www-cmrc.sri.com/CEH/Re . . . urfactantsHouseDetergents.html

http://www.webcom.com/trumpet/TGD/tech.html

http://www.chemistry.co.nz/detergent_class.htm

http://www.chem.wsu.edu/Chem102/102-LipFatSoap.html

http://www.chemistry.co.nz/deterg.htm

[224] SURFACTANTS in Water

http://www.wqa.org/Glossary/suffactants.html

http://www.cchem.berkeley.edu/~cjrgrp/

http://www.igb.fhg.de/Presse/en/Tensioline.en.html

http://www.finishing.com/1400-1599/1410.html

http://www.chemistry.co.nz/propwat.htm

http://gallery.in-tch.com/~rmueller/PUB.htm

http://www.basf.com/businesses . . . ance/html/water_treatment.html

http://www.gwrtac.org/html/tech_eval.html#SURF

[225] SEWAGE IRRIGATION

http://lcweb.loc.gov/lexico/liv/s/sewage_irrigation.html

http://www-dppi.poliba.it/users/sun/water/

http://www.dpws.nsw.gov.au/wc2.html

http://www.ci.long-beach.ca.us/water/sewer.htm

http://www.asic-ne.org/water.html

[226] SPRAY IRRIGATION

http://www.organics.co.uk/sprayirg.htm

http://wwwga.usgs.gov/edu/pictureshtml/irsprayhigh.html

[227] INDUSTRIAL CONTAMINATION of Groundwater

Http://www.geolsoc.org.uk/pubs/books/cat128.htm

http://gw2.cciw.ca/education/gwfacts/d.html

http://gw2.cciw.ca/education/gwfacts/figure09.html

http://gw2.cciw.ca/education/gwfacts/figure10.html

http://gw2.cciw.ca/education/gwfacts/d1.html

http://gw2.cciw.ca/education/gwfacts/d2.html

[228] CALIFORNIA DEPARTMENT OF WATER RESOURCES

http://www.water.ca.gov

http://wwwdwr.water.ca.gov/dir-CA_water_infoR2/
GroundwaterR2.html

[229] METAL WASTES in Groundwater

http://www.epa.gov/gils/records/A00164.html

http://www.doe.gov/html/em52/54856.html

http://es.epa.gov/ncerqa/hsrc/metals/index.html

http://es.epa.gov/ncerqa_abstracts/centers/hsrc/metals/
metal11.html

http://www.greenpeace.org/home/gopher/campaigns/tox-
ics/1991/engdump.txt

[230] MINE WASTES in Groundwater

http://water.wr.usgs.gov/mine/mar/swed.html

http://lcweb.loc.gov/lexico/liv/m/Mine_wastes.html

http://lcweb.loc.gov/lexico/liv/w/Water_pollution.html

http://lcweb.loc.gov/lexico/liv/m/Mine_water.html

http://lcweb.loc.gov/lexico/liv/g/Groundwater.html

http://lcweb.loc.gov/lexico/liv/g/Groundwater_pollution.html

http://www.dem.csiro.au/unrestricted/news/media/
releases/re103.html

[231] PITS AND LAGOONS

http://uts.cc.u.Texas.edu/~mharren/index3.html

http://www.smartref.com/ust/fed-resources.html

http://www.agricycle.com/casefor.html

http://hermes.ecn.purdue.edu:8001/cgi/convertwq?7412

[232] RADIOACTIVE MATERIALS (Wastes)

http://www.dehs.umn.edu/rpdXI/wastesec.html

http://www.ieer.org/ieer/clssroom/r-waste.html

http://www.rw.doe.gov/

http://www.iisd.ca/vol05/0525027e.html

http://www.niehs.nih.gov/odhsb/wasteman/rad/rad8.htm

http://www.westgov.org/wipp/

http://www.env.kyoto-u.ac.jp/english/kumatori/kumatori.html

http://www.sierraclub.org/policy/316.html

http://www.epa.gov/radiation/mixed-waste/mw_pg14.htm

[233] RADON in Water (See also 107)

http://www2.ncsu.edu/bae/programs/extension/publicat/
wqwm/he396.html

http://sedwww.cr.usgs.gov:8080/radon/radonhome.html

http://eande.lbl.gov/IEP/high-radon/hr.html

http://www.epa.gov/iaq/radon/

http://www.nsc.org/ehc/airqual.htm

http://www.sph.umich.edu/~bbusby/radon.htm

[234] INSTITUTION OF CIVIL ENGINEERS

http://www.ice.org.uk

[235] RISK ASSESSMENTS

http://www.riskworld.com [Risk World]

http://ces.soil.ncsu.edu/soilscience/publications/soilfacts/
AG-439-08/index.htm

http://www.csun.edu/~vchsc006/469.html

http://www.sra.org/

http://www.em.doe.gov/emtrain/a3a.html

http://gophisb.biochem.vt.edu/brarg/brasym96/brarg96.html

http://geb.isis.vt.edu/SWAMP/als5984/week8.html

http://www.medscape.com/govmt/...
304.09.lamm/e0304.09.lamm.html

http://risk.lsd.ornl.gov/rap_hp.htm

http://www.eea.dk/Projects/EnvMaST/RiskAss/
Chapter3H.HTML

http://www.usda.gov/agency/oce/oracba/oracba.htm

http://www.neptuneandco.com/950250701.html

http://www.clean-wastewater.com/

[236] MUNICIPAL SLUDGE

http://www2.ncsu.edu/bae/programs/extension/publicat/
wqwn/ag439_3.html

http://www.irim.com/nssh/nsh00147.htm

http://www.env.gov.bc.ca/~cpr/cdd/projects/10203pro.html

http://www.anl.gov/LabDB/Current/Ext/H449-text.001.html

http://www2.ncsu.edu/bae/programs/extension/publicat/
wqwm/ag439_3.html

[237] OIL FIELD BRINES

http://www.ul.cs.cmu.edu/books/groundwater/water033.htm

http://energy.cr.usgs.gov/energy/E%26E/OF97-28/
OF97-28.html

http://www.yournet.com/brine.html

http://eti-geochemistry.com/papers/eti2.htm

http://www.pete.lsu.edu/perttl/research.htm

http://link.tsl.state.tx.us/.dir/fdlp3.dir/a&i_07m.txt

[238] AGRICULTURAL WASTES

http://www/css.orst.edu/Research/Environ/index.html

http://www.jgpress.com/

http://www.metla.fi/conf/iufro95abs/d5pap72.htm

http://www.sarep.ucdavis.edu/SAREP/NEWSLTR/v8n1/
sa-14.htm

http://www.env.duke.edu/faculty/sigmon/agwaste/dialog.html

http://www.env.gov.bc.ca/epd/epdpa/iwhc/faaiw.html

http://www.env.gov.bc.ca/epd/cpr/regs/awcreg.html

http://www.qp.gov.bc.ca/stat_reg/regs/elp/r131_92.htm

http://www.qp.gov.bc.ca/stat_reg/regs/elp/r131_92.htm#links

[239] PESTICIDES—Herbicides, in Groundwater

http://water.wr.usgs.gov/pnsp.gw

http://www.epa.gov/internet/oppts/

http://www.igc.org/panna/

http://www.igc.org/pesticides/

http://agrolink.moa.my/doa/english/laws/rp_pface.html

http://www.epa.gov/pesticides

http://www.ag.ohio-state.edu/~ohioline/b820/

http://ificinfo.health.org/index13.htm

http://www.epa.gov/region4/air/

http://gnv.ifas.ufl.edu/~fairsweb/text/ss/16819.html

http://bluehen.ags.udel.edu/deces/nps/nps-05.html

[240] ANIMAL WASTES and Groundwater

http://www.state.ok.us/osfdocs/nr5797.html

http://hermes.ecn.purdue.edu:8001/server/water/bib/
Waste_Management/Animal.html

http://ss.narc.affrc.go.jp/pro/pro6/indexlab_e.html

http://www.ces.ncsu.edu/whpaper/REactivities.html

http://www.house.gov/resources/105cong/democrat/
press/relea212.html

http://ext.msstate.edu/pubs/pub1878.htm

http://www.niehs.nih.gov/odhsb/wasteman/index.htm

[241] IRRIGATION RETURN FLOWS

http://www.dtsc.ca.gov/rsu/hwrs1c10.htm

http://www.hydrocomp.com/

http://www.hydrocomp.com/hydrolinks.html

http://hermes.ecn.purdue.edu:8001/cgi/convertwq?6531

[242] DISPOSAL AND INJECTION WELLS

http://www.nasda-hq.org/nasdah/nasda/foundation/
state/id/cover.htm

http://gwpc.site.net/classii.htm

http://gwpc.site.net/uicwells.htm

http://gwpc.site.net/factshee.htm

[243] RECHARGE WELLS

    http://www.floridaplants.com/CR/ground.htm

    http://www.floridaplants.com/CR/ground.htm

    http://www.oas.org/EN/PROG/chap1_9.htm

    http://www.state.nv.us/cnr/ndwp/dict-1/WORD_R.htm

    http://water.wr.usgs.gov/projects/ca498.html

[244] SALINE AQUIFERS

    http://www.ce.udel.edu/faculty/cheng/saltnet/saltbib.html

    http://www.wrd.org/broch.htm

    http://www.es.anl.gov/htmls/transport.html

    http://www.brookings.com/bswf/tp8.htm

    http://www.utexas.edu/research/beg/edwards/index.html

    http://twri.tamu.edu/twripubs/NewWaves/vlnl/abstract-3.html

    http://www.amcl.ca/groundwater.htm

    http://www.igb-berlin.de/www/abt1/Galery/galery.htm
      [Java Enabled]

[245] BOREHOLES in Petroleum Industry

    http://www.groundflow.com

    http://twri.tamu.edu/twripubs/NewWaves/v1n1/abstract-3.html

    http://www.august.com/pgeo/pg_cool_wells.html

    http://ladmac.lanl.gov/mgls/95Sym/95Contents.html

    http://www.nrc.gov/NRC/CFR/PART060/part060-0134.html

[246] *IN SITU* Mining

    http://www.wma-minelife.com/uranium/is100006.html

    http://avoca.vicnet.net.au/~seaus/proposed/islnotgood.html

    http://www.bhp.com.au/environment/bhp-env/imi.htm

    http://deq.state.wy.us/lqd/Permits/Ptforms.htm

    http://iridium.nttc.edu/env/tmp/022.html

[247] VOCS AS CONTAMINANTS

    http://iridium.nttc.edu/env/VOC_Arid/VOCA_TOC.html

    http://atsdr1.atsdr.cdc.gov:80...
      0/HAC/PHA/spectron/spe_p2.html

    http://www.exemplar.net/water/report.html

    http://www.epa.ohio.gov/ddagw/voc.html

    http://www.goodwaterco.com/comprob.htm

    http://www.ioc.army.mil/eq/maps/sites/riaa.htm

    http://www.wpi.org/Initiatives/init/feb97/novocs.htm

    http://www.acepump.com/everpure/qc4-voc.html

    http://www.sisweb.com/referenc/applnote/ap8.htm

[248] GEOTHERMAL WELLS

    http://geothermal.id.doe.gov

    http://geothermal.id.doe.gov/geothermal/related.html

    http://www.sandia.gov/geothermal/

    http://www.sandia.gov/geothermal/gdo.htm

    http://wwwrvares.er.usgs.gov/nrp/proj.bib/mariner.html

[249] STORMWATER

    http://www.stormwater-resources.com

    http://www.stormwatermgt.com

    http://www.stormwater-resources.com

    http://www.cabq.gov/flood/swpp.html

    http://www.gatekeeper.com/stormwater/

    http://www.gatekeeper.com/stormwater/
      pollution_abatement/pollution_abatement.html

    http://www.nrdc.org/nrdc/faqs/wastmfic.html

    http://www.state.me.us/dep/blwq/stormwtr/stormwat.htm

    http://www.state.me.us/dep/blwq/stormwtr/index.htm

    http://www.state.me.us/dep/blwq/stormwtr/manage.htm

    http://www.state.me.us/dep/blwq/stormwtr/download.htm

    http://www.state.me.us/dep/blwq/stormwtr/links.htm

[250] U.S. DEPARTMENT OF TRANSPORTATION

    http://www.dot.gov

    http://www.hdrinc.com/otherweb/dot.htm

    http://www.hdrinc.com/hdr_toc.htm

[251] UNDERGROUND STORAGE TANKS

    http://www.epa.gov/oust

    http://www.natlaw.com/pubs/tanks.htm

    http://www.state.in.us/idem/oer/328.html

    http://www.alfaenv.com/ddead.htm

    http://www-emtd.1an1.gov/TD/Tanks.html

    http://www.mckenna-law.com/Envlaw/envlaw.htm

    http://aee.hq.faa.gov/aee-200/FSTQAG/ust.html

    http://denix.cecer.army.mil/denix/Public/Library/Remedy/
      LowryB/lowryb01.html

[252] GROUNDWATER MONITORING Methods

    http://www-cwwr.ucdavis.edu/publications/waterpub.shtml

    http://www.scisoftware.com/lymntr2.htm

    http://www.neosoft.com/internet/paml/groups.G/gwm-1.html

    http://www.sci-sms.com/gw-monitoring.htm

    http://www.emt.com/gw.htm

    http://mpls-wkst00.summite.com/gwmonito.htm

    http://water.nr.state.ky.us/dow/techsvc.htm

    http://water.nr.state.ky.us/dow/monitor.htm

    http://www.ca.blm.gov/GoldenQueen/pub-hyd5.htm

    http://iwrn.ces.fau.edu/steele.htm

[253] ABANDONED TANKS and Groundwater

    http://pasture.ecn.purdue.edu/~agenhtml/agen521/epadir/
      grnwtr/contamination.html

    http://www.swrcb.ca.gov/~cwphome/ust/notice.htm

    http://www.nmenv.state.mn.us/ust/br-aband.html

    http://www.nmenv.state.mn.us/NMED_regs/20nmac5_1.html

    http://www.inspect-ny.com/oiltanks/oiltend.txt

[254] PETROLEUM CONTAMINATED Aquifers

    http://gwrp.cciw.ca/gwrp/studies/lesage/lesage/html

[255] SALTWATER INTRUSION Into Groundwater

    http://www.wrd.org/broch.htm

    http://www.gmssoftware.com/swift-model.htm

    http://www.gwsoftware.com/ifswif.htm

    http://water.wr.usgs.gov/projects/ca429.html

    http://gw2.cciw.ca/gwrp/abstracts/bobba-002.html

    http://www2.hawaii.edu/~nabil/desaltbk.htm

    http://www-sflorida.er.usgs.gov/online_reports/
      wri964285/text.htm

    http://www.regis.berkeley.edu/baydelta.html

    http://enso.unl.edu/ndmc/mitigate/policy/ota/concerns.htm

    http://explorer.scrtec.org/explorer/explorer-db/html/
      836283716-81ED7D4C.html

    http://www.ec.gc.ca/water/en/nature/grdwtr/e_salt.htm

    http://www.nwrc.gov/lessons/saltwate.html

[256] TOXIC CHEMICALS in Groundwater

    http://instruct1.cit.cornell.edu/courses/aben47i

[257] WATERSHED MANAGEMENT

    http://www.watershedthesystem.com

    http://www.watershedthesystem.com/

    http://water.nr.state.ky.us/dow/watrshd.htm

    http://glinda.cnrs.humboldt.edu/wmc/index.html

    http://www.tnrcc.state.tx.us/admin/topdoc/gi/229/toc.html

    http://kyw.ctic.purdue.edu/kyw/kyw.html

    http://www.state.ma.us/dep/brp/wm/wmpubs.htm

    http://watershed.org/wmc/

http://www.ctic.purdue.edu/Catalog/
WatershedManagement.html

http://ice.ucdavis.edu/Califor . . . ver_watershed_
management_plan/

http://www2.nas.edu/wstb/2152.html

http://www.tile.net/listserv/grasslandsl.html

http://www2.ncsu.edu/bae/progr . . . ms/extension/water/
w_shed.html

[258] SAFE DRINKING WATER ACT

htttp://www.citation.com/hpages/sdwa.html

[259] MILITARY TOXICS

http://www4.gue.ch/bci/GreenCrossFamily/gorby/toxic.html

http://www.miltoxproj.org/

http://www.monitor.net/rachel/r227.html

http://www.mapcruzin.com/fotp/links.htm

http://www.gulfweb.org/ngwrc/CaseDU/credits.htm

http://www.mapcruzin.com/fotp/eda.htm

http://ww.ccaej.org/projects/whitpaper/milittoxics.htm

[260] Ecotoxicology

http:/www.apnet.com/www/journal/es.htm

http://www.academicpress.com/www/journal/es.htm

http://fisher.teorckol.lu.se/e . . .
st/1995/Chemical/Chemical.html

http://hplus.harvard.ed/ejournals/ap_ecotoxes.html

http://www.ecotox.lu.se/welcome.html

http://egotox.ecotox.lu.se/

http://www.ramas.com/ecotox.htm

http://www.uga.edu/srel/ecotoxbook.htm

http://192.215.52.3A/www/catalog/es.htm

http://ublib.buffalo.edu/libra . . . es/ejournals/records/ecot.html

http://www.ardeacon.com/field.html

http://www.floridaplants.com/CR/ecotox.htm

http://www.ecotox.lu.se/ecotox/ecotox_theor.html

[261] UNCERTAINTY in Risk Assessment

http://nattie.eh-doe.gov/docs/egm/other/other.0005.txt

http://www.tandfdc.com/Books/toxicology/scijudg.htm

http://www-ep.es.llnl.gov/www-. . .
tompson/SASSFCT96/tsld017.html

http://www.state.nv.us/nucwaste/yucca/gao97.txt

http://www.virginia.edu/~risk/center.html

http://www.kleinfelder.com/health.htm

http://www.ramas.com/

http://data.ctn.nrc.ca/on/content/type3/org72/div1129/
listings/t7373.htm

http://www.stat.washington.edu/NRCSE/events/abstracts/
risksem.html

http://www.ams.med.uni-goettingen.de/mail/biometry/
0419.html

[262] HUMAN RISK Assessment

http://www.cs.auckland.ac.nz/~rajkumar/slideshow/index.htm

http://epawww.ciesin.org.glreis/glnpo/data/arcs/EPA-905-R92-
007/EPA-905-R92-0077.html

http://www.crcpress.com/cgi-
bin/SoftCart.exe/jour/catalog/human.htm?E+storecrc

http://www.pinyonsoftware.com/HomePage.htm

http://www.mrc-cpe.cam.ac.uk/mirrors/llnl/mole_tox/
toxicology.html

[263] AQUIFER RESTORATION

http://offo2.epa.ohio.gov/FERNALD/Aquifer/aquifer.htm

http://www2.ncsu.edu/ncsu/wrri/reports.miller.html

http://offo2.epa.ohio.gov/FERNALD/Aquifer/aquifer.htm

http://www.hullinc.com/documents/tech/ltabst.htm

http://earth1.epa.gov/earth100/records/i10695.html

http://www.clean.rti.org/serdp/CLNP3A.HTM

[264] CERCLA

http://www.em.doe.gov/dd/fctsht2.html

[265] DOD (Department of Defense)

http://disa11.disa.atd.net/index.html

http://www.defenselink.mil/

http://www.denix.osd.mil/

[266] HAZARDOUS WASTE AND TOXIC SUBSTANCES

http://www.rec.hu/poland/wpa/net-hzw.htm

http://ehs.ucsc.edu/hw/hw.html

http://www.hwac.org/

http://www.rec.hu/poland/wpa/net-hzw.htm

http://atsdr1.atsdr.cdc.gov:8080/hazdat.html

[267] VOLATILE COMPOUNDS

http://hudson.cir.tohoku.ac.jp/~sbtu/iinc/volc/vlistC.html

http://www.deltanet.com/sgvw/wqa/vocs.html

http://www.miljoedata.com/voc.htm

http://cpas.mtu.edu/tools/t0026.htm

[268] Agency for Toxic Substances and Disease Registry

http://atsdrl.atsdr.cdc.gov:8080/hazdat.html

[269] THE NATIONAL TOXICOLOGY PROGRAM

http://ntp-server.niehs.nih.gov/

[270] MYELOTOXICITY

http://dir.niehs.nih.gov/dirlep/Webpages/refs.html

[271] HAZARD IDENTIFICATION

http://www.eng.mu.oz.au/eng/Hazard.html

http://www.eng.rpi.edu/dept/chem-eng/Biotech-
Environ/SAFETY/hazard.html

http://cheminform.de/msds3.htm

[272] EXPOSURE ASSESSMENT

ftp://ftp.epa.gov/epa_ceam/wwwhtml/ceamhome.htm

http://www.envsci.rutgers.edu/grad_pgms/exp_asmt/

http://www.eohsi.rutgers.edu/emad/emad.html

http://www.thistlepublishing.com/

[273] DOSE-RESPONSE

http://www.sph.unc.edu/ies/doseresp.htm

http://www.lindsoft.com/

http://www.triumf.ca/safety/tsn/tsn_6_2/section3_2.html

[274] HAZARD RANKING SYSTEM

http://www.erols.com/als/hrshome.html

http://www.nablus.com/hrshome.html

[275] U.S. DOE ORDERS

http://www.et.anl.gov/DOEord.html

[276] ENVIRONMENTAL RESTORATION at DOE

http://www.em.doe.gov/er/

[277] WASTE MANAGEMENT at DOE

http://www.ornl.gov/NSProject/srid/fa-16wm.htm

[278] Nuclear Material

http://www.ca.sandia.gov/NMM/

[279] HAZARDOUS WASTE at DOE Facilities

http://www.hanford.gov/eis/sweis/notice.htm

[280] RCRA

htttp://www.citation.com/hpagess/rcra.html

[281] WASTE TANKS at Hanford

http://www-emtd.lanl.gov/td/Tanks/tanksafety.html

[282] The Galvin Report

http:www.lanl.gov/Internal/News/galvin/

[283] URANIUM
  http://www.uilondon.org/
  http://www.ccnr.org/nfb_uranium_3.html
  http://www.epa.gov/ngispgm3/iris/irisdat/0421.DAT

[284] PLUTONIUM
  http://plutonium-erl.actx.edu/
  http://www.plutonium239.org/
  http://plutonium-erl.actx.edu/
  http://www.pu.org/
  http://www.pu.org/
  http://www.pu.org/main/primer/pu_issue.html

[285] ENVIRONMENTAL REMEDIATION at DOE
  http://www.cnie.org/nle/waste-3.html

[286] WASTE ISOLATION PILOT PLANT
  http://www.em.doe.gov/rtc1994/wipp.html

[287] CLEAN WATER ACT (CWA)
  http://www.webcom.com/~staber/cwa.html

[288] Los Alamos National Laboratory EM Program
  http://www.lanl.gov

[289] TRANSURANIC WASTE at DOE Facilities
  http://www.wipp.carlsbad.nm.us/

[290] LOW-LEVEL Waste
  http://www.em.doe.gov/info/natlowlv.html

[291] MIXED Waste
  http://www.epa.gov/radiation/mixed-waste/

[292] DOE HANFORD SITE
  http://www/hanford.gov

[293] TRITIUM
  http://tritium.lanl.gov/

[294] FERNALD DOE SITE
  http://offo2.epa.Ohio.gov/FERNALD/fernald.htm

[295] TOXIC SUBSTANCES CONTROL ACT
  http://www.r2d3.com/1st100.html

[296] FIFRA
  http://www.epa.gov/pesticides/regleg.htm

[297] SURFACE MINING AND CONTROL AND
  RECLAMATION ACT
  http://www.osmre.gov/welcome.htm

[298] U.S. EPA Groundwater Protection Program
  http://www2.ncsu.edu/bae/programs/extension/publicat/
  wqwm/ag441_5.html

[299] Solid Waste Management Units
  http://www.emo.anl.gov/annrep/1995/chapter2/2_3_9.html

[300] HSWA (Hazardous and Solid Waste Amendment)
  http://www.epa.gov/enviro/html/rcris/rcris_overview.html

[301] RADIOACTIVITY in Groundwater
  http://www.eas.asu.edu/~holbert/gwater-a.html

[302] OCCURRENCE/FATE/TRANSPORT in Groundwater
  http://gwrp.cciw.ca/gwrp/abstracts_f/lesage-009.html

[303] REMEDIATION AND RESTORATION of Groundwater
  http://gwrp.cciw.ca/index.html

[304] RADIOACTIVE Waste
  http://www.nrc.gov/NRC/NUREGS/BR0216/part03.html

[305] HIGH-LEVEL Waste
  http://www.nrc.gov/OPA/gmo/tip/waste.htm

[306] NATIONAL ACADEMY OF SCIENCES
  http://www.nas.edu

[307] YUCCA MOUNTAIN Site, Nevada
  http://www.igc.apc.org/citizenalert/fctshts/yuccal.txt

[308] NATIONAL ENVIRONMENTAL POLICY ACT (NEPA)
  http://ceq.eh.doe.gov/nepa/regs/nepa/nepaeqia.htm
  http://es.epa.gov/oeca/ofa/
  http://tis-nt.eh.doe.gov/nepa/tools/tools.htm
  http://tis-nt.eh.doe.gov/nepa/policy.htm
  http://www.fs.fed.us/forum/nepa/nepaeaseiss.html

[309] MINING Wastes
  http://lu62gw.sds.no/nou/1994-12/ved0521.htm
  http://www.enviromine.com/wetlands/inorganics.htm
  http://www.cais.net/publish/stories/0996wat3.htm
  http://www.epa.gov/epaoswer/other/mining.htm
  http://www.enviromine.com/wetlands/inorganics.htm
  http://www.uswaternews.com/archive/95/waterq/colloids.html

[310] UMTRA
  http://www.em.doe.gov/bemr96/faci.html
  http://www.em.doe.gov/bemr96/bsbo.html
  http://www.osha-slc.gov/OshDoc/Interp_data/I19901228.html
  http://www.em.doe.gov/bemr96/ship.html
  http://www.em.doe.gov/bemr96/grri.html
  http://www.em.doe.gov/bemr96/tostates.html
  http://www.ornl.gov/~jkm/DOEQM/npr0002.html
  http://www.em.doe.gov/bemr96/gius.html
  http://www.em.doe.gov/bemr96/gums.html
  http://www.doegjpo.com/perm-barr/durango.htm
  http://www.em.doe.gov/emprog/winter96/empw16.html
  http://warrior.tubacity.k12.az.us/~man2/winter98/eliza/
  http://www.em.doe.gov/itrd/tubainfo.html
  http://www.em.doe.gov/bemr96/olnc.html
  http://www.em.doe.gov/bemr96/cano.html
  http://www.em.doe.gov/bemr96/mams.html
  http://www.em.doe.gov/bemr96/amla.html

[311] URANIUM MILL TAILINGS RADIATION CONTROL ACT
  http://www.pmei.com/nrc/u/Uran . . .
  ngs_Radiation_Control_Act.html
  http://www.antenna.nl/wise/uranium/ulus.html
  http://www.epa.gov/docs/radiation/radwaste/umt.htm
  http://www.house.gov/commerce_democrats/
  comact04/31396.htm
  http://www.osha-slc.gov/OshDoc/Interp_data/I19901228.html

[312] ALBUQUERQUE DOE OPERATIONS
  http://wwww.doeal.gov/ [Home Page]
  http://www.doeal.gov/qtd/ta.htm [Qualification and Training]
  http://www.doeal.gov/QTD/erc.htm [Resource Center]
  http://tis.eh.doe.gov/does/she/se93spr.0001.txt
  [Safety Connection]
  http://labs.ucop.edu/emc2/Netscape.html
  [Contracting Community]
  http://www.em.doe.gov/closure/final/alb.html
  [Accelerating Cleanup]

[313] NPDES
  http://www.dodson-hydro.com/
  http://www.pugetsound.org/p2/default.html
  http://www.epa.gov/earthlr6/6en/w/sw/home.htm
  http://www.calepa.cahwnet.gov/epadocs/npdes.txt
  http://www.epa.gov/owmitnet/npdes.htm
  http://www.mirsinfo.com/npdesmod.htm
  http://www.epa.gov/owm/tool.htm
  http://ag.arizona.edu/AZWATER/glossary/npdes.html
  http://www.ai.org/idem/owm/npdes/municipal/backgrnd.html

[314] GEOCHEMISTRY

http://www.geo.cornell.edu/geo . . . y/classes/
Geochemweblinks.HTML

http://www-ks.cr.usgs.gov/Kansas/reslab/

http://www.geo.brown.edu/

http://www.maik.rssi.ru/journals/geochem.htm

http://www.ciw.edu/Geo_seismo.html

http://www.undp.org/tcdc/cpr5004.htm

http://www.rgu.ac.uk/schools/egrg/home.htm

http://sedwww.cr.usgs.gov:8080/lillis/

[315] NUCLEAR REPOSITORY

http://www.elsevier.com/inca/p . . . /2/2/2/3/9/
522239.editix.shtml

http://www.glowingcoast.demon.co.uk/bitsbobs/b_nuked.htm

http://loft-gw.zone.org/cgi-bi . . . Num=30&CATEGORY=
[a-z]&STATE=NV

http://www.freeinfo.org/tch/fall96/resource/r7.htm

[316] NAPLS IN GROUNDWATER

http://e3power.com/CRT4.htm

http://www.esm.ucsb.edu/~keller/abstract.html

http://gwrp.cciw.ca/internet/b . . . rem-
archive/1997/msg00531.html

http://www.daienv.com/no_frames/contamin.htm

http://www.ets.uidaho.edu/che470/napl.htm

http://copland.udel.edu/~xf/introduction.html

http://civil.queensu.ca/individ/faculty/kueperp.htm

http://ccs.lbl.gov/TOUGH2/README/READT2VOC.html

[317] TOXIC HEAVY METALS in Groundwater

http://es.epa.gov/ncerqa/rfa/metals.html

http://www.rivm.nl/lib/Reports/719101019.html

http://www.cmst.org/OTD/tech_s . . .
IA/EPA_HSRC/Electr_Sensor.html

http://xre22.brooks.af.mil/estrg/indexes/Cleanup.htm

http://www.udayton.edu/udri/metcer.htm

http://www.anl.gov/LabDB/Current/Ext/H553-text.002.html

http://www.nttc.edu/env/doe/e/einolf.html

http://es.epa.gov/ncerqa_abstr . . .
/sbir/other/rem/tennakoon.html

http://pw2.netcom.com/~lmdmit84/hmm.html

[318] GROUNDWATER MANAGEMENT

http:www.flint.umich.edu/Dep . . . gionalGroundwater/
rgchome.html

http://www.foxcanyongma.org/

http://www.den.doi.gov/wwprac/

http://www.urisaoc.on.ca/prog9602.html

http://www.cumbre-summit.org/cumbre/test/wter0004.htm

http://unhinfo.unh.edu/ur-warm.html

http://www-esd.worldbank.org/envmat/vol2f96/strateg.htm

[319] GROUNDWATER PLANNING

http://www.ait.ac.th/clair/theses/pakistan/pk014.html

http://www.deq.state.mi.us/ogp/

http://www.mt.gov/dnrc/wrd/home.htm

http://www.tuns.ca/wwater/cwrs_ground.html

[320] WATER RESOURCES PLANNING

http://www.asce.org/confconted/wrpm99.html

http://water99.asce.org/Water99/

http://superior.lre.usace.army.mil/planning/planning.html

http://www.oieau.fr/euromed/anglais/ate_1/res/tsiourti.htm

http://cheddar-nyswri.cfe.cornell.edu/wrpc/

http://water.wr.usgs.gov/sw/

http://cosncr1.co.nrcs.usda.gov/wrp.htm

http://www-esd.worldbank.org/envmat/vol2f96/strateg.htm

http://www.netcomuk.co.uk/~jpap/hvidt.htm

http://www.engr.ucdavis.edu/~cedept/grad/water.html

http://www.ce.umanitoba.ca/water/wrr.html

http://www.ca.nrcs.usda.gov/wps/alhambra.html

http://www.pubs.asce.org/journals/marwr.html

http://www.clark.net/wrmi/ascepapr.htm

http://www.uwin.siu.edu/announce/event/1998/event0606.html

[321] GROUNDWATER DEVELOPMENT PLANS

http://lifewater.ca/m_develo.htm

http://www.pu.go.id/publik/pengai~1/html/eng/water.htm

http://www.hwr.arizona.edu/theses.html

http://www.water-ed.org/briefing.html

http://www.ipc.state.id.us/specs/03000/03353.htm

http://watershed.lake-coe.k12.ca.us/lakeinfo/gwater.html

http://www.rivm.nl/lib/Reports/259102011.html

http://www.unicef.org/wwd98/papers/unep.htm

[322] AQUIFER RECHARGE

http://www.epa.gov/r10earth/data/cara.html

http://www.e-aquifer.com/gis/gisnet.htm

http://tx.usgs.gov/program/TX174.html

http://www.charlotte-florida.c . . . Government/
AquiferRecharge.htm

http://txwww.cr.usgs.gov/program/TX169.html

http://twri.tamu.edu/twripubs/Insights/v2n2/article-8.html

http://volusia.org/gis/data/aqr.htm

http://www-dh.lnec.pt/gias/hydrogeology.htm

http://www.uswaternews.com/archive/96/supply/recharge.html

http://tx.usgs.gov/program/TX169.html

http://txwww.cr.usgs.gov./reports/fs/94/048/

http://www.pae.sa.gov.au/html/left_stormwater.htm

http://gw.ehnr.state.nc.us/planb/absrecharge_black_r.htm

http://www.geolsoc.org.uk/pubs/books/cat130.htm

[323] GROUNDWATER ABSTRACTION

http://www.gcrio.org/geo/level.html

http://www.sopac.org.fj/wasp/P . . . ns/Gallery%20Project/
index.htm

http://www.wmo.ch/web/homs/120310.html

http://www.wetlands.demon.co.uk/WaterBal.htm

http://www.cgd.ucar.edu/cas/ACACIA/wilbycv.html

http://isvapcsl.isva.dtu.dk/grc/1994/grcmain.htm

http://www.soloflo.com/

http://www.atlas.co.uk/listons/grndwtr1.htm

[324] CALIFORNIA GROUNDWATER LAW

http://members.aol.com/gdtlrfb/groundwater.htm

http://www.grac.org/

http://www.water-ed.org/briefing.html

http://www.groundh2o.org/

http://www.sdcwa.org/

http://www.ucop.edu/cps/grwater.html

[325] WATER LAW

http://www.waterlaws.com/

http://www.cnr.colostate.educ/CWK/

http://www.uwyo.edu/law/l&wlrev/l&wlrev.htm

http://www.bandersnatch.com/water.htm

http://www.nesarc.org/water.htm

http://www.house.state.mo.us/bills97/bills97/HB288.htm

http://www.cleinternational.com/wywat97.html

http://www.uswaternews.com/archive/97/conserv/texwat3.html
http://www.eluls.org/apr1998_waterlawupdate.html
http://www.bickerstaff.com/waterlawseminar.htm
http://www.cnr.colostate.edu/CWK/aq_wl_lw.htm
http://www.geology.ewu.edu/conf/waterlaw.htm
http://www.cleinternational.com/orlwat97.html
http://www.umkc.edu/umkc/catalog/htmlc/law/c725.html
http://www.cleinternational.com/denwat98.html

[326] KIRTLAND AIR FORCE BASE
http://www.kirtland.af.mil/
http://www.cmc.sandia.gov/about/visit/visit4.htm
http://www.kpc.nm.org/index1.htm
http://library.adelaide.edu.au/hytelnet/us6/us692.html

[327] CITY OF ALBUQUERQUE, NM
http://www.cabq.gov/
http://www.cabq.gov/cip/cipcplan.html
http://www.nmcjnet.org/
http://gisweb.cabq.gov/
http://www.cabq.gov/resources/index.html

[328] HAZARDOUS MATERIALS TRANSPORTATION ACT
http://www.em.doe.gov/emtrain/f3f.html
http://www.citation.com/hpages/hmta2.html
http://homer.hsr.ornl.gov/oepa/law_sum/HMTA.HTM
http://www.claitors.com/prf/catelog/552-070-21739-8.html
http://www.rw.doe.gov/pages/resource/fedreg/180c.htm
http://www.engrng.pitt.edu/~ch . . . s/CHE2983/
    CHE2983-1/sld044.htm
http://www.epa.gov/epaoswer/hazwaste/id/char/append.txt

[329] WATER POLLUTION CONTROL ACT
http://www.epa.gov.tw/english/Laws/wpcact.htm
http://www.usbr.gov/laws/cleanwat.html
http://www.fws.gov/laws/digest/reslaws/fwatrpo.html
http://www.msnbc.com/news/WLD/iframes/WaterQuality.asp
http://www.pmei.com/nrc/f/Fede . . . ter_Pollution_
    Control_Act.html
http://www.spea.iupui.edu/iupu . . .
    97/h320r361/h320env/sld007.htm
http://ucs.byu.edu/bioag/aghort/214pres/watrpoln/tsld004.htm
http://www.engrng.pitt.edu/~ch . . . /CHE2983/
    CHE2983-1/tsld043.htm

[330] ATOMIC ENERGY ACT
http://homer.hsr.ornl.gov/oepa/law_sum/AEA.HTM
http://www.pmei.com/nrc/a/Atomic_Energy_Act.html
http://www.doe.gov/osti/timp.html
http://www.em.doe.gov/emtrain/f3p.html
http://tis-nt.eh.doe.gov/oepa/comments/aea.htm
http://www.hmso.gov.uk/acts/summary/01989007.htm
http://www.nils.com/rupps/706.htm
http://194.128.65.3/acts/summary/01989007.htm

[331] CLEAN AIR ACT
http://earth1.epa.gov/oar/oaqps/peg_caa/pegcaain.html
http://envinfo.com/caalead.html
http://www.epa.gov/swercepp/rmp-imp.html
http://www.epa.gov/acidrain/lawsregs/caaa.html
http://www.dfwinfo.com/envir/bikeped/caa.html
http://www.cnie.org/nle/air-9.html
http://www.citation.com/hpages/caamini.html

[332] OSHA
http://www.osha.gov/
http://www.oshadata.com/

http://www.oshaproof.com/main.htm
http://www.osha-slc.gov/OshStd_toc/OSHA_Std_toc.html
http://www.npr.gov/initiati/common/osha.html
http://www.osha-slc.gov/OCIS/toc_fed_reg.html
http://www.nsi.org/Tips/COMPLAIN.HTM
http://www.osha-slc.gov/html/construction.html
http://www.osha-slc.gov/html/dbsearch.html

[333] NEW MEXICO ENVIRONMENT DEPARTMENT; OVERSIGHT WITH DOE
http://www.nmenv.state.nm.us/ [NMED]
http://www.nmenv.state.nm.us/DOE_Oversight/doetop.html
    [DOE Related]
http://www.clay.net/statag.html [Government Agencies and All
    State's—Main Links]
http://www.nmenv.state.nm.us/DOE_Oversight/speaker.html
    [DOE Related]
http://www.nmenv.state.nm.us/DOE_Oversight/activities.html
    [DOE Related]
http://www.nmenv.state.nm.us/DOE_Oversight/program.html
    [DOE Related]
http://www.nmenv.state.nm.us/DOE_Oversight/techrep.html
    [DOE Related]
http://www.nmenv.state.nm.us/DOE_Oversight/newsletter.html
    [DOE Related]
http://www.nmenv.state.nm.us/DOE_Oversight/annual95.html
    [DOE Related]

[334] HOUSEHOLD HAZARDOUS WASTE
http://www.ae.iastate.edu/water/pm1334g.txt
http://www.curbsideinc.com/ [Home of Hazardous Waste
    Management]
http://www.1800cleanup.org/text/hhwaste/hhwaste.htm
    [Household Hazardous Waste]
http://www.lcswma.org/ [Lancaster Solid Waste Authority,
    Lancaster, PA]
http://www.state.nh.us/des/planning/hhw.htm [New Hampshire
    Department of Environmental Services Program]
http://www.tnrcc.state.tx.us/exec/oppr/hhw/nicadg.html
    [Household Hazardous Waste]
http://www.orcbs.msu.edu/AWARE/pamphlets/list.html
    [Information Pamphlets]
http://www.ci.mpls.mn.us/cityw . . . lic-works/
    solid-waste/hhw.html
http://www.hhw.org/ [Santa Clara Program]
http://www.cswma.org/hhw.htm

[335] OTHER DATABASES AND RELATED LINKS
{1} http://library.dialog.com/bluesheets/html/
    blow.html#WATER [Dialog Databases on Water]
{2} http://earth.agu.org/pubs/inpress.html [Contents and
    Abstracts of Recent American Geophysical Union
    Publications]
{3} http://earth.agu.org/pubs/agu_jour.html [AGU Journals]
{4} http://earth.agu.org/pubs/agu_jourwrr.html [Water
    Resources Research Journal]
{5} http://www.agu.org/wrr/ [Water Resources Research
    Online]
{6} http://www.csa.com/ [Cambridge Scientific Abstracts]
{7} http://www.ei.org/ [Engineering Information Incorporated]
{8} http://www.isinet.com/ Institute for Scientific Information]
{9} http://www.unm.edu/~csel/ [University of New Mexico
    Centennial Science and Engineering Library]
{10} http://www.unm.edu/~csel/electronic_journals.html
    [UNM Electronic Journals]

{11} http://www.pubs.asce.org/ [American Society Civil Engineers Publication Page]

{12} http://www.pubs.asce.org/journals/jlist.html [American Society Engineers—Journals]

{13} http://www.unm.edu/~csel/guides/handouts/electronic_tools.html [UNM Library Tools]

{14} http://www.doe.gov/dra/dra.html [DOE Reports Bibliographic Database]

{15} http://www-sul.stanford.edu/ [Stanford University Library]

{16} http://www.library.yale.edu/ [Yale University Library]

{17} http://cd-rom-guide.com/cdprod1/cdhrec/009/098.shtml [Groundwater and Soil Contamination Database]

{18} http://www.infonordic.se/GROUND.html

{19} http://www.agiweb.org/agi/pubs/newpub.html [Groundwater and Soil Contamination Database]

{20} http://www.silverplatter.com/catalog/gwsc.htm [Groundwater and Soil Contamination Database]

{21} http://lib-www.lanl.gov/ [Los Alamos National Laboratory Research Library]

http://lib-www.lanl.gov/libinfo/libs.htm [Los Alamos National Laboratory Links to Other Libraries]

{22} http://lib-ww.lanl.gov/libinfo/libs.htm#world [Los Alamos National Laboratory Links to World Libraries]

[336] OTHER GROUNDWATER RELATED LINKS

{1} http://www.dnr.state.wi.us/org/water/dwg/gw/ [Groundwater Information]

{2} http://www.ag.state.co.us/DPI/publications/waterbmp.html [Agricultural Chemicals and Groundwater Protection]

{3} http://www.ag.state.co.us/DPI/programs/groundwater.html [Groundwater Publications: pdf]

{4} http://well.water.ca.gov/gwbrochure/ [Groundwater the Hidden Water Supply]

{5} http://www.wqa.org/WQIS/Glossary/A-list.html [WQA Glossary of Water Terms]

{6} http://www.wef.org/docs/wclinkro.html [Water Environment Web-Related Links]

{7} http://www.wef.org/docs/waterenvres.html [Water Environment Research Journal]

{8} http://gwrp.cciw.ca/gwrp/studies/piggott/index.html [Groundwater Geomechanics]

{9} http://www.per.dwr.csiro.au/CGS/research.html

{10} http://ces.soil.ncsu.edu/soilscience/publications/Soilfacts/AG-439-09 [Soil Facts]

{11} http://www.crestech.ca/WaterResources/water.htm [CRESTech Water Resources]

{12} http://ces.soil.ncsu.edu:/soilscience/publications/Soilfacts/AG-439-08/ [Groundwater: Risk Assessment]

{13} http://www.cciw.ca/nwri-e/aerb/groundwater-remediation [Aquatic Ecosystem Restoration]

{14} http://gwintl.gwi.memphis.edu/research.htm [GWI Research Summary]

{15} http://www.clw.csiro.au/research/groundwater [Groundwater Management and Site Remediation]

{16} http://www.uwsp.edu/groundwater [Central Wisconsin Groundwater Center]

{17} http://www.mos.gov.pl/soe/10.htm [Groundwater—Poland]

{18} http://www.library.wisc.edu/libraries/Water_Resources/wgrmp/listing.htm [List of Groundwater Summaries]

{19} http://gw2.cciw.ca/gwrp [Groundwater Remediation Project]

{20} http://clu-in.com/phytotce.htm [Phytoremediation of TCE]

{21} http://gwpc.site.net/mainsite.htm [Groundwater Protection Council]

{22} http://gwrp.cciw.ca/gwrp/publications [USGS California Projects]

{23} http://ndsuext.nodak.edu/extpubs/h2oqual/watgrnd/ae1113w.htm [Improved Pesticide Applications BMPS for Groundwater Protection]

{24} http://water.wr.usgs.gov/projects/ca494.html [USGS California Projects]

{25} http://waterhome.tamu.edu:/texasyst/texasystworkbooks/index.html [TEXASyst]

{26} http://magma.Mines.EDU/igwmc/books [Ground-Modeling Publications]

{27} http://www.sway.com/~pacific [The Pacific Institute for Advanced Studies]

{28} http://sedwww.cr.usgs.gov:8080/radon/rnpubs.html [USGS Publications on Radon]

{29} http://water.wr.usgs.gov/projects/ca474.html [USGS California Projects: Wolf Valley]

{30} http://water.wr.usgs.gov/projects/ca477.html [USGS California Projects]

{31} http://glsun2.gl.rhbnc.ac.uk/what.html [Department of Geology, University of London]

{32} http://www.gwrtac.org:80/html/techdocs.html [GW Technologies Center]

{33} http://www.env.gov.bc.ca/wat/waterbot/gwell-out.html [British Columbia Ministry—Water]

{34} http://www.epa.gov/ORD/WebPubs/pumptreat [Pump and Treat Water Remediation]

{35} http://crs-www.bu.edu [Center for Remote Sensing—Boston]

{36} http://water.wr.usgs.gov/projects/ca493.html [USGS California: San Francisco Bay]

{37} http://vulcan.wr.usgs.gov/Projects/framework.html [USGS Cascades Volcano]

{38} http://gwint1.gwi.memphis.edu/gps/gps.htm [Groundwater Institute—Memphis]

{39} http://www.contaminatedland.co.uk [Contaminated Land in UK and other Links]

{40} http://www.ppc.ubc.ca/sludge.html [University of British Columbia: Wastewater Sludge Management]

{41} http://www.ppc.ubc.ca/bioconv.html [University of British Columbia: Bioconversion of Solid Wastes]

{42} http://www.epa.gov/enviro/html/sdwis/sdwis_ov.html [USEPA Safe Drinking Water Overview]

{43} http://www.nsac.ns.ca/nsdam/pt/projsum/95/pr95r03.htm [Nova Scotia: Evaluation of Wetlands]

{44} http://www.worldbank.org/nipr/comrole.htm [Role of Community in Pollution Control]

{45} http://www.epa.gov/OGWDW/sources/swpguid.html [USEPA: State Source and Water Protection]

{46} http://home.acadia.net/cbm/Rad.html [Radnet: Source Points of Anthropogenic Radioactivity]

{47} http://www.zebra.net/~rctinc [Resource Compliance Technologies, Inc. with Links]

{48} http://eagle.emweb.icx.net/bemr96/itri.html [Inhalation Toxicology Research Institute: Kirtland Air Force Base, NM]

{49} http://www.ipm.ucdavis.edu/PUSE/1993/cc93-sp.01.html [UC, California Pesticide Use Summaries]

{50} http://www.epa.gov/superfund/oerr/impm/products/nplsites/usmap.htm [Superfund]

{51} http://www-ks.cr.usgs.gov/Kansas/reslab/biblio.html [USEPA Superfund Web Site]

{52} http://www.pmac.net/pestenv.htm">http://www.pmac.net/pestenv.htm

{53} http://carpe.gecp.virginia.edu [Central Africa Regional Program for the Environment]

{54} http://ss.niaes.affrc.go.jp/index_e.html [National Institute for Agro-Environmental Sciences, Japan]

{55} http://www.kist.re.kr [Korea Institute of Science and Technology]

{56} http://www.uswaternews.com:80 [US Water News Online]

{57} http://www.hampshire.org [Hampshire Research Institute, New Jersey]

{58} http://www.es.anl.gov/htmls/rma.chem.html [Treating Soil Contaminated by Chemical Warfare Agents]

{59} http://www.sos.state.ia.us/register/r8/r8agri.htm [Iowa, Groundwater Protection]

{60} http://www.aist.go.jp/GSJ/pEQ/eq_top.htm [Earthquake Research and Groundwater, Geological Survey of Japan]

{61} http://www.aist.go.jp/NIRE/index_e.htm [National Institute for Resources and Environment, Japan]

{62} http://water.usgs.gov/software/mocdense.html [USGS Water Resources Application Software, Mocdense]

{63} http://www.ccn.cs.dal.ca/Science/SWCS/INFO/wetlands01 [Literature Citations]

{64} http://biogroup.gzea.com [Bioremediation Discussion Group]

{65} http://water.wr.usgs.gov/gwatlas [USGS Groundwater Atlas, California and Nevada]

{66} http:/www.sws.uiuc.edu [Illinois State Water Survey]

{67} http://www.nwi.fws.gov/values_wais.html [Wetlands Values Database]

{68} http://image.fs.uidaho.edu/center2 [Center for Hazardous Waste Remediation Research]

{69} http://water.usgs.gov/software/hst3d.html [USGS Water Resources Application Software, Hst3d]

{70} http://water.usgs.gov/software/hydrotherm.html [USGS Water Resources Application Software, Hydrotherm]

{71} http://www-ep.es.llnl.gov/www-ep/esd/geochem/geochem.html [Geochemistry Group, LLNL]

{72} http://water.usgs.gov/software/analgwst.html [USGS Water Resources Application Software, Analgwst]

{73} http://www.engg.ksu.edu/HSRC [Rocky Mountain Hazardous Substance Research Center]

{74} http://water.usgs.gov/software/hysep.html [USGS Water Resources Software Applications, Hysep]

{75} http://ianrwww.unl.edu/ianr/wcrec/water/index.htm [Water Research, University of Nebraska]

{76} http://wwwrvares.er.usgs.gov/nawqa/nawqamap.html [USGS—National Water Quality Assessment]

{77} http://wwwrvares.er.usgs.gov/nawqa/nawqa_home.html [USGS—National Water Quality Assessment—Details]

{78} http://hermes.ecn.purdue.edu:8001/server/water/bib/wq.html [Purdue University—Water Quality Materials Bibliography]

{79} http://nj.usgs.gov/delr/index.html [National Water Quality Assessment—Delaware]

{80} http://bowdnhbow.er.usgs.gov/nawqaweb.html [USGS—New England Coastal Basin]

{81} http://www.rtdf.org/phytobib.htm [USEPA Phytoremediation Bibliography]

{82} http://water.usgs.gov/software/radmod.html [USGS Water Resources Application Software, Radmod]

{83} http:/www.clw.csiro.au/research/catchment [Sustainable Catchment Management]

{84} http://www.lib.ttu.edu/playa/og.htm [Playa Lakes and Ogallala Aquifer]

{85} http://ecsask65.innovplace.saskatoon.sk.ca/pages/pub/1995.html [NHRI Publications]

{86} http://water.wr.usgs.gov/fact/b07 [Seawater Intrusion Coastal Aquifer]

{87} http://water.wr.usgs.gov/calbib/index.html [USGS Water Resources Bibliography, California]

{88} http://www.nal.usda.gov/wqic/wgwq/progress.html [USDA Water Quality Report]

{89} http://water.usgs.gov/software/moc3d.html [USGA Water Resources Application Software, Moc3d]

{90} http://water.usgs.gov/software/modpath.html [USGS Water Resources Application Software, Modpath]

{91} http://wwwsd.cr.usgs.gov/nawqa/vocns [USGS National Assessment of Volatile Organics]

{92} http://gwrp.cciw.ca/gwrp/studies/piggott/index.html [Groundwater Geomechanics]

{93} http://gwrp.cciw.ca:80/gwrp [Groundwater Remediation Project]

{94} http://ces.soil.ncsu.edu/soilscience/publications/Soilfacts/AG-439-09 [Soil Management Protects Groundwater]

{95} http://wrri.nmsu.edu/ [New Mexico Water Resources Research Institute]

{96} http://wrri.nmsu.edu/publish/techrpt/techrpt.html [NMWRRI Publications]

{97} http://www.r3-bardos.demon.co.uk/NATO/natoreps.html [NATO/CCMS]

{98} http://arl.cni.org/scomm/copyright/uses.html [Fair Use in Electronic Age]

{99} http://www.cyberlawcentre.org.uk/ao05000.html [Cyber Law]

{100} http://www.theta.com/trfn/netsurfer.html [Simple Guide to Copyright]

{101} http://www.cup.cam.ac.uk/Journals/JNLSCAT/clj/clj.html [Cambridge Law Journal]

{102} http://www.cyberlaw.com/ [Cyber Law]

{103} http://www.albany.net/allinone/all1www.html [All in One Search Engine]

{104} http://www.metacrawler.com [Search Engine]

{105} http://wombat.doc.ic.ac.uk/foldoc/foldoc.cgi?Uniform+Resource+Locator [URL Def.]

{106} http://wombat.doc.ic.ac.uk/foldoc/index.html [Free Online Dictionary on Computing]

{107} http://wombat.doc.ic.ac.uk/misc.html [Website Lists and Catalogs]

[337] SOME U.S. EPA PUBLICATIONS

{1} SUPERFUND RECORD OF DECISION: MCCOLL SUPERFUND SITE, (GROUNDWATER O.U.), FULLERTON, CA

http://www.epa.gov/ncepihom/Catalog/EPARODR0996154.html

{2} SUPERFUND RECORD OF DECISION: SOUTH-EAST ROCKFORD GROUNDWATER CONTAMI-NATION SITE, ROCKFORD, IL

http://www.epa.gov/ncepihom/Catalog/EPARODR0595277.html

{3} SUPERFUND RECORD OF DECISION: NAVAL AIR ENGINEERING STATION, AREA C SOIL AND GROUNDWATER, LAKEHURST, NEW JERSEY

http://www.epa.gov/ncepihom/Catalog/EPARODR0296270.html

{4} SUPERFUND RECORD OF DECISION: MATERIAL TECHNOLOGY LABORATORY SITE, US ARMY SOILS AND GROUNDWATER O.U., WATERTOWN, MASSACHUSETTS

http://www.epa.gov/ncepihom/Catalog/
EPARODR0196124.html

{5} GROUNDWATER PROTECTION: WATER
QUALITY MANAGEMENT REPORT
http://www.epa.gov/ncepihom/Catalog/
EPAOSW000886.html

{6} SUPERFUND RECORD OF DECISION AMEND-
MENT: KOPPERS COMPANY, INC., SUPERFUND
SITE, (SOIL AND GROUNDWATER OPERABLE
UNIT) (OROVILLE PLANT), OROVILLE, CA
http://www.epa.gov/ncepihom/Catalog/
EPAAMDR0996151.html

{7} PROCEEDINGS: CHESAPEAKE BAY GROUND-
WATER TOXICS LOADING WORKSHOP,
BASINWIDE TOXICS REDUCTION STRATEGY
REEVALUATION REPORT
http://www.epa:gov/ncepihom/Catalog/
EPA903R93010.html

{8} EXPOSURE POINT CONCENTRATIONS IN
GROUNDWATER
http://www.epa.gov/ncepihom/Catalog/
EPA903891002.html

{9} NATIONAL WATER QUALITY INVENTORY:
REPORT TO CONGRESS, GROUNDWATER
CHAPTERS, 1996
http://www.epa.gov/ncepihom/Catalog/
EPA816R98011.html

{10} GROUNDWATER AND LAND USE IN THE WATER
CYCLE {POSTER-LARGER VERSION}
http://www.epa.gov/ncepihom/Catalog/
EPA813H95002.html

{11} GROUNDWATER AND LEACHATE TREATMENT
SYSTEMS {MANUAL}
http://www.epa.gov/ncepihom/Catalog/
EPA625R94005.html

{12} ENVIRONMENTAL RESEARCH BRIEF:
BIOAUGMENTATION WITH BURKHOLDERIA
CEPACIA PR1301 FOR IN SITU BIOREMEDIATION
OF TRICHLOROETHYLENE CONTAMINATED
GROUNDWATER
http://www.epa.gov/ncepihom/Catalog/
EPA600S98001.html

{13} GROUNDWATER AND LEACHATE TREATABIL-
ITY STUDIES AT FOUR SUPERFUND SITES:
PROJECT SUMMARY
http://www.epa.gov/ncepihom/Catalog/
EPA600S286029.html

{14} SURFACE IMPOUNDMENTS AND THEIR
EFFECTS ON GROUNDWATER QUALITY IN THE
UNITED STATES: A PRELIMINARY SURVEY,
JANUARY 1978
http://www.epa.gov/ncepihom/Catalog/
EPA570978005.html

{15} NATO/CCMS PILOT STUDY: EVALUATION OF
DEMONSTRATED AND EMERGING TECHNOLO-
GIES FOR THE TREATMENT OF CONTAMINATED
LAND AND GROUNDWATER (PHASE III)
SPECIAL SESSION: TREATMENT WALLS AND
PERMEABLE REACTIVE BARRIERS
http://www.epa.gov/ncepihom/Catalog/
EPA542R98003.html

{16} NATO/CCMS PILOT STUDY: EVALUATION OF
DEMONSTRATED AND EMERGING TECHNOLO-
GIES FOR THE TREATMENT OF CONTAMINATED
LAND AND GROUNDWATER (PHASE III)
ANNUAL REPORT, 1998
http://www.epa.gov/ncepihom/Catalog/
EPA542R98002.html

{17} EVALUATION OF DEMONSTRATED AND EMERG-
ING TECHNOLOGIES FOR THE TREATMENT AND
CLEANUP OF CONTAMINATED LAND AND
GROUNDWATER, NATO/CCMS PILOT STUDY,
PHASE II, APPENDIX IV, PROJECT SUMMARIES
http://www.epa.gov/ncepihom/Catalog/
EPA542R98001C.html

{18} EVALUATION OF DEMONSTRATED AND
EMERGING TECHNOLOGIES FOR THE TREAT-
MENT AND CLEANUP OF CONTAMINATED
LAND AND GROUNDWATER, NATO/CCMS PILOT
STUDY, PHASE II, OVERVIEW REPORT
http://www.epa.gov/ncepihom/Catalog/
EPA542R98001B.html

{19} EVALUATION OF DEMONSTRATED AND EMERG-
ING TECHNOLOGIES FOR THE TREATMENT AND
CLEANUP OF CONTAMINATED LAND AND
GROUNDWATER (PHASE 2), NATO/CCMS PILOT
STUDY, INTERIM STATUS REPORT, NUMBER 203
http://www.epa.gov/ncepihom/Catalog/
EPA542R95006.html

{20} REMEDIATION CASE STUDIES: GROUNDWATER
TREATMENT
http://www.epa.gov/ncepihom/Catalog/
EPA542R95003.html

{21} GROUNDWATER CURRENTS, ISSUE NUMBER 29,
SEPTEMBER 1998
http://www.epa.gov/ncepihom/Catalog/
EPA542N98008.html

{22} SITE EMERGING TECHNOLOGIES: LASER-
INDUCED PHOTOCHEMICAL OXIDATIVE DE-
STRUCTION OF TOXIC ORGANICS IN
LEACHATES AND GROUNDWATER
http://www.epa.gov/ncepihom/Catalog/
EPA540SR92080.html

{23} INTRODUCTION TO GROUNDWATER
INVESTIGATIONS
http://www.epa.gov/ncepihom/Catalog/
EPA540R95001.html

{24} SEMINAR SERIES: MONITORED NATURAL
ATTENUATION FOR GROUNDWATER
http://www.epa.gov/ncepihom/Catalog/
EPA540F98500.html

{25} APPLICATIONS ANALYSIS REPORT: MEMBRANE
TREATMENT OF WOOD PRESERVING SITE
GROUNDWATER BY SBP TECHNOLOGIES, INC.
http://www.epa.gov/ncepihom/Catalog/
EPA540AR92014.html

{26} APPLICATIONS ANALYSIS REPORT: BIOLOGI-
CAL TREATMENT OF WOOD PRESERVING SITE
GROUNDWATER BY BIOTROL, INC.
http://www.epa.gov/ncepihom/Catalog/
EPA540A591001.html

{27} GUIDE TO PUMP AND TREAT GROUNDWATER
REMEDIATION TECHNOLOGY
http://www.epa.gov/ncepihom/Catalog/
EPA540290018.html

{28} INTERNATIONAL EVALUATION OF IN SITU
BIORESTORATION OF CONTAMINATED SOIL
AND GROUNDWATER
http://www.epa.gov/ncepihom/Catalog/
EPA540290012.html

{29} EVALUATION OF GROUNDWATER EXTRACTION
REMEDIES, VOLUME 3: GENERAL SITE
DATABASE REPORTS INTERIM FINAL
http://www.epa.gov/ncepihom/Catalog/
EPA540289054C.html

{30} EVALUATION OF GROUNDWATER EXTRACTION REMEDIES, VOLUME 2: CASE STUDIES 1–19 INTERIM FINAL

http://www.epa.gov/ncepihom/Catalog/EPA540289054B.html

{31} GROUNDWATER PATHWAY ANALYSIS FOR ALUMINUM POTLINERS (KO88) {DRAFT}

http://www.epa.gov/ncepihom/Catalog/EPA530R97023.html

{32} RCRA, SUPERFUND, AND EPCRA HOTLINE TRAINING MODULE: INTRODUCTION TO: GROUNDWATER MONITORING, 40 CFR PARTS 264/265, SUBPART F, JULY 1996

http://www.epa.gov/ncepihom/Catalog/EPA530R96030.html

{33} RCRA/UST, SUPERFUND, AND EPCRA HOTLINE TRAINING MODULE: INTRODUCTION TO: GROUNDWATER MONITORING, 40 CFR PARTS 264/265, SUBPART F, JULY 1995

http://www.epa.gov/ncepihom/Catalog/EPA530R95065.html

{34} GROUNDWATER PROTECTION STANDARDS FOR INACTIVE URANIUM TAILINGS SITES (40 CFR 192). BACKGROUND INFORMATION FOR FINAL RULE

http://www.epa.gov/ncepihom/Catalog/EPA520188023.html

{35} GROUNDWATER MODELING COMPENDIUM: MODEL FACT SHEETS, DESCRIPTION, APPLICATIONS AND COST GUIDELINES, 2ND EDITION

http://www.epa.gov/ncepihom/Catalog/EPA500B94004.html

{36} ELECTROKINETIC LABORATORY AND FIELD PROCESSES APPLICABLE TO RADIOACTIVE AND HAZARDOUS MIXED WASTE IN SOIL AND GROUNDWATER

http://www.epa.gov/ncepihom/Catalog/EPA402R97006.html

{37} K:\WWWROOT\DOWNLOAD\REMED\GW-TECH.BIB

Summary: This bibliography identifies reports, journal articles, and conference proceedings published from 1990 to 1996 that focus on innovative technologies for the remediation of contaminated groundwater.

http://www.epa.gov:80/swertiol/download/remed/gwbib.pdf

{38} MDA-PPLN

Summary: MONTROSE AND DEL AMO SUPERFUND SITES UNITED STATES ENVIRONMENTAL PROTECTION AGENCY · REGION 9 · SAN FRANCISCO, CA · JUNE 1998: At Montrose and Del Amo Superfund sites, EPA Proposes Groundwater Cleanup Plan (General Fact Sheet Version).

http://www.epa.gov:80/region09/waste/sfund/npl/delamo/document/mda-ppl n.pdf

{39} Agricultural Drainage Wells (5F1)

Summary: Risk Assessment—PDF document

http://www.epa.gov:80/ogwdw000/uic/5fl.pdf

{40} DSS Demonstration Plan

Summary: For each pilot, EPA utilizes the expertise of partner "verification organizations" to design efficient procedures for conducting performance tests of environmental technologies.

http://www.epa.gov:80/etvprgrm/02/dssplan.pdf

{41} ROD.PDF

Summary: Part II Responsiveness Summary: This section presents Environmental Protection Agency's

(EPA's) responses to the written and oral comments received at the public meeting and during the public comment period.

http://www.epa.gov:80/region09/waste/sfund/npl/oii/rod2.pdf

{42} ROD 1.PDF

Summary: Description of the Remedy: This ROD addresses liquids control and contaminated groundwater as well as long-term operation and maintenance of all environmental control facilities at the landfill.

http://www.epa.gov:80/region09/waste/sfund/npl/oii/rod1.pdf

{43} MDA-PPLN

Summary: MONTROSE AND DEL AMO SUPERFUND SITES UNITED STATES ENVIRONMENTAL PROTECTION AGENCY · REGION 9 · SAN FRANCISCO, CA ·JUNE 1998: At Montrose and Del Amo Superfund sites, EPA Proposes Groundwater Cleanup Plan (General Fact Sheet Version).

http://www.epa.gov:80/region09/waste/sfund/npl/delamo/document/mda-ppl n.pdf

{44} 600/SR-94/051

Summary: Potential problem pollutants were identified, based on their mobility through the unsaturated soil zone above groundwater, their abundance in stormwater, and their treatability before discharge.

http://www.epa.gov:80/ordntrnt/ORD/WebPubs/projsum/600sr94051.pdf

{45} OGALLALA GROUNDWATER CONTAMINATION SITE

Summary: The Ogallala Groundwater Contamination Site consists of two properties approximately 15 acres and 1 acre in size, respectively.

http://www.epa.gov/rgytgrnj/programs/spfd/nplfacts/ogallala.html

{46} EPA Region 1—Press Release: EPA ANNOUNCES NO RISKS AT CHESHIRE GROUNDWATER CONTAMINATION SUPERFUND SITE

Summary: To solicit input on the proposal, the EPA will hold a public meeting at 7 p.m., Oct. 24 at the Cheshire Town Hall. In addition, a public comment period will run from Oct. 21 through Nov. 20 for those who would like to send comments to the EPA.

http://www.epa.gov/region01/pr/files/pr1010a.html

{47} Bally Groundwater Contamination

Summary: The Bally Ground Water Contamination site consists of an area of groundwater contamination in and around the Bally Engineered Structures ("BES") plant in the borough of Bally, Pennsylvania. Potential Health Risks are listed.

http://www.epa.gov/reg3hwmd/super/bally/pad.htm

{48} US EPA Region 2: NPL Site Fact Sheets

Summary: The Dover Municipal Well: The Dover Water Commission owns and operates this municipal well field. Further investigations define source areas of contamination or additional areas of groundwater contamination and remediation.

http://www.epa.gov/r02earth/superfnd/site_sum/0200768c.htm

{49} EPA Region 2 Internet [EPA Selects Remedy for Groundwater Contamination at the Goldisc Recordings Superfund Site in Holbrook, Long Island]

Summary: The U.S. Environmental Protection Agency (EPA) has selected a plan to address low-level groundwater contamination at the Goldisc Recordings Superfund site in the Village of Holbrook on Long Island, NY.

http://www.epa.gov/r02earth/epd/98140.htm

{50} Record of Decision (ROD) Abstract

Summary: and extent of groundwater contamination was largely unknown. Road, Eleventh Street and Kishwaukee Street. Primary source of potable water is groundwater.

http://www.epa.gov/oerrpage/superfnd/web/sites/que/rods/r0595277.htm

{51} EPA National Priorities List 0503017n.htm

Summary: The 11-acre Ossineke Groundwater Contamination site resulted from a series of unrelated spills and incidents that contaminated the groundwater of local residents within the LaBell subdivision.

http://www.epa.gov/oerrpage/superfnd/web/sites/nplsites/0503017n.htm

{52} EPA National Priorities List 0500955n.htm

Summary: The Southeast Rockford Groundwater Contamination site covers approximately 4 square miles in Rockford, Illinois.

http://www.epa.gov/oerrpage/superfnd/web/sites/nplsites/0500955n.htm

{53} EPA National Priorities List 0202330n.htm

Summary: The 5-acre Rowe Industries Groundwater Contamination site, located on the eastern side of the Sag Harbor Bridgehampton Turnpike, was owned and operated by Rowe Industries, Inc., from the 1950s through the early 1960s.

http://www.epa.gov/oerrpage/superfnd/web/sites/nplsites/0202330n.htm

{54} Choose a Site by County, City, or Site Name

Summary: 1 10TH STREET SITE 13TH & STOCK-WELL 5 56TH ST GROUNDWATER CONTAMINATION SITE 9 9TH & CALVERT A ABANDONED D O D BOMB FACILITY ADAMS MUNICIPAL WATER SUPPLY WELLS.

http://www.epa.gov/oerrpage/superfnd/web/sites/cursites/toc/nestate.ht m

{55} Choose a Site by County, City, or Site Name

Summary: ALVO FORMER GRAIN STORAGE DOUGLAS FORMER GRAIN STORAGE ADAMS ARMY GUARD WET SITE AYR GROUNDWATER FARMLAND INDUSTRIES INC—HASTINGS CITY LANDFILL—HASTINGS GROUND WATER CONTAMINATION BOX BUTTE ALLIANCE GROUND WATER CONTAMINATION ALLIANCE

http://www.epa.gov/oerrpage/superfnd/web/sites/cursites/toc/necnty.htm

{56} Choose a Site by County, City, or Site Name

Summary: ADAMS ADAMS MUNICIPAL WATER SUPPLY WELLS ALDA GRAND ISLAND DRUMS THREE D INVESTMENTS INC ALLIANCE ALLIANCE GROUND WATER CONTAMINATION ALLIANCE MUNICIPAL LANDFILL ALVO ALVO FORMER GRAIN STORAGE ARAPAHOE

http://www.epa.gov/oerrpage/superfnd/web/sites/cursites/toc/necity.htm

{57} EPA and Superfund Logo

Summary: Site Info | Aliases | Operable Units | Actions | Financial | Choose Another Site 00

SITEWIDE 01 EXTRACTION WELL 02 FINAL REMEDY

http://www.epa.gov/oerrpage/superfnd/web/sites/cursites/c3wi/o505186.htm

{58} EPA and Superfund Logo

Summary: Site Info | Aliases | Operable Units | Actions | Financial |

http://www.epa.gov/oerrpage/superfnd/web/sites/cursites/c3tn/o404230.htm

{59} EPA and Superfund Logo

Summary: Site Info | Aliases | Operable Units | Financial | Choose Another Site Alias ID Alias Name/Address Alias Latitude/Longitude

http://www.epa.gov/oerrpage/superfnd/web/sites/cursites/c3tn/1404230.htm

{60} EPA and Superfund Logo

Summary: Site Info | Aliases | Operable Units | Actions | Financial | Choose Another Site 00

SITEWIDE 01 GROUNDWATER CONTAMINATION 02 03

http://www.epa.gov/oerrpage/superfnd/web/sites/cursites/c3pr/o203140.htm

{61} EPA and Superfund Logo

Summary: Site Info | Aliases | Operable Units | Actions | Financial | Choose Another Site 00

SITEWIDE 01 02 03 GROUNDWATER CONTAMINATION

http://www.epa.gov/oerrpage/superfnd/web/sites/cursites/c3pa/o301227.htm

{62} EPA and Superfund Logo

Summary: LEBANON GROUNDWATER CONTAMINATION Site Info | Aliases | Operable Units | Actions | Financial |

http://www.epa.gov/oerrpage/superfnd/web/sites/cursites/c3or/o001756.htm

{63} EPA and Superfund Logo

Summary: SMITHTOWN GROUNDWATER CONTAMINATION Site Info | Aliases | Operable Units | Actions | Financial |

http://www.epa.gov/oerrpage/superfnd/web/sites/cursites/c3ny/o204148.htm

{64} EPA and Superfund Logo

Summary: EXETER GROUNDWATER CONTAMINATION Site Info | Aliases | Operable Units | Actions | Financial |

http://www.epa.gov/oerrpage/superfnd/web/sites/cursites/c3ne/o703127.htm

{65} EPA and Superfund Logo

Summary: WAYNE HWY 15 GROUNDWATER CONTAMINATION Site Info | Aliases | Operable Units | Actions | Financial |

http://www.epa.gov/oerrpage/superfnd/web/sites/cursites/c3ne/o703087.htm

{66} EPA and Superfund Logo

Summary: ROSCOE HWY 30 GROUNDWATER CONTAMINATION Site Info | Aliases | Operable Units | Actions | Financial |

http://www.epa.gov/oerrpage/superfnd/web/sites/cursites/c3ne/o703086.htm

{67} EPA and Superfund Logo

Summary: CRAIG GROUNDWATER CONTAMINATION Site Info | Aliases | Operable Units | Actions | Financial | http://www.epa.gov/oerrpage/superfnd/web/sites/cursites/c3ne/o703083.htm

{68} EPA and Superfund Logo

Summary: HERMAN GROUNDWATER CONTAMINATION Site Info | Aliases | Operable Units | Actions | Financial |

http://www.epa.gov/oerrpage/superfnd/web/sites/cursites/c3ne/o703082.htm

{69} EPA and Superfund Logo

Summary: LAWRENCE GROUNDWATER CONTA-MINATION Site Info | Aliases | Operable Units | Actions | Financial |

http://www.epa.gov/oerrpage/superfnd/web/sites/cursites/c3ne/o703046.htm

{70} EPA and Superfund Logo

Summary: LESHARA GROUNDWATER CONTAMI-NATION SITE Site Info | Aliases | Operable Units | Actions | Financial |

http://www.epa.gov/oerrpage/superfnd/web/sites/cursites/c3ne/o703033.htm

{71} EPA and Superfund Logo

Summary: GURLEY GROUNDWATER CONTAMI-NATION SITE; Site Info | Aliases | Operable Units | Actions | Financial |

http://www.epa.gov/oerrpage/superfnd/web/sites/cursites/c3ne/o703032.htm

{72} EPA and Superfund Logo

Summary: MILFORD GROUNDWATER CONTAMI-NATION Site Info | Aliases | Operable Units | Actions | Financial |

http://www.epa.gov/oerrpage/superfnd/web/sites/cursites/c3ne/o703020.htm

{73} C:\~BIRUTE\COVER.FIN

Summary: Page ES-1 EXECUTIVE SUMMARY The U.S. Environmental Protection Agency (EPA), Office of Solid Waste has investigated potential gaps in the current hazardous waste characteristics promulgated.

http://www.epa.gov:80/epaoswer/hazwaste/id/char/scopingp.pdf

{74} ROD.PDF

Summary: Part II Responsiveness Summary: This section presents Environmental Protection Agency's (EPA's) responses to the written and oral comments received at the public meeting and during the public comment period.

http://www.epa.gov:80/region09/waste/sfund/npl/oii/rod2.pdf

{75} ROD 1.PDF

Summary: Description of the Remedy: This ROD addresses liquids control and contaminated ground-water as well as long-term operation and maintenance of all environmental control facilities at the landfill.

http://www.epa.gov:80/region09/waste/sfund/npl/oii/rod1.pdf

{76} MDA-PPLN

Summary: MONTROSE AND DEL AMO SUPER-FUND SITES UNITED STATES ENVIRONMEN-TAL PROTECTION AGENCY · REGION 9 · SAN FRANCISCO,CA · JUNE 1998: At Montrose and Del Amo Superfund sites, EPA Proposes Groundwater Cleanup Plan (General Fact Sheet Version).

http://www.epa.gov:80/region09/waste/sfund/npl/delamo/document/mda-ppl n.pdf

{77} 600/SR-94/051

Summary: Potential problem pollutants were identified, based on their mobility through the unsaturated soil zone above groundwater, their abundance in storm-water, and their treatability before discharge.

http://www.epa.gov:80/ordntrnt/ORD/WebPubs/projsum/600sr94051.pdf

{78} OGALLALA GROUNDWATER CONTAMINATION SITE

Summary: The Ogallala Groundwater Contamination Site consists of two properties approximately 15 acres and 1 acre in size, respectively.

http://www.epa.gov/rgytgrnj/programs/spfd/nplfacts/ogallala.html

{79} EPA Region 1 - Press Release: EPA ANNOUNCES NO RISKS AT CHESHIRE GROUNDWATER CONTAMINATION SUPERFUND SITE

Summary: To solicit input on the proposal, the EPA will hold a public meeting at 7 p.m., Oct. 24 at the Cheshire Town Hall. In addition, a public comment period will run from Oct. 21 through Nov. 20 for those who would like to send comments to the EPA.

http://www.epa.gov/region01/pr/files/pr1010a.html

{80} Bally Groundwater Contamination

Summary: The Bally Ground Water Contamination site consists of an area of ground water contamination in and around the Bally Engineered Structures ("BES") plant in the borough of Bally, Pennsylvania. Potential Health Risks listed.

http://www.epa.gov/reg3hwmd/super/bally/pad.htm

{81} US EPA Region 2: NPL Site Fact Sheets

Summary: The Dover Municipal Well No. The Dover Water Commission owns and operates this municipal well field. Further investigations define source areas of contamination or additional areas of groundwater contamination and remediation.

http://www.epa.gov/r02earth/superfnd/site_sum/0200768c.htm

{82} EPA Region 2 Internet: [EPA Selects Remedy for Groundwater Contamination at the Goldisc Recordings Superfund Site in Holbrook, Long Island]

Summary:—The U.S. Environmental Protection Agency (EPA) has selected a plan to address low-level ground-water contamination at the Goldisc Recordings Super-fund site in the Village of Holbrook on Long Island.

http://www.epa.gov/r02earth/epd/98140.htm

{83} Record of Decision (ROD) Abstract

Summary: and extent of groundwater contamination was largely unknown. Road, Eleventh Street and Kishwaukee Street. primary source of potable water is groundwater.

http://www.epa.gov/oerrpage/superfnd/web/sites/query/rods/r0595277.htm

{84} EPA National Priorities List 0503017n.htm

Summary: The 11-acre Ossineke Groundwater Contami-nation site resulted from a series of unrelated spills and incidents that contaminated the groundwater of local residents within the LaBell subdivision.

http://www.epa.gov/oerrpage/superfnd/web/sites/nplsites/0503017n.htm

{85} EPA National Priorities List 0500955n.htm

Summary: The Southeast Rockford Groundwater Conta-mination site covers approximately 4 square miles in Rockford, Illinois.

http://www.epa.gov/oerrpage/superfnd/web/sites/nplsites/0500955n.htm

{86} EPA National Priorities List 0202330n.htm

Summary: The 5-acre Rowe Industries Groundwater Contamination site, located on the eastern side of the Sag Harbor Bridgehampton Turnpike, was owned and operated by Rowe Industries, Inc. from the 1950s through the early 1960s.

http://www.epa.gov/oerrpage/superfnd/web/sites/nplsites/0202330n.htm

{87} Choose a Site by County, City, or Site Name

Summary: 1 10TH STREET SITE 13TH & STOCK-WELL 5 56TH ST GROUNDWATER CONTAMI-

NATION SITE 9 9TH & CALVERT A ABAN-
DONED D O D BOMB FACILITY ADAMS
MUNICIPAL WATER SUPPLY WELLS.
http://www.epa.gov/oerrpage/superfnd/web/sites/
cursites/toc/nestate.htm

{88} Choose a Site by County, City, or Site Name
Summary: ALVO FORMER GRAIN STORAGE
DOUGLAS FORMER GRAIN STORAGE ADAMS
ARMY GUARD WET SITE AYR GROUNDWA-
TER FARMLAND INDUSTRIES INC - HASTINGS
HASTINGS CITY LANDFILL HASTINGS
GROUND WATER CONTAMINATION BOX
BUTTE ALLIANCE GROUND WATER
CONTAMINATION ALLIANCE MUNICIPAL
http://www.epa.gov/oerrpage/superfnd/web/sites/
cursites/toc/necnty.htm

{89} Choose a Site by County, City, or Site Name
Summary: ADAMS ADAMS MUNICIPAL WATER
SUPPLY WELLS ALDA GRAND ISLAND DRUMS
THREE D INVESTMENTS INC ALLIANCE
ALLIANCE GROUND WATER CONTAMINA-
TION ALLIANCE MUNICIPAL LANDFILL ALVO
ALVO FORMER GRAIN STORAGE ARAPAHOE
http://www.epa.gov/oerrpage/superfnd/web/sites/
cursites/toc/necity.htm

{90} EPA and Superfund Logo
Summary: Site Info | Aliases | Operable Units | Actions |
Financial | Choose Another Site 00
SITEWIDE 01 EXTRACTION WELL 02 FINAL
REMEDY
http://www.epa.gov/oerrpage/superfnd/web/sites/
cursites/c3wi/o505186.htm

{91} EPA and Superfund Logo
Summary: Site Info | Aliases | Operable Units | Actions |
Financial |
http://www.epa.gov/oerrpage/superfnd/web/sites/
cursites/c3tn/o404230.htm

{92} EPA and Superfund Logo
Summary: Site Info | Aliases | Operable Units | Actions |
Financial | Choose Another Site Alias ID Alias
Name/Address Alias Latitude/Longitude
http://www.epa.gov/oerrpage/superfnd/web/sites/
cursites/c3tn/l404230.htm

{93} EPA and Superfund Logo
Summary: Site Info | Aliases | Operable Units | Actions |
Financial | Choose Another Site 00
SITEWIDE 01 GROUNDWATER
CONTAMINATION 02
http://www.epa.gov/oerrpage/superfnd/web/sites/
cursites/c3pr/o203140.htm

{94} EPA and Superfund Logo
Summary: Site Info | Aliases | Operable Units | Actions |
Financial | Choose Another Site 00
SITEWIDE 01 02 03 GROUNDWATER
CONTAMINATION
http://www.epa.gov/oerrpage/superfnd/web/sites/
cursites/c3pa/o301227.htm

{95} EPA and Superfund Logo
Summary: LEBANON GROUNDWATER CONTAMI-
NATION Site Info | Aliases | Operable Units | Actions
| Financial |
http://www.epa.gov/oerrpage/superfnd/web/sites/
cursites/c3or/o001756.htm

{96} EPA and Superfund Logo
Summary: SMITHTOWN GROUNDWATER CONTA-
MINATION Site Info | Aliases | Operable Units |
Actions | Financial |

http://www.epa.gov/oerrpage/superfnd/web/sites/
cursites/c3ny/o204148.htm

{97} EPA and Superfund Logo
Summary: EXETER GROUNDWATER CONTAMI-
NATION Site Info | Aliases | Operable Units | Actions
| Financial | Choose Another Site
http://www.epa.gov/oerrpage/superfnd/web/sites/
cursites/c3ne/o703127.htm

{98} EPA and Superfund Logo
Summary: WAYNE HWY 15 GROUNDWATER CON-
TAMINATION Site Info | Aliases | Operable Units |
Actions | Financial | Choose Another Site
http://www.epa.gov/oerrpage/superfnd/web/sites/
cursites/c3ne/o703087.htm

{99} EPA and Superfund Logo
Summary: ROSCOE HWY 30 GROUNDWATER
CONTAMINATION Site Info | Aliases | Operable
Units | Actions | Financial |
http://www.epa.gov/oerrpage/superfnd/web/sites/
cursites/c3ne/o703086.htm

{100} EPA and Superfund Logo
Summary: CRAIG GROUNDWATER CONTAMINA-
TION Site Info | Aliases | Operable Units | Actions |
Financial |
http://www.epa.gov/oerrpage/superfnd/web/sites/
cursites/c3ne/o703083.htm

{101} EPA and Superfund Logo
Summary: HERMAN GROUNDWATER CONTAMI-
NATION Site Info | Aliases | Operable Units | Actions
| Financial |
http://www.epa.gov/oerrpage/superfnd/web/sites/
cursites/c3ne/o703082.htm

{102} EPA and Superfund Logo
Summary: LAWRENCE GROUNDWATER CONTA-
MINATION Site Info | Aliases | Operable Units |
Actions | Financial |
http://www.epa.gov/oerrpage/superfnd/web/sites/
cursites/c3ne/o703046.htm

{103} EPA and Superfund Logo
Summary: LESHARA GROUNDWATER CONTAMI-
NATION SITE Site Info | Aliases | Operable Units |
Actions | Financial |
http://www.epa.gov/oerrpage/superfnd/web/sites/
cursites/c3ne/o703033.htm

{104} EPA and Superfund Logo
Summary: GURLEY GROUNDWATER CONTAMI-
NATION SITE Site Info | Aliases | Operable Units |
Actions | Financial |
http://www.epa.gov/oerrpage/superfnd/web/sites/
cursites/c3ne/o703032.htm

{105} EPA and Superfund Logo
Summary: MILFORD GROUNDWATER CONTAMI-
NATION Site Info | Aliases | Operable Units | Actions
| Financial |
http://www.epa.gov/oerrpage/superfnd/web/sites/
cursites/c3ne/o703020.htm

{106} EPA and Superfund Logo
Summary: ASHLAND GROUNDWATER CONTAMI-
NATION Site Info | Aliases | Operable Units | Actions
| Financial |
http://www.epa.gov/oerrpage/superfnd/web/sites/
cursites/c3ne/o702978.htm

{107} EPA and Superfund Logo
Summary: RAYMOND GROUNDWATER CONTAM-
INATION Site Info | Aliases | Operable Units |
Actions | Financial |

http://www.epa.gov/oerrpage/superfnd/web/sites/
cursites/c3ne/o702977.htm

{108} Archive (NFRAP) Site Information

Summary: Site Information: Site Name: TORONTO
GROUNDWATER CONTAMINATION

SITE Address: MAIN ST TORONTO, KS 66777 EPA
ID: KSD985014034 EPA Region: 07

County: 207 WOODSON Congressional District: 02
Metro Statistical Area: Action: OU Action Type
Action

http://www.epa.gov/oerrpage/superfnd/web/sites/
arcsites/reg07/a0702748.htm

{109} Archive (NFRAP) Site Information

Summary: Site Information: Site Name: VALENTINE
GROUNDWATER CONTAMINATION

Address: HWY 83 & 20 JCT VALENTINE, NE 69201
EPA ID: NED986386993 EPA Region: 07

County: 031 CHERRY Congressional District: 03 Metro
Statistical Area: Action: OU Action Type

http://www.epa.gov/oerrpage/superfnd/web/sites/
arcsites/reg07/a0702732.htm

{110} SUPERFUND RECORD OF DECISION: SOUTH-
EAST ROCKFORD GROUNDWATER CONTAMI-
NATION SITE, ROCKFORD, IL

Summary: Title: SUPERFUND RECORD OF DECI-
SION: SOUTHEAST ROCKFORD GROUND-
WATER CONTAMINATION SITE, ROCKFORD, IL
Department of Commerce National Technical Infor-
mation Service 5285 Port Royal Rd Springfield, VA
22151 Phone Number: 800-553-6847

http://www.epa.gov/ncepihom/Catalog/
EPARODR0595277.html

{111} EDRI Project: 04911-07 HEALTH HAZARDS FROM
GROUNDWATER CONTAMINATION

Summary: Description: Research is directed toward a
better understanding of potential health hazards arising from contamination of groundwater and soils with
chemicals commonly found at hazardous waste sites.

http://www.epa.gov/edrlupvx/inventory/NIEH-018.html

{112} IN00032 TITLE

Summary: IN00032 TITLE: Cleaning up explosives
contamination at Army munitions plants using incin-
eration. PUBLISHER: Hazardous Materials Control
Resources, Inc., Greenbelt, Md.

PUBLICATION DATE: 1989. DESCRIPTION: Journal
article: 9 pages, figures, tables.

http://www.epa.gov/bbsnrmrl/attic/a2/IN00032.html

{113} Agricultural Drainage Wells (5Fl)

Summary: The RTC defines abandoned drinking water
wells as abandoned or improperly plugged wells that
have become waste receptacles, whether the waste
disposal is intentional or unintentional.

http://www.epa.gov:80/ogwdw000/uic/5flold.pdf

{114} Stormwater Drainage Wells (5D2)

Summary: Draft—For EPA Workgroup Review Only—
Do Not Cite or Distribute — Draft TABLE OF
CONTENTS SUMMARY OF INFORMATION OF
STORMWATER DRAINAGE WELLS FOR
WORKGROUP REVIEW I. Introduction.

http://www.epa.gov:80/ogwdw000/uic/5d2.pdf

{115} Stormwater Drainage Wells (5D2)

Summary: According to the 1987 Report to Congress
(RTC), municipalities with limited stormwater sewer
systems or those experiencing rapid growth and increased impervious surface may experience floods.

http://www.epa.gov:80/ogwdw000/uic/5d2old.pdf

{116} 124//Wednesday, June 26, 1996//Proposed Rules ENVI-
RONMENTAL PROTECTION AGENCY

Summary: 124//Wednesday, June 26, 1996//Proposed
Rules ENVIRONMENTAL PROTECTION
AGENCY 40 CFR Parts 152 and 156 [OPPP±36190;
FRL±4981±9]RIN 2070±AC46 Pesticides and
Ground Water State Management Plan Regulation
AGENCY: Environmental Protection Agency
(EPA).ACTION

http://www.epa.gov:80/fedrgstr/EPA-PEST/1996/June/
Day-26/pr-768DIR/pr-768.pdf

{117} US EPA Region 2: NPL Site Fact Sheets

Summary: The earliest actions are focusing on removal
actions to address off-site groundwater contamination,
sources of groundwater contamination and highly
contaminated soil.

http://www.epa.gov/r02earth/superfnd/sitesum/
0202841c.htm

{118} Preliminary Study

Summary: Despite major improvements in these and
other steel mill streams, state agencies have identified
40 iron and steel mills with discharges to impaired
water bodies.

http://www.epa.gov:80/ostwater/ironsteel/pdf/
prelim3.pdf

{119} Summary: This document addresses all primary transportation (highway, rail, aviation, and maritime transport)
and all environmental media (air, water, and land resources), and covers the full "life-cycle" of transportation.

http://www.epa.gov:80/oppetptr/indicall.pdf

{120} FOREWORD Soil vapor extraction (SVE) has been
used at many sites:

Summary: FOREWORD Soil vapor extraction (SVE)
has been used at many sites to remove volatile organic
compounds (VOC) from soil in the vadose zone.

http://www.epa.gov:80/swertio1/download/remed/
sveenhmt.pdf

{121} Summary: ABSTRACT The EEAC addressed the design, conduct, and results of the contingent valuation
study (undertaken for the EPA Office of Solid Waste).
There is little doubt that this study represents a substan-
tive contribution, extending our understanding.

http://www.epa.gov:80/science1/eeac9401.pdf

{122} SEL.PDF

Summary: Criteria for Selection of Environmental Data
Sets Decision Support Software

Demonstration: One objective of the demonstration program is to test the capability of the selected Decision
Support Software (DSS).

http://www.epa.gov:80/etvprgrm/02/dss.pdf

[338] Some U.S. Department of Energy R&D Project Summaries

{1} Evaluation of in situ Bioremediation of BTEX Groundwater Plumes http://www.doe.gov/rnd/data/11324.html

{2} Partnership in Computational Science (PICS): Software
Support for Groundwater Transport and Remediation
http://www.doe. gov/rnd/data/12404.html

{3} Analytical and Numerical Methods
http://www.doe.gov/rnd/data/12411.html

{4} Fundamental Research in the Geochemistry of Geothermal
Systems
http://www.doe.gov/rnd/data/12795.html

{5} Thermodynamic Mixing Properties of C-O-H-N Fluids
http://www.doe.gov/rnd/data/12797.html

{6} Expedited Site Characterization: Application and Continued Development of Rapid, Focused Site Characterization
Methodology for Federal Facilities http://www.doe.
gov/rnd/data/13748.html

{7} In situ Groundwater Treatment Using Magnetic Separation
http://www.doe.gov/rnd/data/13749.html

{8} Development & Demonstration of In-Well Sonication for in situ Removal of Organic Contaminants from Groundwater http://www.doe.gov/rnd/data/13764.html

{9} Fast, Easily Applied Scale-Up Techniques for Optimizing Reservoir Production
http://www.doe.gov/rnd/data/13901.html

{10} Materials Science: Engineering Chemistry—Aqueous Corrosion   http://www.doe.gov/rnd/data/14054.html

{11} Environmental Research: Percolation Modeling of Microbial Transport in the Subsurface
http://www.doe.gov/rnd/data/14107.html

{12} Radionuclide Speciation in Groundwater Systems
http://www.doe.gov/rnd/data/14123.html

{13} SITE CHARACTERIZATION OF GROUNDWATER FLOW
http://www.doe.gov/rnd/data/17893.html

{14} FASCHEM: VERIFCATION, APPLICATION, AND SENSITIVITY
http://www.doe.gov/rnd/data/18102.html

{15} DETERMINATION OF SOLUBILITIES AND COMPLEXATION
http://www.doe.gov/rnd/data/18229.html

{16} KESTERON RESERVIOR RESEARCH PROGRAM
http://www.doe.gov/rnd/data/26067.html

{17} ULTIMATE FATE OF HAZARDOUS WASTE INJECTION STUDY
http://www.doe.gov/rnd/data/26069.html

{18} SITE CHARACTERIZATION OF GROUNDWATER FLOW & TRANSPORT IN FRACTURED ROCK SYSTEM http://www.doe.gov/rnd/data/26069.html

{19} Research and Development Monitoring http://www.doe.gov/rnd/data/27484.html

{20} CARBON METABOLISM IN SYMBIOTIC NITROGEN FIXATION http:www.doe.gov/rnd/data/27879.html

{21} Site Remediation Technologies: In situ Bioremediation of Organic Contaminants
http://www.doe.gov/rnd/data/28084.html

{22} Remote Chemical Sensor Development
http://www.doe.gov/rnd/data/28084.html

{23} COMPARATIVE ANALYSES OF SUBSURFACE BACTERIAL COMMUNITY STRUCTURE: CORRELATION BETWEEN COMMUNITY COMPOSITION, ORIGIN AND ENVIRONMENTAL PARAMETERS http://www.doe.gov/rnd/data/28202.html

{24} Research Titled "Acoustically Enhanced Remediation of Contaminated Soils and Groundwater"
http://www.doe.gov/rnd/data/28257.html

{25} BIOPOLYMER BARRIER TO GROUNDWATER CONTAMINANTS
http://www.doe.gov/rnd/data/28451.html

{26} ENERGY CROP CHEMICAL FATE AND NUTRIENT CYCLING STUDY
http://www.doe.gov/rnd/data/28692.html

{27} DEVELOPMENT OF A DATA MANAGEMENT SYSTEM FOR ASSISTANCE IN CONDUCTING AREAS OF REVIEW (AORS) IN CALIFORNIA
http://www.doe.gov/rnd/data/28737.html

{28} DEVELOP DATA MANAGEMENT SYSTEM FOR ASSISTANCE IN CONDUCTING AREA OF REVIEW (AORS) IN KANSAS http://www.doe.gov/rnd/data/28752.html

{29} MODELING CONTROL COSTS AND LAKE ACIDIFICATION EFFECTS FOR THE TRACKING ANALYSIS FRAMEWORK: A PROPOSAL http://www.doe.gov/rnd/data/28838.html

{30} A MULTINUCLEAR MAGNETIC RESONANCE STUDY OF THE INTERACTIONS OF POLLUTANTS WITH MAJOR SOIL COMPONENTS http://www.doe.gov/rnd/data/28855.html

{31} PARTNERSHIP IN COMPUTATIONAL SCIENCE
http://www.doe.gov/rnd/data/28885.html

{32} THE PROPOSAL RESEARCH ADDRESSES THE PROBLEM OF THE NUMERICAL MODELING OF GROUNDWATER http://www.doe.gov/rnd/data/28890.html

{33} TECHNICAL SUPPORT FOR THE COLEMAN DATA
http://www.doe.gov/rnd/data/30717.html

{34} CHARACTERIZATION & MONITORING FOR THE MAGNE
http://www.doe.gov/rnd/data/30721.html

{35} IN SITU INORGANIC REMEDIATION OF GROUNDWATER
http://www.doe.gov/rnd/data/30722.html

{36} MAG*SEP Process Chemistry Support
http://www.doe.gov/rnd/data/30723.html

{37} OFF-GAS TREATMENT SAMPLING AND ANALYSIS
http://www.doe.gov/rnd/data/30729.html

{38} PENETROMETER FOR SITE CHARACTERIZATION
http:www.doe.gov/rnd/data/30730.html

{39} STATE AND NATIONAL ENERGY AND ENVIRONMENTAL RISK ANALYSIS SYSTEMS FOR UNDERGROUND INJECTION CONTROL
http://www.doe.gov/rnd/data/30744.html

{40} ANAEROBIC METABOLISM OF AROMATIC COMPOUNDS BY PHOTOTROPHIC BACTERIA
http://www.doe.gov/rnd/data/32256.html

{41} THE DETERMINATION OF 222RN FLUX FROM SOILS BASED ON 210PB AND 226RA DISEQUILIBRIUM http://www.doe.gov/rnd/data/32317.html

{42} Rocky Mountain 1 (RM1) Underground Coal Gasification (UCG) Project
http://www.doe.gov/rnd/data/32343.html

{43} Groundwater Colloids: Their Mobilization from Subsurface Deposits http://www.doe.gov/rnd/data/32532.html

{44} Surface Chemistry Investigation of Colloid Transport in Packed Beds
http://www.doe.gov/rnd/data/32534.html

{45} Factors Affecting Transport of Bacterial Cells in Porous Media
http://www.doe.gov/rnd/data/32540.html

{46} The Influence of Interfacial Properties on Two-Phase Liquid Flow of Organic Contaminants in Groundwater
http://www.doe.gov/rnd/data/32544.html

{47} MECHANISM CONTROLLING PRODUCTION & TRANSPORT OF METHANE CARBON DIOXIDE & DISSOLVED SOLUTES WITHIN A LARGE BOREAL PEAT BASIN
http://www.doe.gov/rnd/data/32616.html

{48} MODEL OF ACIDIFICATION OF GROUNDWATER IN CATCHMENTS & TESTING OF THE REVISED MODEL USING SOURCES
http://www.doe.gov/rnd/data/32813.html

[339] OTHER RELATED MISCELLANEOUS REFERENCES

{1} Optimal Groundwater Management: 2. Application of Simulated Annealing to a Field-Scale Contamination Site,

R. A. Marryott, D. E. Dougherty, and R. L. Stollar. Water Resources Research WRERAQ, Vol. 29, No. 4, p 847-860, April 1993. 11 fig, 5 tab, 22 ref. See Review at: http://www2.uwin.siu.edu:4001/usr/local/data/wrsic/1993.txt_13634590_1993.

{2} Geraghty and Miller's Groundwater Bibliography. Water Information Center, Inc., Plainview, NY. Fifth Edition. 1991. 507p. Compiled and Edited by Frits van der Leeden. See Review at:

http://www2.uwin.siu.edu:4001/usr/local/data/wrsic/1993.txt_13887969_1993

{3} Groundwater Contamination Risk Assessment: A Guide to Understanding and Managing Uncertainties, E. Reichard, C. Cranor, R. Raucher, and G. Zapponi.IAHS Publication No. 196, 1990. International Association of Hydrological Sciences, Wallingford, England. 204p, See Review at:

http://www2.uwin.siu.edu:4001/usr/local/data/wrsic/1993.txt_1059803_1993

{4} Groundwater Contamination in the United States, D. W. Moody. Journal of Soil and Water Conservation JSWCA3, Vol. 45, No. 2, p 170–179, 1990. 28 ref. See Review at:

http://www2.uwin.siu.edu:4001/usr/local/data/wrsic/1991.txt_19626226_1991

{5} Geochemistry of Natural Waters (3rd ed.) by J.I. Drever

Reviewed by Carl Bowser, Department of Geology & Geophysics, University of Wisconsin-Madison, 1215 W. Dayton St., Madison, Wisconsin 53706. *May–June 1998, vol. 36, no. 3, p. 391* See Review at: http://www.ngwa.org/publication/98book.html#geochemistry

{6} Groundwater Geochemistry, Fundamentals and Applications to Contamination by William J. Deutsch

Reviewed by G. L. Macpherson, Dept. of Geology, 120 Lindley Hall, University of Kansas, Lawrence, KS 66045. *May–June 1998, v. 36, no. 3: 392–393.* See Review at:

http://www.ngwa.org/publication/98book.html#geochemistry

{7} U.S. Geological Survey: Open-File Report 97-219

Title : PRELIMINARY HYDROGEOLOGIC ASSESSMENT OF A GROUNDWATER CONTAMINATION AREA IN WOLCOTT, CONNECTICUT

Author: Janet Radway Stone, George D. Casey, Remo A. Mondazzi, and Timothy W. Frick, Availability: U.S. Geological Survey Earth Science Information Center, Open-File Reports Section, Box 25286, MS 517, Denver Federal Center, Denver, CO 80225, USGS OFR 97-219, 29 p., 6 figs., 8 pl.

See Abstract at: http://water.usgs.gov/public/swra/help.html

{8} U.S. Geological Survey: Water-Resources Investigation 97-4003

Title: TRANSPORT AND TRANSFORMATIONS OF CHLORINATED- SOLVENT CONTAMINATION IN A SAPROLITE AND FRACTURED ROCK AQUIFER NEAR A FORMER WASTEWATER-TREATMENT PLANT, GREENVILLE, SOUTH CAROLINA

Author: Don A. Vroblesky, Paul M. Bradley, John W. Lane, Jr., and J. Frederick Robertson, Availability: Copies are available for inspection at the U.S. Geological Survey office in Columbia, S.C. (Stephenson Center-Suite 129, 720 Gracern Road, Columbia, SC 29210-7651) and at most large libraries in South Carolina. Paper and microfiche copies can be purchased at cost from U.S. Geological Survey, Branch of Information Services, Open-File Reports Section, Box 25286,

MS 517, Denver Federal Center, Denver, CO 80225, USGS Water-Resources Investigations Report 97-4003, 76p., 11 figs., 6 tabs., and 56 refs.

See Abstract at: http://water.usgs.gov/lookup/getabstract?WRI974003

{9} U.S. Geological Survey: Water-Resources Investigation 96-4116

Title: ANALYSIS OF GROUNDWATER DATA FOR SELECTED WELLS NEAR HOLLOMAN AIR FORCE BASE, NEW MEXICO, 1950–95

Author: G. F. Huff, Availability: Available from USGS Branch of Information Services, Denver Federal Center, Box 25286, Denver, CO 80225, USGS Water-Resources Investigations Report 96-4116, 37 p, 24 fig.

See Abstract at: http://water.usgs.gov/lookup/getabstract?WRI964116

{10} Water-Resources Investigation 95-4091

Title: GEOHYDROLOGY AND SIMULATION OF GROUNDWATER FLOW NEAR LOS ALAMOS, NORTH-CENTRAL NEW MEXICO

Author: P. F. Frenzel, Availability: Available from USGS, Earth Science Information Center, Open-File Reports Section, Box 25286, MS 517, Denver Federal Center, Denver, CO 80225, USGS Water-Resources Investigations Report 95-4091, 92 p, 31 fig.

See Abstract at: http://water.usgs.gov/lookup/getabstract?WRI954091

{11} Water-Resources Investigation 94-4251

Title: SIMULATION OF GROUNDWATER FLOW IN THE ALBUQUERQUE BASIN, CENTRAL NEW MEXICO, 1901–1994, WITH PROJECTIONS TO 2020

Author: J. M. Kemodle, D. P. McAda, and C. R. Thorn, Availability: Available from USGS, Earth Science Information Center, Open-File Reports Section, Box 25286, MS 517, Denver, CO 80225, USGS Water-Resources Investigations Report 94-4251, 1995, 114 p, 1 plate, 65 fig, 8 tab, 49 ref.

See Abstract at: http://water.usgs.gov/lookup/getabstract?WRI944251

{12} U.S. Geological Survey: Open-File Report 95-773

Title: Chemical analyses of groundwater samples from file Rio Grande Valley in the vicinity of Albuquerque, New Mexico, October 1993 through January 1994

Author: D. W. Wilkins, J. L. Schlottmann, and D. M. Ferree, Availability: Available from USGS, Earth Science Information Center, Open-File Reports Section, Box 25286, MS 517, Denver Federal Center, Denver, CO 80225, USGS Open-File Report 95-773, 27 p, 11 fig.

See Abstract at: http://water.usgs.gov/lookup/getabstract?OFR95773

{13} U.S. Geological Survey: Water-Resources Investigation 96-4006

Title: PLAN OF STUDY TO QUANTIFY THE HYDROLOGIC RELATIONS BETWEEN THE RIO GRANDE AND THE SANTA FE GROUP AQUIFER SYSTEM NEAR ALBUQUERQUE, CENTRAL NEW MEXICO

Author: D. P. McAda, Availability: Available from USGS, Earth Science Information Center, Open-File Reports Section, Box 25286, MS 517, Denver Federal Center, Denver, CO 80225, USGS Water-Resources Investigations Report 96-4006, 58 p, 5 fig.

See Abstract at: http://water.usgs.gov/lookup/getabstract?WRI964006

{14} U.S. Geological Survey: Open-File Report 95-768

Title: LISTINGS OF MODEL INPUT AND SELECTED OUTPUT VALUES FOR THE SIMULATION OF

GROUNDWATER FLOW NEAR LOS ALAMOS, NORTH-CENTRAL NEW MEXICO—SUPPLEMENT TO WATER-RESOURCES INVESTIGATIONS REPORT 95-4091

Author: P. F. Frenzel, Availability: Available from USGS, Earth Science Information Center, Open-File Reports Section, Box 25286, MS 517, Denver Federal Center, Denver, CO 80225, USGS Open-File Report 95-768, 3 p, 1 diskette.

See Abstract at: http://water.usgs.gov/lookup/getabstract?OFR95768

{15} U.S. Geological Survey: Open-File Report 95-385

Title: GROUNDWATER-QUALITY AND GROUND-WATER-LEVEL DATA, BERNALILLO COUNTY, CENTRAL NEW MEXICO, 1990-93

Author: G. E. Kues, and B. M. Garcia Availability: Available from USGS, Earth Science Information Center, Open-File Reports Section, Box 25286, MS 517, Denver Federal Center, Denver, CO 80225, USGS Open-File Report 95-385, 76 p, 24 fig.

See Abstract at: http://water.usgs.gov/lookup/getabstract?OFR95385

{16} Quest for Water: http://www.anglia.ac.uk/~trochford/persgard/qanatlr.htm

The Qanat Project: http://www.pangea.org/org/unesco/wat14/Sca.html

Water in the Middle East: http://ah.soas.ac.uk/Centres/IslamicLaw/WaterIntro.html

{17} World Health Organization: http://www.who.org/

http://www.who.org/home/map_ht.html

http://www.who.org/home/map_ht.html#Diseases:Communicable/Infectious

http://www.who.int/emc/

http://www.who.int/peh/ [Environmental Health]

http://www.who.int/peh/specprg.htm#Water supply and sanitation in human s

{18} NSF Standard for Performance: http://www.culligan-systems.com/nsf.html

{19} Natural Resources Defense Council: http://www.nrdc.org/

http://www.nrdc.org/sitings/fslink.html

{20} U.S. Forest Service: http://www.fs.fed.us/

{21} Uniform Plumbing Code: http://www.iapmo.org/

{22} State of Oregon Department of Water Resources: http://www.wrd.state.or.us/

http://www.wrd.state.or.us/groundwater/index.html [Groundwater and Wells]

{23} University of California: http://www.ucop.edu/

{24} American Petroleum Institute: http://www.api.org/

{25} Creosote Waste and Chemistry:

http://bordeaux.uwaterloo.ca/biol447/assignment1/creosote.html

http://www.ec.gc.ca/library/elias/bibrec/3020711B.html

http://bordeaux.uwaterloo.ca/biol447/assignment1/creo2.html

http://toxics.usgs.gov/toxics/bib/bib-field.shtml

http://toxics.usgs.gov/toxics/bib/bib-pensa.shtml

http://clu-in.com/Sewood.htm

http:/www.isva.dtu.dk/grc/1994/ass-e4.htm

{26} U.S. Fish and Wildlife Service: http://www.fws.gov/

{27} Reclamation's Laws and Regulations: http://www.usbr.gov/laws/chronol.html

http://www.usbr.gov/laws/mines.html [Geothermal Steam Act 1970]

{28} Bureau of Reclamation: http://www.lc.usbr.gov/

http://www.usbr.gov/main/index.html [Managing Water in the West]

http://www.usbr.gov/rsmg/wsi/ [Water Supply Information]

{29} American Gas Association: http:www.osha-slc.gov/SLTC/nrtl/aga.html

http://www.osha-slc.gov/SLTC/nrtl/aga.html

{30} Office of Pipeline Safety: http://ops.dot.gov/home.htm [Home]

http:/www.bts.gov/smart/cat/rspahsty.html [Research and Special Programs]

http://www.viadata.com/search2.htm [Pipeline Safety Federal Regulation for Windows]

http://www.dps.state.nm.us/pipeline/ [Minnesota Office of Pipeline Safety]

http://www.dot.gov/affairs/rspa2298.htm [DOT National Pipeline Safety Mapping]

http:/www.law.cornell.edu/uscode/33/1232a.shtml [Ports and Waterways Programs]

http://www.ornl.gov/etd/etdsect.htm [Nuclear Related]

http://www.nas.edu/trb/directory/db_usdot.html [DOT Databases]

http://www.bts.gov/ntl/DOCS/ng_year.html [Natural Gas Pipeline Summary]

http://www.cycla.com/opsiswc/wc.dll?sii~toppage [System Integrity Inspection Program]

http://www.state.nj.us/infobank/circular/eow13.htm [Natural Gas Pipeline Explosion]

{31} Applied Statistics

http://www.carfax.co.uk/jas-ad.htm [Journal of Applied Statistics]

http://www.cas.lancs.ac.uk/cas.html [Center for Applied Statistics, Lancaster, England]

http://franz.stat.wisc.edu/pub/MASS2 [Modem Applied Statistics]

http://www.math.mun.ca/ [University of Newfoundland]

http://www.math.mun.ca/remote.html [Links to Mathematical Sites]

http://euclid.math.fsu.edu/Science/math.html [Florida State University, Virtual Math Library]

http://euclid.math.fsu.edu/Science/Journals.html [Math Electronic Journals]

{32} Air & Waste Management Association

http://www.awma.org/ [Home]

http://online.awma.org/journal/ [Journal Online]

http://online.awma.org/journal/_vti_bin/shtml.dll/search.htm [Search]

{33} Aberdeen Proving Ground Related

http://quake.wes.army.mil/FS_APG.html [Hydrogeology—Environmental Management]

http://www.apg.army.mil/AboutAPG.html [About]

http://www.apg.army.mil/GAR-SUP.html [Garrison Services]

{34} Nuclear Regulatory Commission

http://www.shef.ac.uk/uni/academic/A-C/chem/web-elements/nofr-uses/Th.html

http:/www.nrc.gov/NRC/contract.html

{35} Thorium

http://www.shef.ac.uk/~chem/web-elements/Th.html

http://micronmetals.com/90.htm

http://www.shef.ac.uk/uni/academic/A-C/chem/web-elements/nofr-uses/Th.html

{36} Asbestos
http://www.asbestos-institute.ca/index.html [Asbestos Institute Online]
http://www.louisville.edu/admin/dehs/hsasbes.htm [Health and Safety]

{37} Environmental Audits
http://www.rpi.edu/~breyms/audit/reports/audit_explanation.html
http://www.wheatonma.edu/Admin/environmental/audit.HTML
http://www.airportnet.org/depts/environ/audtform.htm
http://www.ucalgary.ca/UofC/departments/UP/0-919813/0-919813-92-5.html
http://www.fcx.com/fcx/envaudit.htm

{38} Department of the Interior
http://www.doi.gov/
http://library.doi.gov/

{39} National Science Foundation
http://www.nsf.gov/ [Home]
http://www.nsf.gov/home/geo/start.htm [Geosciences]

{40} New Mexico Bureau of Mines and Mineral Resources
http://geoinfo.mnt.edu/

{41} Indian Nations and Tribes
http://www.azcentral.com/depts/dest/interests/indianres.shtml [Arizona]
http://www.doi.gov/bia/aitoday/q_and_a.html
http://law.house.gov/31.htm [Indian Nations and Tribes]
http://www.encarta.com/index/concisein-dex/42/04291000.htm

{42} New Mexico Net
http://www.state.nm.us/ [NM Government]
http://www.bncinc.com/nmnet/all.html [All Listings]
http://www.state.mn.us/state/city_county.html [City, County Government]

{43} A Call for Accountability, Competency, and Ethics in DOE.
http://www.mindspring.com/~jpcarson/

{44} U.S. DOE Whistleblower Protection Program
http://www.oha.doe.gov/whistle1.htm

{45} Whistleblower Protection Resources
http://altavista.looksmart.com/r?comefrom=avsearch-e74444&izf&e74444

{46} Government Accountability Project
http://www.whistleblower.org/

{47} Safety Culture Meltdown at Hanford
http://www.whistleblower.org/www/safetymelt.htm

{48} Security Concerns Ignored at Rocky Flats
http://www.whistleblower.org/www/graf.htm

{49} Nuclear Warhead Transportation
http://www.whistleblower.org/www/TSD.htm

{50} Government Accountability Project—Program Descriptions
http://www/whistleblower.org/www/program.htm

{51} Nuclear Weapons Program
http://www.whistleblower.org/www/nuclear.htm

{52} Hanford Nuclear Site
http://www.whistleblower.org/www/hanford.htm

{53} Report of Hanford Tank Vadose Zone—Baseline Characterization
http://www.whistleblower.org/www/Brodeur%20Report/index.html

{54} Hanford Whistleblower Fired Again
http://www.whistleblower.org/www/taylorrcipr.htm

{55} Slide Presentation on Hanford Contamination
http://www.whistleblower.org/images/Buske/buskeslidel.htm

{56} Hanford Reach Protection, a No Brainer
http://www.whistleblower.org/www/spain.htm

{57} Hanford Salmon Face New Risk—Strontium-90
http://www.whistleblower.org/www/hotfishpr.htm

{58} Strontium-90 Adjacent to Fall Chinook Salmon Redds at H-Reactor Area
http://www.whistleblower.org/www/hotfishreport.htm

{59} Hanford Whistleblower Files $240 Million Fraud Claim
http://www.whistleblower.org/www/fraud.htm

{60} U.S. DOE Secret Meetings
http://www.whistleblower.org/www/fftfnerac.htm

{61} Radioactive Fruit Flies Attack Hanford
http://www.whistleblower.org/www/fruitfly.htm

{62} HOT WATER—Groundwater Contamination at the Hanford Nuclear Reservation a Report of the Previous and Current Problems (Adobe pdf document)
http://www.whsitleblower.org/www/hotwater.pdf

{63} Judge Orders Contractors to Pay Hanford Whistle-blowers
http://www.whistleblower.org/www/pipefitvic.htm

{64} Labor Department Orders Punitive Damages in Hanford Retaliation Case
http://www.whistleblower.org/www/Ruuddecpr.htm

{65} WHITE PAPER: Blowing Off Safety Concerns at Hanford Tank Farms
http://www.whistleblower.org/www/103cpr.htm

{66} Batelle Memorial Institute Settles Suit with Whistle-blower
http://www.whistleblower.org/www/Laulset.htm
http://www.whistleblower.org/www/spokanerev.htm

{67} Scientist Prevails in Whistleblower Complaint
http://www.whistleblower.org/www/cornetwin.htm

{68} Hanford Tank Cover-Up Alleged
http://www.whistleblower.org/www/twrspr.htm
{69} Serial Blacklisting
http://www.whistleblower.org/www/hanford.htm

{70} Links for Whistleblowers—Legal Resources and Case Law
http://www.whistleblower.org/www/Connections.htm

{71} Whistleblowers in the News
http://www.whistleblower.org/www/wbnews.htm

{72} Whistleblower Wins EG&G Appeal
http://www.whistleblower.org/www/Jonespr.htm

{73} Lawrence Livermore National Laboratory Retaliates Against Whistleblower
http://www.whistleblower.org/www/Lappapr.htm

{74} Pantex Article
http://www.whistleblower.org.www/nprarticle.htm

{75} Going Critical at Pantex
http://texasobserver.org/subjects/enviro/LD.10.12.7.html

{76} Indecent Exposures—Health and Safety at the Nation's Bomb Complex
http://www.whistleblower.org/www/indecent.htm

{77} Flour Daniel Northwest Investigation by U.S. Department of Labor OSHA
http://www.whistleblower.org/www/OSHApipe.htm

{78} Safety Meltdown at Hanford—Sham Investigation
http://www.whistleblower.org/www/safetymeltart.htm

{79} Hartford's Self-Investigation of Plutonium Plant Explosion Emergency Response Extends Cover-Up
http://www.whistleblower.org/www/hoablast.htm

{80} Review of the Federal Management of the Tank Waste Remediation System (TWRS) Project
http://www.whistleblower.org/www/60dayrev.htm

{81} U.S. Department of Labor and Westinghouse Hanford Company
http://www.whistleblower.org/www/ruuddec.htm

{82} Committee Receives GAO Report on Brookhaven
http://www.pubaf.bnl.gov/pr/hscpr111397.html

{83} Information on the Tritium Leak and Contractor Dismissal at Brookhaven National Laboratory (Adobe pdf Document)
http://www.pubaf.bnl.gov/pr/gao111397.pdf

{84} Los Alamos Study Group, Case No. VFA-0346, November 19, 1997

Los Alamos Study Group, Case No. VFA-0298, June 19, 1997 and Other Cases

http://www.oha.doe.gov/cases/foia/vfa0298.htm
http://www.oha.doe.gov/cases/foia/vfa0346.htm
http://www.oha.doe.gov/cases/foia/vfa0316.htm
http://www.oha.doe.gov/cases/foia/vfa0463.htm
http://www.oha.doe.gov/cases/foia/vfa0481.htm

{85} Los Alamos National Laboratory—Howard T. Uhal, Case No. VFA-0160, May 31, 1996
http://www.oha.doe.gov/cases/foia/vfa0160.htm

{86} Los Alamos National Laboratory—Homesteaders Association of the Pajarito, Case No. VFA-0355, December 22, 1997
http://www.oha.doe.gov/cases/foia/vfa0355.htm

{87} Los Alamos—Cases Received by the Office of Hearings and Appeals
http://www.oha.doe.gov/reports/submission/1999/sub0301.htm

{88} Los Alamos—Decisions and Orders January 4 through January 8, 1999
http://www.oha.doe.gov/reports/decisionlist/1999/d&o0104.htm

{89} Los Alamos—William H. Keenan, Case No. VFA-0521 September 30, 1999
http://www.oha.doe.gov/cases/foia/vfa0521.htm

{90} Los Alamos National Laboratory—Michael J. Ravnitzky, Case No. VFA-0188, June 22, 1998
http://www.oha.doe.gov/cases/foia/vfa0188.htm

{91} Los Alamos National Laboratory—William Payne, Case No. VFA-0436, September 3, 1998
http://www.oha.doe.gov/cases/foia/vfa0436.htm

{92} Los Alamos National Laboratory—Natural Resources Defense Council, Case No. VFA-0338, October 31, 1997
http://www.oha.doe.gov/foia/vfa0338.htm

{93} Los Alamos National Laboratory—Johnson Controls, Gary Roybal, Case No. VBU-0016
http://www.oha.doe.gov/cases/whistle/vbu0016.htm

{94} Los Alamos National Laboratory—Charles Montano, Case No. VBZ-0016
http://www.oha.doe.gov/cases/whistle/vbz0016.htm

{95} Los Alamos National Laboratory—University of California, Case No. VWZ-0017
http://www.oha.doe.gov/cases/whistle/vwz0017.htm

{96} Contractor Employee Protection Cases—Whistleblower Protection Cases
http://www.oha.doe.gov/whistlec.htm

{97} Richard W. Gallegos, Case No. VWA-0004 (H. O. Klurfeld)
http://www.oha.doe.gov/cases/whistle/vwa0004.htm

{98} Los Alamos National Laboratory—Edward J. Seawalt, Case No. VBU-0039
http://www.oha.doe.gov/cases/whistle/vbu0039.htm

{99} Oak Ridge National Laboratory—Dr. Nicolas Dominguez, Case No. VFA-0386, April 2, 1998
http://www.oha.doe.gov/cases/foia/vfa0386.htm

{100} Sandia National Laboratory—William H. Payne, Case No. VFA-0128, March 26, 1996
http://www.oha.doe.gov/cases/foia/vfa0128.htm

{101} Benton County, Washington
http://www.oha.doe.gov/cases/pett/lpa0001.htm

1

1, 1, 1-Trichloroethane 13, 28, 30,
1, 2-Dichlorobenzene, 25
1, 1, 1-Trichloroethylene, 28
1, 1-Dichloroethane, 28
1, 1-Dichloroethylene 5, 28, 30
1, 1-Dichloroethylene, 28
1, 2-Dichloroethane, 30
1, 2-Dichloropropane, 29
^{129}I, 34, 45, 66
^{14}C, 34, 46, 66
^{137}Cs, 56, 64, 71

2

2, 4-Dichlorophenol, 24
2, 4-Dichloroaniline, 24
^{222}Rn, 66
^{226}Ra, 59, 66
^{228}Ra, 59, 66
^{235}U, 56, 66, 40
^{238}Pu, 64
^{238}U, 40, 66
^{239}Pu, 64, 68
^{240}Pu, 64, 68
^{241}Am, 64-65

3

3, 4-Dichloroaniline, 24
^{32}Si, 66
^{36}Cl, 66
^{39}Ar, 66
^{3}H, 66

4

4-Chlorophenol, 24

5

5-Nucleotidase, 29

6

^{60}Co, 71
^{63}Ni, 34, 46

8

^{85}Kr, 66

9

^{90}Sr, 33-34, 45-46, 55-56, 64, 71
^{99}Tc, 33-34, 45-46
^{99}Technetium, 45-46

A

Abandoned hazardous waste sites, 12
Abandoned wells, 12, 18, 114
Abstract water, 7
Abstraction, 6
Abstractions and consequences, 3
Accessible environment, 23
Accidental surface spills, 78
Accumulation of salts, 7
Acetone, 28, 30
Actinides, 65
Action plan to implement groundwater
  protection
  NM, 111
Actual risk, 23
Acute
  and chronic sensitivities, 24
  contamination events, 11
  exposure in humans
  radioactively related, 55
  from toxic chemicals, 26
  to chronic effects, 24
  chronic relationship log noec, 24
Adherence of contaminants to subsurface
  materials, 76
Adverse
  environmental impacts, 8
  health, social, environmental, and economic
  impacts, 20
Agence financiere de bassin rhine-meuse, 4
Agency for toxic substances and disease
  registry, 28
Agricultural
  activities, 8
  chemicals, 13
Agriculture, 4
Air
  quality, 26
  sparging, 75

stripping
  as remediation technique, 79
  stripping and adsorption technologies, 76
Alanine amino transferase, 28-29
Albuquerque
  basin, 54, 81, 84
Aldicarb, 29
Algae, 24
Alkaline phosphatase, 29
Alternative groundwater development
  activities, 8
Am, 68
American National Standards Institute, 94
American Petroleum Institute, 94
American Society
  for Testing and Materials, 94
  of Chemical Engineers, 2
  of Civil Engineers, 1
Americium, 57, 64, 65
Ammonium nitrate, 29
Ammonium sulfa-mate, 24
Amphibia, 24
Analyses of transfer pathways, 22
Analytical chemistry of mixtures, 30
Animal
  feedlots, 13
  tissue waste
    Los Alamos, 58
  waste disposal, 13
Anions, 34, 46
Anthropogenic, 2
  substances, 34
    identification of, 46
Applicable or relevant and appropriate,
  requirements (ARARS), 26
Approaches to risk assessment, 25
Appropriative Doctrine, 19
Aquifer
  cleanup project, 11
  groundwater classification program, 11
  proper management, 1
  protection controls, 10
  restoration, 11
    achievability, 80
ARARS, 26

Area of body exposed
   as related to radionuclides, 55
Argon, 66
Argonne National Laboratory, Illinois, 57
Arid
   and semiarid countries, 4
   regions, 4
Arizona, 10
Arkoses, 66
Arochlor (1260), 28, 30
Arroyo de Coyote fan
   NM, 85
Arroyos
   temporary rivers, 7
Arsenic, 13, 24
   (III) 90, 30
   trioxide, 28
Artificial
   outflow, 7
   recharge, 6-7
Asbestos, 44
Asia, 10
Assessment
   drinking water needs, 10
   risks, 25
   and cleanup of inactive doe sites, 55
   of potential risks from exposure, 32
   risk management, 31
Assumptions and uncertainties, 21
Atazine, 29
Atmospheric
   radiation, 66
   transport mechanisms, 60
Atomic Energy Act, 38, 59, 91, 117
   NM, 94
Atomic Energy Commission, 37, 57

B

Bacteria, 8, 11, 13, 24
Bacteriological/virological contaminants, 12
Bandelier tuft, 64
Bangkok, 8
Barrier, and source 1 and source 2
   computer models, 60
Baseline DOE 1995 report, 38
Basin yield, 9
Belfield, ND (UMTRA), 62
Benzene, 11, 13, 25, 28, 30, 125
Bernalillo County, 81, 83, 96, 98
Beryllium-containing weapons, 44
Best Management Practices, 10, 11
Bioavailability and toxicity of these metals
   Savannah River Site, 47
Biochemistry, 32
Biological
   integration, 24
   organization, 24
   transformations of organic chemicals, 15
Biometry, 32
Biostatics, 32
Biota, 22
Birmingham, 8
Bismuth, 58
BMP, 10

Bone marrow
   cellularity, 30
   parameters, 30
   progenitors, 30
   sarcoma and head carcinoma, 31
Boron concentrations, 86
Brine/salinity, 13
Brines, 1
Brookhaven National Laboratory, 41
BTEX, 76
BUN and BUN/CREAT, 29
BUN/CREATININE RATIO, 29

C

Ca and alkalinity gradients, 48
Cadmium, 24, 28
   (II) 510, 30
   acetate, 28
Calcium, 25, 81
Calculated
   health risk, 23
   risk, 23
   risk reduction, 27
California, 9, 10, 23, 29, 31
California water law, 10
California's new groundwater legislation, 9
Cancer
   risk, 27, 31
   risk due to groundwater contamination, 21
Canonsburg, PA, 62
Capacity of wells and well fields, 2
Cape Cod, 10
Carbon adsorption, 79
Carbon dioxide, 76
Carbon tetrachloride, 4, 28, 30, 33-34, 45-46
Carbon tetrachloride (ct), 77
Carbon tetrachloride toxicity, 29
Carcinogenic
   potency of metabolized PCE dose, 23
   risks with guidelines, 26
Carcinogens, 21, 25-26
Cartography, 32
Case history models
   doe, 53
Cations, 34, 46
Cause-effect relationships, 3
CCl_4, 29
CCl_4 hepatotoxicity, 29
Cd, 48
CERCLA, 43, 48, 49, 61, 68, 73, 117
Certification programs, 15
Cesium, 44, 56, 137
Cesium-137, 64
Characteristics of waste, 32
Characterization
   relevant to SNL/KAFB, 83
Chelate formation of radionuclides, 71
Chelating agents on the migration, 71
   of radionuclides, 71
Chemical
   plants, 12
   precipitating treatment plants
     Los Alamos, 64
   separation of fission products, 40

separation wastes, 40
   treatment sludge
     Los Alamos, 58
   waste landfill (CWL)
     SNL, 84
Chemistry (analytical and organic), 32
Chernobyl
   accident, 66
   evacuation zone, 66
Chi-square, 71
Chlorendic acid, 24
Chloride, 81
Chloride salts
   from molten-salt extraction, 63
Chlorides, 17
Chlorinated
   ethylenes, 79
   solvents, 47
Chlorobenzene, 28
Chlorobenzene, 30
Chloroform, 28, 30, 79
Chromate solutions, 59
Chromium, 33-34, 39, 45
Chromium
   (III) 360, 30
   chloride, 28
Chronic
   aquifer contamination, 11
   exposure
     compared to acute exposure, 55
     effects of, 55
   hazard index, 26
   spills and leaks, 12
Cibola National Forest
   NM, 85
City of Albuquerque, 81
   Public Works Department, 17
Classical toxicology and ecotoxicology
   compared, 22
Clean Air Act, 91
   NM, 94
Clean Water Act, 11-12, 42, 49, 50
Cleanup cost overruns, 53
Clitellata, 24
Closing the circle
   DOE, 57
Coefficients
   of radioactive nuclides
     diffusion related, 71
Coelenterata, 24
Cold War, 37-38, 40, 53
Cold War Mortgage, 37
Colloid formation
   and transport mechanism, 70
   and transport mechanisms, 70
Colorado, 10, 54, 57
   Department of Health, 53
Columbia River, 34, 46
   (Hanford Site), 57
Commercial
   LLRW disposal sites, 60
   commercial/industrial water supplies, 12
Community planning, 10
Compilation of historical water quality data,
   17

Complexities and uncertainties, 8
Comprehensive Environmental Response,
    Compensation, and Liability Act, 13, 26, 28,
    33, 42, 46, 49.
Comprehensive
    groundwater monitoring databases, 16
    groundwater protection
        Bernallilo County
            NM, 100
    planning programs, 1
    state groundwater protection program
        (csgwpp), 11
Computer
    models, 18
    models for data summarization, 16
    systems and software, 19
Computerized database information system, 18
Conceptual
    hydrogeological model (CM)
        SNL/KAFB, 86
    model (CM)
        SNL/KAFB, 86
Configuration of the system, 2
Confined and unconfined aquifers, 7
Congressional office of technology assessment,
    53
Connecticut, 10, 11, 12
Conserving water
    NM, 110
Containment
    at DOE sites, 41
Contaminant characterization activities
    SNL, 84
Contaminant
    transport
        factors controlling, 70
    transport and fate
        difficulties and complexities, 69
            subsurface issues, 69
    transport processes, 69
Contaminants
    at Hanford, 45
    found in groundwater, 12
    in groundwater, 20
Contaminated groundwater emanating from a
    chemical plant, 21
Contamination of groundwater by organic and
        inorganic chemicals, 20
    potential, 17
    problems
        as acute or chronic, 11
Continuous monitoring
    for tritium, 75
Control
    of groundwater development, 5
    strategies, 21
Copper, 24
Corps of Engineers, 17
Corrective activities program
    Los Alamos, 63
    programs, 14
Cosmic rays, 66
Cosmogenic
    decay, 66
    radiation, 66

Cost
    analysis, 3, 76
    components of remedial investigations, 80
    of remediation
        DOE, 57
Costa de Hermosillo, 9
Council On Environmental Quality, 59
Coyote Canyon
    NM, 86
Cresols, 25
Crustacea, 24
Cu, 48
Curium, 57
Current federal laws, 15
CWA, 43
Cyanide, 13, 33-34, 39. 45-46
Cytogenetics, 28

D

Damage to wildlife, crops, vegetation, and
        physical structures
    by exposure to waste constituents, 35
Daphnias, 23
Database
    and data extraction system, 16
        for Rio Grande Basin, 83
    system, 18
Datura innoxia (Angel's Trumpet), 79-80
Decommissioning facilities
    Los Alamos, 45
Decontamination and Decommissioning (D&D),
    61
Degreasing agents, 77
Delaware, 37
Demand
    and supply management, 1
    management, 2
Demonstrating alternative on-site systems and
        other liquid waste disposal options
    NM, 112
Demonstration projects, 15
Denver, 54
Department of Defense, 26, 83
Department of Energy (DOE), 73
Department of Energy's Applied Research,
    Development, Demonstration, Testing,
    and Evaluation (RDDT&E), 79
Desiccation of existing agricultural lands, 7
Designing contaminated groundwater
        remediation systems
    by use of pump-and-treat strategy, 79
Detection
    of contaminated groundwater areas, 20
    correction, prevention, and stabilization, 20
Development
    of programs and plans, 19
    testing, and evaluation plan, 61
Di(2-ethylhexyl)-phthalate, 28
Dibromochloropropane, 29
Dichloro-ethane 14, 30
Dielectric mineral oil
    Los Alamos, 63
Difficulties in characterizing the subsurface, 76
Diffusion of contaminants into inaccessible
        regions, 76

Dilution attenuation factors (DAFS), 27
Direct natural recharge, 6
Distribution coefficient (Kd), 67
District of Columbia, 37
DNA damage, 55
DNAPLs, 74, 76
DOE
    Albuquerque Operations (DOE/AL), 83
    baseline program, 41
    environmental management program, 37
    environmental restoration program, 57
    nuclear weapons laboratories, 54
    Order 5400.3
        hazardous and mixed waste program, 59
    Order 5480, 14, 33, 45
    subcontractor nuclear facilities, 37
DOE'S
    CERCLA response actions, 49
    cleanup efforts, 39
    department's office of civilian radioactive
        waste management, 42
Dose
    of radiation, 55
    related cfu-gm difference, 30
    response analysis, 24
    response assessment, 32
    response models, 23
Drawdown, 9
Drinking Water Standards, 23
Drought phenomena, 7
Drum storage area, 26
Durango, CO (UMTRA), 62

E

E1-Dabaa (the site of the future Egyptian power
    reactor), 71
Ecological, 3
    receptors, 25
    risk assessment, 21, 24
    systems
        potential damage, 43
Ecologically significant groundwater, 12
Ecology, 26, 32
Ecosystem
    definition of, 22
Ecosystems, 20, 22, 23, 24
Ecotoxicological
    assessment, 22
    concepts and principles, 35
    danger, 22
    testing, 23
Ecotoxicology, 22, 32
    definition of, 22
Edgemont, SD (UMTRA), 62
Effectiveness of federal, state, and local
    programs, 14
Effluent waste monitors (TEWMS), 75
EG&G Mound Plant, OH, 62
Eh-Ph diagrams, 65, 67
EIA
    procedures, 8-9
    process, 8
Electro-refining
    Los Alamos, 63
Emergency Management Act, 94

English Channel, 4
Enriched uranium, 39, 56
Environmental
    advisory board, 39
    cleanup at DOE facilities, 39
    contaminants, 22
    costs of nuclear weapons production, 40
    guidelines, 5
    health, 32
    health and safety, 23
    health department, 17
    health scientists, 32
    impact assessment (EIA), 8
    impacts, 6, 20
    Improvement Act, 94
    management program, 37-38, 40
    pollutant, 22
    protection, 24
    Protection Agency, 88
    protection laws and regulations
        as related to radionuclides, 55
    regulations, 21
    remediation
        DOE, 56
    remediation/restoration
        in United States, 72
    restoration, 37
        SNL/KAFB, 83
    restoration activity
        Los Alamos, 44
    restoration and remedial actions, 61
    restoration program at Los Alamos, 44
    restoration within the DOE, 41
    reviews of existing projects, 8
    risk assessment, 25
    services gross receipts tax acts
        nm, 93, 96
    toxicology, 32, 35
Environmentally
    sensitive freshwater plant species, 9
    sound groundwater management, 7
EPA, 39, 43, 49, 50, 60
Epidemiological studies, 27, 32
Epidemiologists, 27
Epidemiology, 32
Epididymis, 31
Equilibrium, 6
ES&H vulnerabilities
    Los Alamos, 63
Estimating the Cold War Mortgage, 56, 57
ETEX, 27
Ethylbenzene, 3, 25, 28, 30
Ethylene dibromide, 29, 77
Ethylenediaminetetraacetic acid, 71
Euphrates, 4
Euthanatized, 30
Evaluating the sources of uncertainty, 23
Evaluation
    of a system of permits, 6
    of risk factors, 22
Evaporation, 7
Evapotranspiration, 43
Exogenous inputs, 2
Experimental reactors
    Los Alamos, 44

Exploratory studies facility (esf)
    DOE, 56
Exposure
    assessment, 25, 32
    related fate, 22
Extending water and sewer services
    as a groundwater protection policy
    NM, 113
Extraction procedure (EP), 39
Extraction projects, 10

F

F_0 and F_1 mice and rats, 28
Facility investigation/corrective measures study
    under RCRA, 34
Falls City, TX (UMTRA), 62
Fate
    and transport, 27
    of toxic chemicals
        from hazardouos waste sites, 51
Fate/transport, 65
Fe, 48
Feasible controllable measures, 2
Federal
    and state government, 15
    facility compliance act, 42
    facility compliance agreement, 46
    framework, 14
    insecticide, fungicide and rodenticide act, 14,
        49, 50
    support to the states, 15
Feed Material Production Center (FMPC), 46
Feedback to the planning agency, 4
Feedlots, 18
Fernald, 62
    environmental management project (FEMP),
        46, 73
    groundwater concerns
        DOE site, 46
Field Units
    Los Alamos, 44
FIFRA, 50, 68, 91
    NM, 94
Fire and explosion containment, 32
Fire code
    NM, 96
Fish, 20, 23, 25
Flammable liquids statute
    NM, 93, 95
Flexible groundwater models, 5
Florida, 10
Flow
    model, 9
    paths through the vadose zone, 87
Fluorides, 13
Food chain, 25, 55
Fossil water, 7
France, 4
FSVE modeling assumptions and accuracy,
    75
Fuel
    and oil leaks, 40
    element wastes, 66
    rod assemblies, 56
Fulvic and humic acids, 71

Fumigants, 77
Fundamentals of groundwater hydrology, 10

G

Galleries, 7
Galvin Report, 39
Gamma scans, 34
Gasoline, 13
Geochemistry
    radionuclide, 65-66
Geographic information system (GIS), 5, 82
Geographical component, 21
Geohydrological regime, 2
Geologic mapping of the Travertine Hills
    SNL, 84
Geological
    repository, 42
    sampling, 5
Geophysical
    borehole logging, 5
    logs, 81
GIS, 5
    geographic information, 5
Grand Junction, CO (UMTRA), 62
Granulocyte-macrophage
    colony formations, 27
    progenitor cells, 30
Gravity-segregated vertical equilibrium
    (GSVE), 74
Green River, UT (UMTRA), 62
Green-fields concept, 57
Gross alpha, gross beta, and gamma scans, 34,
    46
Groundwater
    abstraction, 3, 5, 6, 9
    classification system, 12
    cleanup, 23
    computer models, 19
    contamination
        at DOE sites, 41
    contamination and prevention models, 5
    contamination and prevention plan, 5
    contamination and risk factors, 22
    contamination cleanup, 10
    contamination evaluation program, 16
    contamination prevention plan, 6
    contamination protection plan, 16
    development plan, 3, 4, 5
    flow
        direction of, 35
    flow modeling investigations
        for Rio Grande basin, 83
    flow velocities, 25
    hydrogeology, 11
    legislation, 9
    management, 6, 8
        criteria, 1
    modeling, 10, 26, 43
        plan, 8-9
    monitoring, 43
        program at Hanford, 46
        related data, 18
    planning cycle, 3
    planning elements, 10
    protection act

NM, 93
protection management plan (GWPMPP)
  Los Alamos, 58, 88
protection policy
  Albuquerque
    NM, 98
protection policy and action plan
  NM, 97
protection programs and strategies, 11, 19
protection/control/stabilization plan, 18
quality management programs, 13
quality standards, 11
remediation
  using simulated annealing algorithm, 78
resources, 10, 20
resources management strategies, 2
resources plan, 3
resources planning processes, 2
utilization, 2
withdrawal permits, 12
GSGWPP process, 11-12
Guidelines
  and framework for project selection, 5
  for addressing contaminants, 15
  in setting priorities, 15
  to assist in conducting reliable hydrogeologic
    investigations, 15
  to protect groundwater, 15
GWPMPP
  Los Alamos, 89
GWSCREEB, BLT
  computer models, 60

H

Half-life
  definitions, 54
Handling spent fuel, 39
Hanford, 33-34, 39, 42, 45, 62, 64
  groundwater, 34, 46
  nuclear reservation, 47, 56-57
  potentially explosive tanks, 39
  site-wide groundwater monitoring, 45
  tanks, 39
Hastings groundwater contamination site, 77
Hazard
  analysis, 20
  assessment, 24-25
  communication act, 92
  evaluations, 23
  identification, 24, 32
  ranking system (hrs), 32
Hazardous
  and Solid Waste Amendments (HSWA), 85
  chemicals, 43
  Materials Transportation Act, 91
  waste, 37
    definition of, 38
  waste act
    NM, 93
  waste at DOE facilities, 38
  waste management and risk assessment
    processes, 32
  waste management units, 38
  waste site investigations, 51
  waste sites, 21

waste/groundwater contamination concerns
  DOE, 53
wastes
  under RCRA, 38
Hazards from the groundwater pathway at
  hanford, 33
Head waters, 7
Health
  and environmental protection standards for
    uranium mill tailings, 59
  hazard
    radon, 59
  risk, 65
    potential, 43
  risk to the workers, 22
  risks
    by exposure to waste constituents, 35
    of remediation concerns, 72
  safety, and environmental concerns, 6
  safety planning
    remediation process, 80
Health-risk-based approach, 23
Heavy metals, 30, 40, 57
Hematological parameters, 30
Hematopoiesis in mice, 30
Hepatic parenchymal damage, 29
Hepatotoxicity, 29
  of CCl$_4$, 29
Herbicides, 20, 44
High explosives, 44
  in abondoned buildings
    Los Alamos, 45
High-dose, 21
High-level radioactive waste
  defense related, 38, 42, 56, 67
  radionuclides, 72
High-level waste and spent nuclear fuel, 42
High-quality groundwater discharge, 12
Historical
  maps and aerial photographs, 18
  waste disposal practices, 22
HLW storage tanks
  leaking tanks at hanford, 56
Holland, 4
Hot-cell waste
  Los Alamos, 58
HRS groundwater route, 33, 45
HSWA permit, 89
Hubbell Spring
  NM, 85-87
Human activities, 8
Human health and the environment, 49
Hydraulic conductivity
  related to cleansing of zones, 72
Hydraulic head, 7-8
Hydraulic jetting of clay materials, 76
Hydrodynamic isolation, 69
Hydrogeologic
  conceptual model
    SNL/KAFB, 86
  conditions
    related to containment of groundwater, 73
  investigations, 14
Hydrogeological
  and environmental information, 6

characteristics, 35, 43
consequences, 3
  of simulated scenarios, 3
information, 5
Hydrogeologists, 50
Hydrographs, 5
Hydrologic
  analysis, 17
  balance, 18
  parameters of the aquifer, 18
Hydrostratigraphic architecture, 87

I

Idaho, 39, 56
  chemical processing plant, 42
  national engineering laboratory, 42
  national engineering laboratory (INEL), 56
Immune
  system, 28
  system of humans, 28
Immunodeficiency, 27
Implement rational groundwater management
  plans, 9
Improper groundwater management, 8
In situ
  air sparging, 78
  biodegradation, 75
  bioremediation, 78
  chemical treatment, 11, 78
  colloid remediation process, 74
  leaching, 58
  remediations, 72
  soil vapor extraction, 78
  technologies, 11, 79
Inadequate sewage treatment, 8
Increased salinity, 8
Indian reservations, 16
Indirect
  ionizing radiation such as gamma or x-rays
    and neutrons, 54
  recharge, 7
Individual sensitivity
  as related to radionuclides, 55
Industrial
  and commercial facilities, 12
  hygiene, 32
  landfills, 13
  waste discharge, 1
  waste disposal sites, 13
  wastes, 18
  water use, 3
Infrared
  from sunlight, 7
  photography, 7
Inhalation Toxicology Research Institute (ITRI),
  84
Injection
  of chemicals into saturated zones, 75
  of hazardous waste
    into drinking water, 48
Injection wells, 14, 18, 49
  including u/c class v, 12
Insecta, 24
Installation of flow meters, 5
Institution of civil engineers, 53

Integrated
  approach, 1, 3
    groundwater management, 1
    to hazardous waste regulations and
      programs, 35
  conceptual model (cm)
    SNL/KAFB, 86
  finite difference method, 9
  groundwater resources planning, 3
  plan of activities, 3
  water resources management, 1
Integration
  of consumers, 3
  of disciplines, 3
  of issues
    concept of, 3
  of supply and demand, 3
International atomic energy agency, 53
Interstate stream commission, 16
Ionizing radiation (alpha particles, beta
  particles), 54
Iraq, 4
Iron, 25
Iron hydroxides
  precipitation, 64
Irrigated
  areas, 7
  fields, 7
Irrigation
  and removal of contaminants, 77
  districts, 9
Isleta Pueblo Indian Reservation, 83, 84, 86

J
Jemez
  geothermal reservoir, 81
  mountains
    New Mexico, 44
Jordan, 4

K
KAFB, 83, 85
Kansas, 10
Kansas City Plant, KS
  DOE, 62
Kentucky, 12
Key contamination sources, 13
Kirtland Air Force Base (KAFB), 83, 85

L
Lagoons, 18
Lakeview, or (UMTRA), 62
Land
  application treatment agricultural activities,
    12
  disposal restrictions (LDR), 48
  leveling of wells, 5
  use plans, 10
  use/zoning, 33
Landfills, 8, 12, 18
  contaminated with chemicals, 40
Land-use concerns

related to remediation, 72
  related to remediation concerns, 72
LANL, 63, 89
LANL material disposal areas
  Los Alamos, 57
Las Vegas
  Nevada, 56
Latrrop Wells Cinder Cone, 56
Lawrence Livermore Laboratory, California, 54,
  57
Laws and regulations, 2
LC_{50} value, 23
Leachate collection systems, 33
  groundwater transport pathway, 27
Leaching, 31
  of chemicals, 8
Lead, 28, 45, 58
  acetate, 28
  shielding
    from radionuclides, 59
Leaking
  canals, 7
  underground fuel tanks, 1
    Bernalillo County, 99
Legislative mandate to protect groundwater
  quality, 11
Lethal
  concentration (LC_{50}), 24
  toxics, 22
Leukocyte or erythrocyte levels, 30
Levels, and other information from wells, 18
Licensing or permitting system, 5
Linear programming techniques
  related to flow and transport, 73
Liquid-liquid partitioning, 69
Little skull mountain, 56
Livermore, CA, 54
Livestock holding pens, 22
LLRW, 55, 60
LLRWPA, 43
Local ordinances
  NM, 96
London, 8
Long-term
  allocation of groundwater resources, 4
  risks to the hematopoietic system, 30
Los Alamos
  environmental restoration program, 44
  meson physics facility (LAMPF), 63
  municipal sanitary landfill, 44
  National Laboratory, 37, 42, 43, 44, 53, 54,
    55-57, 60, 62, 64, 68, 81
  case history review
    environmental survey 1987, 62
    groundwater protection plan, 88
    disposal of radioactive wastes, 64
Louisiana, 12
Low-level radioactive waste, 38, 4245, 54, 56
  disposal aspects, 60
  solidification aspects, 71
  policy act (LLRWPA), 42
  policy act of 1990, 60
  (LLRW), 60
  measurement, 54

M
Magic bullet, 21
Magnesium, 25, 81
Magnitude of risk that is acceptable, 25
Management of household hazardous waste
  promote concerns, 103
Management strategies
  groundwater, 1
Managing the residual risk, 21
Manganese, 13
Manhattan Project
  Los Alamos, 58
Manufacture of nuclear weapons
  Fernald, 46
Manzanita Mountain, 85, 87
Manzano (or Manzanita) Mountains, 85
Marrow granulocyte macrophage progenitors
  (CFU-GM), 30
Maryland, 12
Massachusetts and cape cod, 10
Mathematical modeling, 32
Mathematical models, 10, 25
  application to fate and transport studies, 69
Maximum Concentration Limit (MCL), 34,
  49, 85
Mechanisms for preventing contamination, 16
Mercuric chloride, 28
Mercury, 24, 28, 39
Mercury (II), 30
Metagranites, 66
Metals, 13
Meteorological investigations, 32
Methylene chloride, 28, 30
Mexico, 9
Microbial ecology, 69
Microbiological principles
  biorestoration, 69
Microbiology, 32
Microorganisms, 20
Micropollutants in water, 79
Micro-purge low-flow sampling
  at Fernald, 73
Middle east, 4
Migration
  of contaminated groundwater, 9
  of radionuclides, 71
  containment routes, 32
Military defense initiative, 20
Mill tailings, 40
Mining
  activities, 12
  and mill tailings waste, 58
  and milling, 39
  and milling uranium
    process explained, 58
Mixed waste, 45
  definition of, 59
  landfill
    SNL, 84
Mn, 48
Modeling and repository siting, 67
Modeling
  strategy, 10
  subsurface contaminant transport and fate, 69

Mollusca, 24
Monitoring
and control, 4
and surveillance network plans and programs, 16
groundwater quality
NM, 115
of groundwater levels and storage, 10
of groundwater quality, 19
of the flow system, 15
of water levels, 6
programs (air, water, and other), 32
strategies
to evaluate data, 51
systems
post closure, 61
wells, 13
surveillance, protection, control, or
stabilization programs, 16
surveillance and protection plans, 19
Monte Carlo Simulation Method (MCS), 21
Mortandad Canyon
Los Alamos, 64, 68, 74
Motor Carrier Act
NM, 95
Mound Facility, Ohio, 57
Multicriteria analysis, 3
Multidisciplinary approaches, 32
Multifactor nonideality models
development and evaluation, 70
Multiparty agreements
DOE, 61
Multipoint-source model for landfill, 25
Municipal
Health Act
NM, 93, 95
landfills, 12
water supply wells, 17
Myelotoxic, 30
Myelotoxic agents, 30
Myelotoxicity, 30

N

NAPLs, 72, 74, 76
NAPL-contaminated sites, 74
National
Academy of Sciences, 56
contingency plan (NCP), 50
Environmental Policy Act (NEPA), 89
fire code
NM, 94
hazardous substances and oil contingency
plan, 47
Laboratory Complexes
DOE, 53
legislative mandate to protect groundwater,
42
policies and priorities, 4
policies related to the protection of
groundwater, 14
priority list, 33, 45
research council, 10, 11, 24
toxicology program, 28

Natural
attenuation, 41
at DOE sites, 41
ecosystems, 22
gas field exploration activities, 18
outflow of groundwater from an aquifer, 7
Naturally
occurring
metals, 13
radioactive elements, 12
Naval reactors, 38
NCSN method, 71
Nebraska Department of Environmental
Quality, 77
NEPA, 61, 92, 117
Neurobehavioral signs, 28
Nevada, 12
test site, 42
New Jersey, 10, 11, 12
New Mexico, 16, 17, 19
Bureau of Mines and Mineral Resources, 82, 98
Conservation Division, 16
Environment Department, 88
Environment Department (NMED), 91, 98
Game and Fish Department, 16
Hazardous Waste Management Society
NM, 101
Mining Act, 94
New York, 10, 42
Newton's (NCSN) Method, 71
Ni, 48
Nickel 28, 68, 30
Nickel acetate, 28
Nile, 4
Nitrate ion, 34, 46
Nitrates, 8, 11, 13, 29, 39
Nitrogen, 66
NM Environmental Improvement Agency, 16
NMED groundwater discharge plan permit
permit requirements
NM, 107
No Further Action (NFA) status, 27
Nodal domain integration technique, 9
Nonaqueous phase liquid (NAPL), 72
Noncarcinogens, 25-26
Nonideality models
application of single factor, 70
Nonlinear chi-square, 71
North Africa, 7
NPDES, 62
NRC, 57, 60
Nuclear
criticality, 44
devices, 40
environmental legacy, 40
fuel, 56
fuel cycle, 56
magnetic resonance spectrometry, 79
Waste Policy Act (NWPA), 42, 43
Waste Policy Act of 1982, 56
weapons, 37
production, 40
weapons sites
in United States, 72

O

Oak Ridge, 64
National Laboratory, Tennessee, 57
Reservation, 42
Oases, 7
Objectives for applying an EIA, 8
Occupational Safety and Health Act, 92
Occurrence
fate, and transport, 65
fate/transport
of groundwater, 53
Office
of DOE Environment, Safety and Health
Environmental Audit, 62
of Environmental Management at DOE, 37
of Environmental Restoration and Waste
Management
DOE, 60
of Environmental Restoration and Waste
Management (EM), 61
of Solid Waste and Emergency Response
(OSWER)
EPA, 48
of Technology Assessment, 15, 53
of Technology Development
DOE, 54
Ohio, 46
Oil
and Gas Act
NM, 95
Oil
and gas brine pits, 12
Oil
exploration and production activities, 12
Oils, 59
Old plutonium processing facility
Los Alamos, 44
Omega West Reactor
Los Alamos, 45
On-site
industrial landfills, 12
sewage disposal systems
alternatives to
NM, 99, 101
Open space, 18
Oregon, 62
Organic
chemicals, 11, 55
complexes, 65
compounds, 25
contaminant plumes, 51
contaminants in plant remediation strategies
fate of, 77
solvents, 59
volatiles, 30
Organically complexed ions, 48
Savannah River Site, 48
OSHA, 91-92
Overcharging, 7
of aquifers, 6
Overexploitation
and irrational use, 8
of aquifers, 6

Over-pumping and overcharging concerns, 6
Overview of risk assessment, 31
Oxidation of TCE and/or PCE
    by OH radicals, 79
Oxide minerals, 66
Ozone, 79

P

Pajarito Plateau, 44
    Los Alamos, 44
Pantex Plant, TX, 62
Paradigm
    for contaminant transport
        within porous media, 70
Particle methods to reliable identification
    and pollution sources, 74
Pathogenesis of disease, 27
Pathway exposure factors (PEFS), 23
PCB transformers, 44
    and capacitors
        Los Alamos, 63
PCE, 79
    concentrations available in large public water
        supplies, 23
Perceived risk, 23
Perception of risk, 21
Perchloroethylene (PCE), 79
Permatid production by the testis, 31
Permeability enhancement
    as related to diffusion, 76
Pesticide
    and fertilizer contamination of groundwater,
        29
    and fertilizer toxicity, 29
    Control Act
        NM, 93, 95
Pesticides, 13-14, 20-21, 44, 50, 55
Petroleum, 11, 21
    products, 13
pH, 71
Pharmacokinetic profile, 29
Phase separator pit, 45
Phenol, 25, 28, 30, 31
Phosphates, 8
Photocatalytic process
    for water and wastewater treatment, 76
Photofission experiments, 44
Photoreactors, 76, 77
Phreatic water, 7
Physical
    geography, 32
    heterogeneity, 76
Physicochemical
    properties of prevailing complexes, 67
    reactions, 71
Pipeline leaks, 12
Pipeline safety act
    NM, 93, 95
Pisces, 24
Planning
    and decision-making processes, 2
    and implementation of groundwater project, 8
    and preventive programs, 10
    processes, 2

Plants in remediation
    use of, 77
Plume control remediation technologies, 75
Plutonium, 39, 40, 43, 45, 56, 57, 63, 64, 65,
        73
    Finishing Plant (PFP)
        at Hanford Site Washington State, 34, 73
    metal production activities
        Los Alamos, 45, 63
    processing waste, 22
    removal from low-level wastewater effluents
        at Hanford Site, 73
Point-source contamination areas, 14
Political
    indecisiveness or derivitives thereof, 14
    judgments concerning the role of the federal
        government, 14
Polonium, 58
Polychlorinated biphenyls (PCBs), 40, 44
Polyelectrolyte catfloc
    to stop colloid migration, 74
Ponds, 18
Population at risk, 25
Potable water supplies, 12
Potassium permanganate, 24
Potential
    groundwater contaminants, 14
    risks to human health, 25
    water problems, 18
Potentially responsible parties (PRPS)
    as related to spills, 78
Precipitation of radionuclides, 66
Predicted effects from monitoring networks, 5
Predictive
    accuracy, 24
    capacity, 32
Prepare and implement the plans and programs,
        16
Presence of nonaqueous-phase liquids (NAPLs),
        76
Preservation of natural habitat, 18
PRESIDENT TRUMAN, 83
Pretreatment
    on the site, 14
    programs, 1
Prevention
    based approach, 19
    is still the best fix
        as a proverb, 80
    planning, and emergency response, 10
Primary pathways of concern from inactive
        waste sites
    at Hanford, 45
Princeton Plasma Physics Laboratory, 41
Principles of Health Risk Assessment, 32
Priorities for cleanups, 26
Priority
    pesticides, 22
    pollutants, 17
Private sector and semigovernmental agencies,
        4-5
    water supplies, 12
Proactive significance
    groundwater management strategies, 1

Process of Human Health Risk Assessment
    (HRA), 24
Production and plutonium, 39
Projected
    demands on the aquifer, 18
    pumping activities, 18
Prostate, 31
Protecting groundwater resources, 10
Protection
    and remediation programs, 12
    measures
        for prevention of groundwater
            contamination
            NM, 104
Protozoa, 24
Pthalate esters, 25
Pt-tio$_2$, 76
Pu, 68
Public health, 4
    concerns, 20
    impact, 26
        radioactive wastes, 54
    service, 28
Public involvement
    remediation programs, 72
Public
    nuisance provision
        nm, 93, 95
    sector, 5
    water supplies, 3, 12, 65
    works, 4
Pulsing, 75
    as related to modeling, 75
Pump and treat operations
    at doe environmental restoration sites, 49
Pump-and-treat, 73
    strategies
        simulation models, 79
    systems, 78
Pumping and treating
    at doe sites, 41

Q

Qanats, 7
Qualitative and quantitative estimates, 20
Quality assurance, 32
Quantify colloid migration
    in porous media, 71
Quantitative
    data, 22
    mathematical models, 32
    risk assessment, 21

R

Radiation
    on hematopoiesis, 30
    protection act
        NM, 93, 95
Radioactive
    and chemical substances, 34
    and chemical waste, 37
    colloid transport, 70
    colloids

or radiocolloids, 74
contaminants, 44
decay, 66
landfill
Los Alamos, 45
material, 13, 40, 43-44
radium and thorium, 40
waste and spent fuel from nine weapons
production reactors, 39
waste repositories, 71
wastes, 22, 38, 54
definition of, 54
generated at Los Alamos, 58
various types, 55
Radioactivity, 33, 37, 53
Radiocolloids
containing plutonium and americium, 74
migration of, 71
Radionuclide behavior
predictions of, 67
Radionuclide contamination, 46
Radionuclide
disposal
site consideration, 68
geochemistry, 66, 68
inventories, 34
inventories and radionuclide mobility, 46
migration, 68, 70
to groundwater, 66, 71
mobility, 34
assessing, 67
modeling
limitations, 67
transport
through biosphere, 60
transport models, 67
Radionuclides, 2, 20, 44, 57, 65
long-lived, 41
mobility actions, 60
transport
problem evaluation, 67
Radium, 40, 58
Radon, 59, 68
Rainfall, 6
Ranking System (RS), 18
Rate
at which groundwater is withdrawn, 43
of abstraction, 7
of groundwater abstraction, 8
of groundwater flow, 43
of recharge, 7
Rating factor system (RFS), 33
Rationale
prevent pollution of groundwater
NM, 104
RCRA, 38-39, 42-43, 48-50, 59, 61, 63, 68, 85,
91, 99, 117
corrective and CERCLA response actions, 49
Part B Permit, 60
Subtitle C Regulation, 39
Reactor
fuel element, 39
irradiation wastes, 40
Recharge

aquifers, 6
areas and zones, 18
Reclamation/restoration, 32
Reduce
or eliminate uncertainties, 2
or manage the risk, 21
waste volume, 14
Reducing
salt loads, 1
uncertainty, 23
Refuse collection ordinance
NM, 96
Regional
aquifer water steering group, 16
groundwater monitoring
NM, 116
plan, 4
Regression and correlation analyses, 24
Regulated hazardous waste sites, 12
Reinjection of treated groundwater, 75
Reliable techniques for conducting
hydrogeologic investigations, 15
Remedial
Action Master Plan (RAMP), 26
actions
DOE, 61
activities, 22
investigation/feasibility study under
CERCLA, 34
Remediating contaminated sites
soil and groundwater, 79
Remediation of the five operable units
Hanford, 47
program at SRS, 47
restoration, 72
Remote sensing methods, 7
Remote-well sensing equipment, 18
Removal of plutonium from low-level process
wastewaters
by absorption, 73
Renal histopathological alterations, 29
Repository
development, 42
environments
and transport mechanisms, 70
low permeability aspects, 67
Research
and development, 15
and development activities, 15
on development of water quality standards,
15
on environmental and economic impacts of
contamination, 15
on the behavior of contaminants in
groundwater, 15
on toxicology and adverse health effects, 15
testing wastes, 40
Reserve basin water quality, 9
Resource Conservation and Recovery Act
(RCRA), 26, 42, 49
Restoration
costs, 10
of groundwater quality, 10
Restoration/remediation groundwater

contamination concerns, 37
restrictions in wellhead protection areas
restrict activities damaging to groundwater
NM, 105
RFS, 33
Rhine System, 4
Rhizosphere, 77
microbial communities
biodegradable aspects, 77
Rhode Island, 37
RI/FS work plan, 46
Rio Grande
Basin, 82
River
New Mexico, 44
Valley, 82
Risk
low-level radioactive wastes to public health,
60
reduction to human health and environment,
61
to public and environment, 77
Risk analysis
definition of, 10, 21
Risk assessment, 10, 11, 20-21 28, 32
contractor's, 26
and evaluation, 32
evaluations, 26
guidance, 31
paradigm, 24
process, 31
strategies, 21, 24, 34
technique, 22, 26, 31
work plan addendum (RAWPA), 47
with toxic chemicals, 26
Risk factors
and radium, 31
actual hazardous events, 24
Risk models, 24
Risk-based
priority system
DOE, 55
strategy
DOE, 58
Risk-characterization, 32
Risks
public fears of exposure, 65
Rivers in arid regions, 4
Road salting, 12
Rocky Flats, CO, 53, 57, 62
Role
of regional groundwater planning bodies, 4
Runoff
and infiltration, 43

S

Safe Drinking Water Act, 11, 17, 42, 49, 92
Safe yield, 2, 18
Safety, 32
Safety/environmental health interactions, 35
Sahara, 7
Salinity, 1, 6, 7
Salinization, 6
Salt content, 6

Salt flats
  as outlet areas for groundwater, 7
Salt lakes, 7
Saltwater intrusion, 1, 12
  control of, 9
Sampling and analysis, 32
  data interpretation, 15
San Mateo basin, 9
Sandia Mountains, 85-86
Sandia National Laboratories
  characterization project, 83
Sandia National Laboratory, 54, 81, 83, 85
Sandia/Hubbell fault line, 86
Sanitary landfills, 13
Sanitary wastewater consolidation system
  (swcs), 62
Santa Fe Group, 81, 82, 86, 87
Santo Domingo basin, 81
SARA, 91, 99
SAS (statistical analyses system), 19
SAS institute, 17
Saturated zone processes
  SNL/KAFB, 87
Savannah River (Savannah River Site), 42, 56-
  57, 64
Savannah River Plant (SR), 47-48, 56-57
  monitoring for tritium, 75
  west valley, 39
Scarcity phenomenon
  of water resources, 3
Scientific
  uncertainties, 21
Scope
  of environmental restoration/remediation, 61
Screening, 3
SDWA, 43, 50, 68
Seepage
  from storage reservoirs, 7
Semiarid
  areas, 2
  to arid climates, 7
Seminal vesicles, 31
Septage
  disposal of, 18
Septic tank systems
  Los Alamos, 63
Septic tanks, 12
  and leachfields, 18
Septic, sewage, and wastewater treatment
  sludge, 13
Sewer and wastewater ordinance
  NM, 97
Sewer-service expansion
  NM, 109
Sharp-interface model
  for assessing NAPL, 74
Silica gel
  as related to photocatalyst, 76
Simazine, 29
Simulated chemical mixture of groundwater
  contaminants, 30
Single-value estimate of risk, 21
Sink-holes, 12
Site conceptual model, 21

Site-specific problems, 14
Site-wide hydrogeologic characterization (swhc)
  project, 85
Slurry wall containment, 76
SNL mixed waste landfill
  characterization activities, 84
SNL/KAFB, 85, 86, 87
  hydrologic modeling, 88
SNL/NM environmental restoration (er)
  program, 85
Snowmelt, 6
Social aspects, 3
Socioeconomic factors, 2
Sodium, 81
  chlorate, 24
Soil
  adsorption of radioactive wastes
    Los Alamos, 64
  contaminated with high-explosive waste, 40
  permeability, 33
  salinization, 7
  vapor extraction, 75
  venting
    as remediation technique, 79
Solid Waste Act
  NM, 93, 95
Solidification/stabilization, 21
Solid-phase cd, 48
Solvents, 40, 77
  metals, and salts, 38
Sorption-desorption, 69
Source of elimination
  of cantaminants, 41
Sources of contaminants, 12
South Carolina, 47, 56
South Valley Site, Ambrosio Lake, NM
  (UMTRA), 62
Southwest Alluvial Basins Regional Aquifer-
  systems Analysis (U .S. Geological
  Durvey, 1988), 81
Specific policies
  GPPAP Mission
    NM, 102
Spent fuel, 56
  reprocessing, 56
  nuclear fuel, 38, 42
Spermatogenesis in b6c3fi mice, 30
Spill prevention control plan
  Los Alamos, 63
Spill-time histories, 74
Sprague-dawley rats, 28
Springs or seeps, 7
Sprinkler irrigation treatment
  for remediation of groundwater, 77
SRM (Site Rating Methodology), 33
SRS groundwater remediation program, 47
Stabilize
  groundwater contamination, 16
  known contaminated areas, 11
Stabilizing groundwater levels, 1
Standards for drinking water and groundwater
  quality, 11
State and Federal
  efforts with groundwater protection/

    control/stabilization, 19
  programs, 14
  programs that protect groundwater quality, 11
  regulatory agencies, 10
State
  and local groundwater programs, 11
  liquid waste disposal regulations
    NM, 95
  regulations
    NM, 92
  water quality control commission (wqcc)
    regulations
      NM, 117
State, regional, or local comprehensive plans, 18
State-of-the-art
  continuous systematic monitoring and
    surveillance, 5
  information exchange, 15
Statistical measurements of margins of
  uncertainty, 24
Steel tank institute
  NM, 94
Stochastic-deterministic models
  use of, 70
Storage coefficient, 7
Storm runoff, 1
Stormwater runoff drainage, 18
Strontium, 44, 56, 64, 90
Styrene, 25
Subchronic ingestion, 28
Subdivision
  Act
    NM, 93, 96
  ordinances
    NM, 97
Subhumid areas, 6
Subsurface
  contamination from volatile organic
    compounds (VOCs), 78
  drains, 7
  saltwater intrusion, 9
Sulfate, 13, 81
Sun fuels groundwater remediation, 76
Superfund, 25, 80, 92, 99
  amendment and reauthorization act, 46, 49
  legislation
    and congress, 39
  NPL site, 26
  program, 48
  sites, 61
Surface impoundments, 12
Surface Mining Act
  NM, 93, 96
Surface
  mining and control and reclamation act, 50
  mining control and reclamation act
    (SMCRA), 49
  water law, 10
  water resources, 4
Sustainable groundwater management program,
  9
Synergistic or antagonistic interactions
  as related to transport mechanisms, 70
Syria, 4

Systematic update of a groundwater development plan, 5

T

Tank farms at the hanford site, 39
Target
  organisms, 22
  organs, 28
Tc(IV), 67
Tc(VII), 67
TCLP, 39
Technetium, 34
Technical assistance to the states by the federal government, 15
Technical evaluations of groundwater mid Rio Grande area of New Mexico, 81
Technique for screening pesticides, 22
Tetrachioroethylene (pce), 47
Tetrachloroethylene, 24, 28, 30, 34
Tetrachloroethylene perchloroethylene (pce), 23
The Clean Water Act of 1987, 13
The Emergency Management Act, 93
The Environmental Improvement Act
  NM, 93
The New Mexico Mining Act
  NM, 93
Thermonuclear bomb testing, 66
Thorium, 40, 66
Threatened or endangered bird species, 9
Three mile island nuclear plant accident, 41
Tigris, 4
Tijeras Arroyo, 85
  NM, 84
Time-history of toxicant concentrations, 25
Toluene, 13, 25, 28, 30, 70
Tonopah test range, 62
Total capacity of an aquifer, 2
Total dissolved solids, 86
Total dose
  as related to radionuclides, 55
Tourism, 4
Toxic
  chemicals
    behavior of in hazardous waste sites, 51
  effects of chemicals, 22
  endpoints, 28
  heavy metals, 40
  recovery aspects, 80
  recovery from contaminated groundwater, 79
Toxic materials, 14
Toxic metals in water and bottom sediments savannah river site, 47
Toxic potency, 23
  associated with the delivered dose, 23
Toxic substances
  physical containment, 33
Toxic substances control act (TSCA), 14, 49-50
Toxic substances
  with high exposure, 25
Toxicity
  metals analysis, 25
Toxicological program, 29
Toxicology, 32
TPSB, 24

Trace element distribution in various phases Savannah River Site, 47
Trace metals, 34, 46
Transport and fate, 25
  assessments, 69
  in subsurface areas, 68-69
Transport codes
  radionuclides. 67
Transport parameters
  SNL/KAFB, 88
Transport processes
  radionuclides, 68
Transuranic (TRU) wastes, 38, 42, 45, 56-57
Transuranic radionuclides ^{239}pu, ^{238}pu, or ^{241}am, 58
Treating water contaminated from multiple sources, 16
Trichloroethane (TCA), 77
Trichloroethylene, 13, 21, 24, 28, 47, 77, 79
Tritium, 34, 40-41, 44, 64
TRU wastes, 57
TSCA, 91-92
Tuba City, AZ (UMTRA), 62
Tubellaria, 24
Two-dimensional finite element model, 9
Types of radiation, 54

U

U.S. Air Force, 27
U.S. Department of the Interior, 50
U.S. Department of Energy (DOE), 33, 37, 38, 45-46, 53, 60, 80
  Albuquerque Operations Office, 62
  baseline environmental management reports, 37
  Hanford Site, 33, 45
  FUSRAP (Formerly Utilized Sites Remedial Action Program), 72
  National Laboratories, 81
U.S. Department of The Interior. 13
U.S. District Court of Tennessee, 38 (DOE National Laboratories), 40
U.S. Environmental Protection Agency, 9, 12, 27, 32, 33, 38, 45-46, 117
  strategy, 19
U.S. Forest Service, 17, 45, 83,85
U.S. General Accounting Office (GAO), 58
U.S. Geological Survey, 15, 16, 17, 80, 81, 82
U.S. National Policy Options, 14
U.S. Nuclear Complex, 37
U.S. Regulatory Commission (NRC), 56
Uk, 8
Ukraine, 66
UMTRA project, 59
Uncertainties, 8, 14, 20-22, 26, 31
  in conceptual models, 87
  in radionuclide complexes, 67
  model input, 70
  surface water and vadose zone, 87
  transport and fate mechanisms, 69
  with colloid migrations in groundwater, 74
Unconfined aquifers underlying large irrigated areas, 6
Uncoupled flow and transport equations, 9

Underground
  nuclear tests, 40
  pipeline evaluations, 35
  repository
    site aspects, 68
  storage tank, 12-13 34, 61
  NM, 108
Underground tests
  nuclear, 40
Uniform and National Fire Codes, 94
Uniform Fire Code
  NM, 94
United Kingdom, 4
United States, 22, 40, 55, 57, 80
University of Nebraska-Lincoln (UNL), 77
Unpermitted disposal, 13
Unquantifiable uncertainty, 31
Unsound waste disposal practices, 8
Upconing of more saline water, 6
Update of the groundwater development plan, 4
Upper Rio Grande
  NM, 98
Uranium, 33-34, 43, 45-46, 57, 66-67
  enriched, 40
Uranium (enriched, depleted, normal, or $_{238}$U), 58
  components, 44
  hexafluoride gas, 40
  mill tailings radiation control act (UMTRA), 59
  mill tailings remedial action, 59, 61
  mill tailings sites, 59
  mining and milling, 66
  mining and milling wastes, 40
  ore, 58
  plume
    at Fernald, 73
  targets, 40
Urban
  and industrial site runoff, 13
  use of fertilizers and pesticides, 12
Urea nitrogen (BUN), 29
Used oil burned for energy recovery, 39
USFS, 83
UST regulations and the ground water protection act
  NM, 93
Utah, 12

V

Vacuum extraction technology, 47
Vadose zone characteristics
  SNL/KAFB, 85
Vaginal cytology evaluations (SMVCE), 28
Van't Hoff Law in Chemistry, 6
Vapor phase migration, 31
Viral, 8
VOCS, 13, 78
Volatile
  and semivolatile organics, 44
  compounds, 27
  organic chemical, 23
  organic components
    cleanup, 62

Volatile (continued)
 organic compounds, 41, 57, 77
 organics, 34, 46
Volatiles, 30
Volatility, 33

W

Wallerian degeneration, 28
Warhead components, 40
Wartime activities
 Los Alamos, 63
Washington State Department of Ecology
 Report, 47
Waste dissolution in a nuclear waste repository,
 71
Waste Isolation Pilot Plant (WIPP), 42, 58
Waste
 management, 37
 minimization
  NM, 108
 recycling, 15
 repositories, 71
 tanks
  Savannah River Site, 39
Wastewater collection and treatment
 Bernallilo County
  NM, 106
Wastewater treatment and septic systems
 Los Alamos, 44
Water
 abstraction, 3
 balance, 6, 43
 code, 10
 conservation, 1, 2-3, 116
  implement techniques in city, 103
 detention facilities, 18

 for irrigation, 7
Water Law, 5, 18-19
Water Pollution Control Act
 Clean Water Act, 92
Water Quality, 16
 Act
  NM, 93
 data, 14
 distribution and affects on water use, 5
 monitoring plan, 19
Water quantity, 16
Water recycling, 10
Water resources, 26
 development plans, 4
 resources planners, 3
 resources planning, 4
 rights, 7, 18
 sampling and analysis, 5
 supply wells, 13
 use patterns, 2
Waterlogging, 7
 in semiarid regions, 6
 of irrigated fields, 6
Watershed protection areas, 18
Weapons
 assembly and maintenance wastes, 40
 complex (DOE)
  near Denver, CO, 54
 production, 37
Well
 abandonment and well destruction program, 9
  administration aspects, 9
 construction policies, 10
 drilling tools, 18
 probes, 18
 testing, 5

Wellfields
 in remediation efforts, 72
Wellhead protection areas
 grants, 11
 identification and management, 9
 programs, 11-12
 strategies, 12
West valley demonstration project in new york,
 42
West Virginia, 12
Westinghouse
 Hanford Company, 47
 Savannah River Company, 47
Wetlands, 7
Wildlife, 43
WIPP, 42
Wisconsin, 10
World War II, 37, 44

X

Xylene, 16, 28, 30

Y

Yellowcake
 radioactive concentrate, 58
Yucca Mountain
 Nevada, 56
  project in Nevada, 58

Z

Zinc, 24
Zn, 48
Zoning ordinances
 NM, 97

T - #0260 - 071024 - C0 - 279/216/11 - PB - 9780367398699 - Gloss Lamination